Organizational Behaviour in Construction

Anthony Walker
BBS, MSc, PhD, FRICS
Emeritus Professor of Real Estate and Construction
University of Hong Kong

A John Wiley & Sons, Ltd., Publication

This edition first published 2011 by Blackwell Publishing Ltd.

Blackwell Publishing was acquired by John Wiley & Sons in February 2007.
Blackwell's publishing program has been merged with Wiley's global Scientific, Technical and Medical business to form Wiley-Blackwell.

Registered Office
John Wiley & Sons Ltd, The Atrium, Southern Gate, Chichester, West Sussex, PO19 8SQ, UK

Editorial Offices
9600 Garsington Road, Oxford, OX4 2DQ, UK
The Atrium, Southern Gate, Chichester, West Sussex, PO19 8SQ, UK
2121 State Avenue, Ames, Iowa 50014–8300, USA

For details of our global editorial offices, for customer services and for information about how to apply for permission to reuse the copyright material in this book please see our website at www.wiley.com/wiley-blackwell.

Library of Congress Cataloging-in-Publication Data

Walker, Anthony, 1939–
Organizational behaviour in construction / Anthony Walker.
p. cm.
Includes bibliographical references and index.
ISBN 978-1-4051-8957-6 (pbk. : alk. paper)
1. Construction industry–Management. 2. Organizational behavior. I. Title.
HD9715.A2W268 2011
624.068′3–dc22

2010048241

A catalogue record for this book is available from the British Library.

Set in 10/12pt Minion by SPi Publisher Services, Pondicherry, India
Printed and bound in Malaysia by Vivar Printing Sdn Bhd

1 2011

Contents

Preface

In the preface to the first edition of my book on construction project management twenty-six years ago I said that it did not pretend that it covered the whole field and that 'techniques and procedures etc. are well covered elsewhere and behavioural aspects need a separate treatment'. Six editions later the former continues to be well covered and behavioural aspects have developed considerably, particularly in the mainstream management literature but also in construction publications. My project management book is still essentially about organizational structures for construction project management based on the systems approach, and although its later editions touched on organizational behaviour it certainly does not do so comprehensively. The increase in the interest of the effects of behaviour on construction activities at the professional and managerial level has been substantial, influenced to a large extent by the more complex nature of construction projects arising from greater specialisation, more complex clients and stakeholder activities, by the advent of partnering and other relational arrangements, and by government and other reports aimed at improving the performance and reputation of the construction industry; all of which generate more intense focus on people. This has been evidenced by the surge in publications in the academic and professional press on matters such as motivation, trust, culture and power amongst many others. So the need for a separate treatment of behavioural aspects relating to construction remains and has intensified and this book hopes to contribute to fulfilling this need.

The field of organizational behaviour itself is vast and has over the years expanded from a focus on the behaviour of people in organizations to encompass all aspects of organizations, including structural issues, so that many organizational behaviour books now run to more than five hundred pages. However, the focus of this book is firmly on the behaviour of people in construction organizations, as publications on the structures of construction projects and firms are provided elsewhere. Also the topics of this book are selective as some areas are seen to be more relevant to construction than others. The reasons for defining this book in this way is to make its size manageable, because in order to examine the relevance and application of organizational behaviour topics to construction it is necessary not only to present the topics with a critical edge but then to add and analyse the construction situations in which they do or do not apply.

In order to apply organizational behaviour to the field of construction I have relied essentially on a number of splendid writers for knowledge of organizational behaviour. In particular, I have drawn substantially on the work of the following authors (cited in the text) enshrined in the following books:

Huczynski, A. and Buchanan, D. (2007) *Organizational Behaviour: An Introductory Text.* Harlow England; New York: Prentice Hall/Financial Times.

Robbins, S. and Judge, T. (2008) *Essentials of Organizational Behavior.* Upper Saddle River, NJ: Pearson Educational.

McShane, S. and Von Glinow, M. (2003) *Organizational Behavior.* New York: McGraw-Hill.

Fincham, R. and Rhodes, P. (2005) *Principles of Organizational Behaviour.* Oxford: OUP.

Moorhead, G. and Griffin, R. (2001) *Organizational Behaviour: Managing People and Organizations.* Boston MA: Houghton Miffin.

I am extremely grateful for the accessibility of their writing and the understanding they provided and I recommend my readers to their books for the depth of treatment I am unable to provide here.

My hope is that this book makes some small contribution to the understanding of the construction process because the process is about people and organizational behaviour is about people. Projects do not go wrong by themselves but by how people behave in developing and constructing them. Traditionally, construction people have tended to treat construction as a mechanical process but increasingly the realization has emerged that this is not so. However, construction people believe that they should be self-reliant so have not been inclined to look to other fields in seeking to improve their understanding and performance. One such significant field is organizational behaviour and there are encouraging signs in some – but not all – construction publications that authors are now underpinning their work with the fundamentals of organizational behaviour. This book aims to bring together the elements of organizational behaviour relevant to construction and illustrate their application so that practitioners and academics can gain a foundation from which to better understand the context of their work and from which to read further on organizational behaviour with greater understanding of its relevance to construction.

I have been extremely fortunate in my working life to have had colleagues in both academia and practice who stimulated and sustained my interest in the management of construction processes. In particular, my gratitude goes to Anita Liu, Richard Fellows and Steve Rowlinson of the University of Hong Kong for their advice and assistance over many years. I give special thanks to Ray Thomas, a most successful practitioner of the art of project management, who diligently read the draft of this book to prevent me making errors in my representation of practice in the construction industry. I must also record my thanks to the many colleagues, and to those not known to me personally, who publish in the field of the management and economics of construction to which is now added those in the field of organizational behaviour and associated fields whom I have relied on for my knowledge and without whom this book could not have been written. Nevertheless, I remain solely responsible for any faults that remain. And finally, but certainly not least, my sincere thanks once again to my wife for her enduring tolerance and encouragement.

Anthony Walker
Hong Kong

1 Introduction

Until relatively recently, people working for companies associated with the construction industry – whether as directors, managers, professionals or technicians, and whether for firms providing professional services or the many types of development and construction companies – have not generally recognized the wealth of mainstream management literature as having much relevance to their working lives. However, with growth in the complexity of the industry and its firms and its projects this has changed. There has been a significant expansion of research and the number of books related to the softer side of project and construction management. But even the recent surge in 'softer' management books has been mainly on the organizational aspects of management which have generally adopted a rational/ systems approach to the subject. These books have contributed much but there remains a significant gap in the management literature relating to construction. Interestingly, this gap relates to what is the most significant aspect of management but also that which is the most difficult to deal with; people. Firms and projects do not go wrong by themselves, people make them go wrong. So an understanding of how people work in organizations is paramount to creating effective organizations as much in construction as in any other industry if not more so.

The study of people in organizations generally has for many years been recognized as important by mainstream management under the title of organizational behaviour. In construction, however, recognition of the way in which individuals' personal characteristics affect the smooth running of firms and projects, principally as a result of the need for the collaborative style of working necessary to produce successful construction projects, has only been recognized implicitly and this knowledge has not been formalized. Whilst mainstream management literature stresses that *managers* need to develop people skills and that an understanding of organizational behaviour can play a vital role in managerial work (Moorhead and Griffin, 2001), the construction industry requires that *all* its professional and technical members, not just managers, need such skills. Nevertheless, this recognition has been no more than implicit and has only rarely been the subject of research; one exception was an investigation of the importance of design consultants' soft skills in design-build projects

Organizational Behaviour in Construction, First Edition. Anthony Walker.

(Ling et al., 2000) which considered conscientiousness, initiative, social skills, controllability and commitment, all of which were supported as being important. With increasing complexity of the industry and its firms and increasing self-awareness of the industry's members, it is necessary to make explicit the issues surrounding the way in which members behave if the management of firms and projects is to become more effective. The more that can be done to prevent misunderstandings between people, which lead to problems within firms and projects, the better.

No matter what the state of the labour market, high calibre employees are in short supply and high demand. Retaining such employees depends not just on the level of salary and fringe benefits but on the conditions within which they work, particularly the way they relate to their colleagues and hence the sense of supportiveness in their working environment. So working with people with good interpersonal skills is likely to make the workplace more amenable, which, in turn, makes it easier to hire and keep qualified people and enhances productivity (Robbins and Judge, 2008). Creation of a supportive working environment needs an understanding of one's own and others' behavioural characteristics by all members of an organization. Though this may realistically be viewed as the 'Holy Grail', nevertheless, the more one can understand one's own behaviour and the behaviour of others the greater chance there is of creating an appropriate organization setting. Construction organizations present a great challenge in this respect due to the range of characteristics required by the diversity of the contributing professions and technologies. It is somewhat surprising, therefore, that the field of organizational behaviour has not until relatively recently been taken seriously by members of the construction industry. This book intends to go some way to redressing the balance, beginning with consideration of what is meant by organizational behaviour.

1.1 What is organizational behaviour?

From our very early years, each of us has been influenced by the behaviour of others and, in turn, has influenced others by our own behaviour. As we grew up, this occurred in social and organizational settings such as family, friends, school and sports clubs. We continually asked why he or she did this or that, seeking explanations of others' behaviour, and they have been asking similar questions about us. When we enter the world of work the number of times such questions are asked seems to multiply relative to the complexity of the settings. As a consequence, the academic field of organizational behaviour has developed to try to rationalize our understanding of our and other's behaviour. The term 'organizational behaviour' is in reality a convenient but inaccurate term. Organizations do not behave; only people behave. The term is shorthand for how people behave within an organizational context, similarly to the way the term 'management of organizations' is used; organizations are not managed, people are.

There is actually no such study as behaviour of organizations, only study of the behaviour of individuals in organizations (Naylor et al., 1980).

Organizational behaviour (OB) has still not yet been clearly defined but a current working definition can be taken as 'the study of human behaviour in organizational settings, of the interface between human behaviour and the organization, and of the organization itself' (Moorhead and Griffin, 2001).[1] They also say that 'we can focus on any one of these three areas, ... [but] all three are ultimately necessary for a comprehensive understanding of organizational behaviour'. The breadth of this definition is common to many writers on OB and encompasses organizational structure as well as behaviour. Whilst many claim this is to enable understanding of the context of behaviour within the organizational structure, many go to great lengths in describing structural issues. Hence many OB books tend to be extremely long as they interpret OB as having the widest possible boundary (or even no boundary at all!). This book aims to avoid focusing on organizational theory and organizational structure; rather, they are seen as the context of organizational behaviour and there are many books dealing with these topics, including those relating to construction. Hence, it only incorporates descriptions of organizational theory and structure where they are necessary to illustrate aspects of behaviour directly. In other cases they will be referenced to the literature. In this way this book seeks to avoid duplication and will keep to a manageable length and focus.

The field of OB cannot claim to be cohesive and integrated. Rather, it comprises a range of topics that aim to help us understand the behaviour of individuals at work. The topics include, amongst others, personality, perception, motivation, communication, group processes, emotions and stress. In the case of construction, some specific aspects are more important than others and this is reflected in the topics of this book. Huczynski and Buchanan (2007)[2] reflect the orientation of many organizational behaviour scholars in stating that: 'organizational behaviour enjoys a controversial relationship with management practice'. It seems that some organizational behaviour academics believe that considering management in conjunction with OB somehow distorts a study of OB. But a study of OB independently of its implication for management seems to render the subject arid, particularly in relation to construction. This book tries to avoid this problem and application to the management of construction projects will feature strongly.

The complexity of situations which OB seeks to address is reflected in the multidisciplinary nature of the knowledge, theories and skills needed to help us understand the behaviour of people in organizations. The disciplines which make major contributions to the field are:

[1] From Moorhead and Griffin, *Organizational Behaviour, 6th Edition.* © 2001, South-Western, a part of Cengage Learning, Inc. Reproduced by permission. www.cengage.com/permissions.

[2] From Huczynski and Buchanan, *Organizational Behaviour: An Introductory Text, 6th Edition.* © 2007, Prentice Hall/Financial Times: Harlow, England.

Psychology: the science which seeks to measure, explain and sometimes change the behaviour of humans.
Sociology: the study of people in relation to their social environment or culture.
Social Psychology: a blend of concepts from both psychology and sociology. It focuses on people's influence on one another.

Even psychoanalysis has been held to offer a great contribution to organizational phenomena (Yannis, 1999). An outline illustration of the contributions of each discipline to OB within the wrapper of the context of OB is shown in Figure 1.1.

The complexity of the field of OB is caused by the multidisciplinary nature of the subject and the minefield which is human emotions.

Adverse reactions to problems which arise on construction projects and within firms which contribute to projects are shown in employees' behaviour – uncooperativeness, avoidance of responsibility, delay in making a contribution and indecision can be symptoms of a variety of underlying causes. Given the

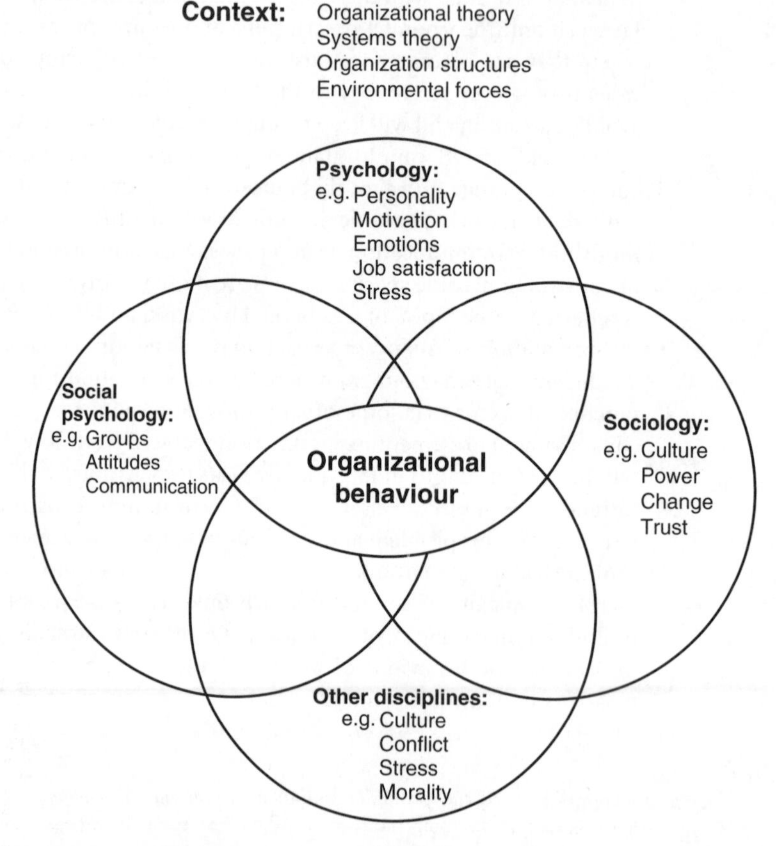

Figure 1.1 The disciplines constituting the field of OB with examples of their contributions.

complexity of construction projects and the complexity of the human emotions produced by the characteristics of individuals and their backgrounds (architect, engineer, etc.), there are numerous potential causes for an inappropriate response; these can include pressure due to, for example, workload leading to long hours and the resulting fatigue; uncertainty surrounding proposed organizational changes at work; poor work–life balance; low motivation (due to a number of reasons); a serious disagreement with the boss. Whilst many such incidents are soon overcome with no serious effect, the outcome could be more significant. For example, if the quantity surveyor asks the building services engineer for clarification and information about an item of equipment which is essential for the quantity surveyor to complete contract documentation and meet the deadline for distributing tender documents to bidders, and the building services engineer behaves in a disruptive manner, is obtuse and does not act conscientiously in providing what the quantity surveyor requires, the progress of the project is delayed if funding deadlines are missed. Of course, the vast majority of firms are managed effectively and projects progress smoothly with little disruption but even on such projects personnel issues do emerge and need management in order to maintain the cohesiveness of firms and projects and to further develop the effectiveness of the organization.

When the complexity of the construction industry is added to the complexity of the subject, it is perhaps not difficult to imagine why the industry has failed to face up to such complex issues earlier. OB can be seen as a complex subject well worthy of study by construction personnel as it is particularly relevant as the interdependency of projects require intense, continual and sensitive relationships between members of project teams and firms.

1.2 Critiques of organizational behaviour

Some textbooks and other publications on OB are criticized on the basis that they are bland and unquestioning. One critic is Wilson (1999) whose book is 'designed to challenge what constitutes organizational behaviour'. She reinforces the point made earlier by stating that the term OB is far from clear. For example she asks: 'Is it behaviour that occurs in some specified place and not in others or behaviour controlled by an organization?' Whilst she says her book is about 'the particular ways that individuals' dispositions are expressed in organizational settings', she points out the broader base of behaviour by saying that: 'What happens in rest and play, both inside and outside the organization, impacts on organizational life.' She is critical of the organizational psychology base of the discipline, which she says is focused on a 'highly empirical–analytical form of science, rooted in the natural sciences'. She considers that such an approach is used to justify a belief that there is no problem with the status of knowledge relating to OB which must be true as scientists are neutral agents, resulting in a very tidy and sanitized view of what goes on in organizations and an unwillingness to self-reflect and critique the discipline.

Ackroyd and Thompson (1999) agree with this view in saying that standard textbooks on organizational behaviour 'suggest, as much by implication as direct assertion, that behaviour in organizations is 'almost exclusively, conforming and dutiful'. It is difficult to deny their view that this assumption simply does not square with most people's experience of working in organizations. They develop their theme by focusing on what they term 'organizational misbehaviour'. They include within misbehaviour 'failure to work very hard or conscientiously, through not working at all, deliberate output restriction, practical joking, pilferage, sabotage and sexual misconduct' and define it as 'anything you do at work you are not supposed to do'. They also recognize that 'the behaviour of managers is much the same as that of the ordinary employees' and that 'managers are also capable of recognizing that there is not a precise correspondence between their own interests and those of their employing company'.

Business organizations are usually portrayed as tidy, coherent entities working towards a common agreed goal because this is how the bosses want it to be seen. But underlying this image is the reality of uncertainty, confusion and disorganization. This is particularly so in construction because of the size and complexity of its projects, the complexity of its clients and the multi-disciplinary nature of its constituent professions and firms. What is remarkable is the ability of all members of business organizations – managers and workers – to overcome these phenomena. Issues such as conflict, power, politics, gender, class, race and ethics amongst others, are endemic to organizations and lead to inequality, domination, subordination and manipulation. These issues need recognizing and treating in an open and critical manner. The mutual dependency demanded by the interrelationships necessary to successfully complete construction projects make it essential for members of construction teams and firms to find their way through the uncertainty and confusion. They will not do this by taking a bland view of OB but by developing a critical understanding which helps them to creatively resolve problems with working relationships.

1.3 Working in construction

The diversity of the characteristics and the skill bases of people working in construction is probably as wide as in any industry. A listing of the professions and talents involved – town planners; architects; quantity surveyors; structural, civil and building services engineers; contractors; subcontractors; and all the specialisms within each group – covers a vast range of psychological profiles drawn from a combination of an individual's personality and the attitudes ingrained by a person's profession. And whilst there are preconceived notions of the type of person who will fit which category (for example designers are 'airy-fairy' and not practical, engineers are pragmatic, quantity surveyors have no vision and are only concerned with the small print and contractors are avaricious), they represent a hugely over-simplified stereotyping. Each category will contain a wide range of personality types, each of which will bring their own personal attitude to their work.

But before taking this further we should ask the question, why do people work? Obviously, the prime reason most people work is to earn money to provide the basic essentials for living, then to buy extras for a better standard of living and, continuing the financial argument, to gain financial independence. Another set of reasons is also significant for many people (Rose, 1994). These are the 'expressive' reasons which are about the satisfaction derived from being able to demonstrate a skill, or the ability to do something well which is recognized by others, creating self respect, status and recognition leading to personal fulfilment. Alongside this group of reasons are the social benefits which include the camaraderie of working and a sense of belonging. So the reasons for working are complex and help to explain why, on retirement, people may have feelings of dislocation, and why some lottery winners decide to continue working.

These considerations underpin much of the theme of this book and help us to respond to questions such as: 'Where does work in the managerial and professional end of the construction industry fit into these ideas?' Certainly, in the higher echelons, careers are satisfying and economically rewarding. At the pinnacle, being involved in world-famous projects is immensely satisfying and, even at the more humble end, a small housing development or a refurbishment can also be very pleasing, particularly on a local level. Frustrating though construction can be, very few move out of it through lack of satisfaction. It could be thought that some professions would be more likely to gain satisfaction than others – architects would be the one to spring to a layman's mind as having the greatest potential for achieving high satisfaction – but is this not also likely to be the case for a contractor having built a good project which could be equally satisfying in their terms?

The point of this explanation is to show that there is great scope for satisfaction for those working in construction. There is also great scope for the behavioural characteristics of those involved to become significant in the success of firms and projects as a successful construction project requires high level of collaboration and communication. Inappropriate behaviour can have a serious effect on the smooth running of projects. If people do not understand their own behaviour and the behaviour of others, their chance of contributing appropriately within the myriad of relationships which are generated by projects is likely to be limited.

1.4 An illustration

It seems appropriate when beginning a book about OB to illustrate some behavioural issues by describing those which arise when someone begins a new job or joins a new project team. Practically all of us will have started a new job and can remember vividly the emotions of the first few days: feelings of excitement, apprehension and tension as we try to make a good impression, find our way around the workplace, and weigh up our co-workers. They in turn are trying to weigh us up. First impressions tend to be more important than they should be, so the initial behaviour of everyone involved is likely to colour our impressions for quite a time. These impressions affect the way in which we deal with the people we have met and the way they deal with us. In construction this scenario is particularly complex as it plays out not just when joining a firm but every time a new project team is formed and members from other firms become involved. Fortunately, some members of project teams will have worked together before and this eases the 'getting to know you' phase. But some will be new to each other and if many of the others have worked together before then the newcomers may feel isolated and find it harder to integrate. In construction collaboration is all important, so the way in which we deal with people is crucial if firms and projects are to be successful.

When applying for a job, an early behavioural facet occurs when candidates decide how to present themselves for interview. They will have researched the company and will try to represent themselves in a way which they believe will appeal to the company. But this may paint a false picture. The picture the company paints may also not truly reflect what the company is like from the inside. Nevertheless, the candidate will, for example, decide the degree of formality to adopt, both in dress and presentation and the interviewers will present the company in the way in which they wish it to be seen. Whilst supporting documentation, for example references, experience, etc., is no doubt considered seriously, the interview is supposed to 'get to the bottom of things' but the likelihood is that both sides are posturing to some extent.

Appointment processes are notoriously difficult in spite of the veneer of rigorous assessment. Most of us will know of poor appointments being made by senior members of companies. Appointees are blamed if they are not able to do the job but should not the blame lie with the appointers for making a poor appointment? But formal appointment processes offer reassurance that a good decision has been made even when it has not and enable those making the decision to hide behind the process. The interview is the most notorious element of the appointment process. How often do we hear comments such as 'we should not rely on the interview as it is subjective and not sufficiently probing to judge the candidate' but how often do we find that the interviewing panel disregard the documented information on the candidate and go by their 'gut feeling' or other imprecise factor in deciding whether or not to appoint a particular candidate.

Appointment processes, particularly for senior posts in complex organizations, are open to political manipulation. Invariably, even in small organizations, decisions on appointments are made by a selection panel. It is unlikely that every member of the panel will act entirely in the interest of the firm when deciding which candidate to support. Some members of the panel could have a 'personal agenda' which will influence their choice of candidate. Even those outside the formal selection process may seek to influence the choice by attempting to manipulate those on the selection panel. Selection of senior members of organizations can be seen to be one of the most politically charged events in a company and may be even more pronounced when internal candidates are involved rather than solely external candidates. We have all heard of cases where a full section process has taken place even though it has been known all along by the selectors that it has already been decided who is to be appointed.

An example could be of an interdisciplinary construction consultancy firm which is making an appointment of a senior associate on the design side of the firm. The selection panel may comprise top executives from the range of skills offered by the company. The member representing the cost control department of the firm may favour an architect with strong understanding of the need for and the discipline of cost management as this would make his or her life easier but the company's overriding need is really for a top design architect. If the cost management representative fights successfully for the person he or she wants it may be to the detriment of the firm as a whole. If the selection panel has a number of members who are self interested in this way, there will be great difficulty in making an appropriate appointment and an uncomfortable compromise may result. A danger in the appointment process is that of always appointing people who are similar to those on the selection panel or those who are similar to those whom the panel have previously appointed. Whilst there is great merit in developing a company culture, if change does not take place within the company, it may fail to keep up with its competitors. If there is a preponderance of members of the panel who do not welcome change, or a preponderance that do, an inappropriate decision may be made. There can be many other underlying factors which influence the way in which members of selection panels behave – spite, the desire to settle old scores, envy, arrogance. The behavioural characteristics of an interview panel's members are fundamental to making appropriate appointments so the person (or people) who decides who shall make up the selection panel has a great responsibility to ensure its members are sound. To achieve this they have to be cognizant with the behavioural characteristics of those they select for panel membership and should attempt to balance them in a way that enables the candidate who best fits the needs of the company to be appointed.

This brief account of appointing someone to a company is a microcosm of the range of behaviour to be found within many aspects of organizational life and confirms the contents of this book. The behaviour exposed during an appointment process is likely to be present in all aspects of organizations' and projects' activities which shows the need to treat these matters seriously. They include such issues as personal characteristics, emotions, ethics, motivation, communication, politics, culture, team working, decision making, conflict, leadership, organizational learning and change.

1.5 The development of organizational behaviour

The first formal approach to management in the modern era appeared in the early years of the 20th century through the ideas and writing of Taylor (1911) and Fayol (1949 trans.). Their work was the foundation of management practice up to about 1950, although elements of their ideas continue to the present day.

Taylor laid the foundation of 'scientific management'. The approach believed that it was possible to scientifically analyse and structure the tasks to be performed so that the maximum output could be obtained with the minimum input. People appeared to be perceived as machines and efficiency was the sole criteria of success. The outcome of such an approach leads to increasing specialization of the workforce. Managers were also seen to be governed by set processes and procedures as much as the workers.

Fayol, whose work did not become widely available in English until 1949, was nevertheless influential before then as others in the US had developed his ideas. Fayol's ideas were embedded in the so-called 'principles of management' which were concerned with such things as pyramidal structure, unity of command, line and staff, the scalar chain and span of control. The primary element was the pyramidal organization structure and the idea that authority was delegated downwards. Division of labour was advocated so that the sub-goals of the various units would add up to the overall organizational goals and coordination would be handled through the management hierarchy. The principles emphasized formalization and specialization and were in this way complementary to and supportive of Taylor's scientific approach.

What emerged from this classical view of management, and hence of organizations, was a deterministic perception. The principles were held to be universal truths about how management should be undertaken and the only way to manage business activities or processes. Hence, an extremely rigid view of how to organize emerged which paid scant attention to the needs, feelings and reactions of people working in such rigidly structured organizations.

Of great importance at this early stage in the development of thinking on organizations was the work of the German sociologist Max Weber (1968 trans.). He worked independently of Taylor and Fayol and adopted a different stance. Rather than focusing on how to improve organizations he took a far more academic approach by seeking to describe the characteristics of newly emerging bureaucratic structures. The characteristics he identified were generally compatible with other ideas of the time such as specialisation, hierarchy, etc., but were developed in much more depth. His primary focus was on organizations as power structures in which control is achieved through an organization hierarchy. Discipline was the keyword which required the exact execution of orders from above. Whilst it could be said that Weber recognized behavioural characteristics in workers, it appears that he recognized them only in so far as they should be controlled by a hierarchy.

The classical approach to management was therefore seen as essentially rigid and about structure not people, leading Bennis (1959) to describe the classical approach as representing 'organizations without people'. The classical approaches originated from military and church models which strongly influenced the way early managers performed. It did not make explicit the behavioural component. It may be that in those days the construction professions and industry recognized through their experience that such approaches were not really appropriate to the management problems of the industry and its projects and, for this reason, did not fully develop management in their training at an early date.

Whilst some serious attempts were made in the early 20th century to recognize the importance of individual and social processes in organizations, these processes were not widely recognized until the 1950s. In the early 1900s Robert Owen, a British industrialist, attempted to better the condition of industrial workers; Hugo Munsterberg, a German psychologist, argued that industrial psychology could provide valuable insights into areas such as motivation; and Mary Follett argued that organizations should strive harder to accommodate their employees' human needs and become more democratic in their dealings with employees (Moorhead and Griffin, 2001). At this time theirs were minority views in a field that believed that workers should be treated in the same way as machines.

It took until the late 1950s before OB began to emerge as a field for serious study but earlier the significant Hawthorne experiments, conducted between 1927 and 1932, provided the foundation of the forerunner of OB which was known as the human relations movement. The experiments took place at Western Electric's Chicago plant. Research was instigated by Western Electric's Employee Relations Research Department and subsequently included Harvard Business School researchers. The studies revealed that the human element in the workplace had a greater effect on organizations than had previously been believed. As Scott (1992) states:

> The experiments served to call into question the simple motivational assumptions on which the prevailing rational models rested. Individual workers do not behave as 'rational' economic actors but as complex beings with multiple motives and values; they are driven as much by feelings and sentiments as by facts and interests; and they do not behave as individual, isolated actors but as members of social groups exhibiting commitments and loyalties stronger than their individualistic self-interest.

Whilst there has subsequently been criticism of the study, these were about the researchers' interpretation of the results, not about the significance of the behaviour of the subjects of the research. For example:

> If the workers had in fact had the kind of trust in management's good intentions that Mayo claims, would they have found it necessary to resist the experiments so actively in this period? The picture we get instead is of a group of rather wary workers engaged in a continuing skirmish with management and determined not to be taken advantage of. Rather than become a part of the 'company' team, they became a team of their own, rather coolly looking out for their own economic interests in an adversarial relationship with management...' (Bramel and Friend, 1981).

Such criticism only serves to reinforce the way in which the Hawthorne experiments exposed the impact of individual and group behaviour on the performance of organizations. As a result of Hawthorne, some sociologists and social psychologists devoted their attention to examining how people in organizations actually behaved, how they actually related to their superiors, subordinates and peers and what were the factors which motivated members of organizations. Around this time Barnard's (1938) work was also significant in these respects. He stressed that organizations were cooperative ventures which integrate the contributions of their members. He also dealt significantly with authority in organizations which is reflected in his view that goals are imposed from the top down whilst their attainment depends on the bottom up.

Subsequent significant work by Maslow (1954) focused on the needs which motivated people. He identified basic physiological and safety needs at the lower level and psychological needs for self-esteem and self-realization at the higher level. McGregor (1960) then encapsulated much of what had gone before by contrasting the human relations approach with the classical approach in his now famous 'Theory X' and 'Theory Y' assumptions about how people behave in organizations. A Theory X manager assumes that people are idle and wish to avoid responsibility but in contrast a Theory Y manager takes a positive view of the people he manages by believing that people are self motivated and seek responsibility and enjoy it.

The change of name of the field from human relations to organizational behaviour is well stated by Huczynski and Buchanan (2007):

> Jack Wood (1995) notes that the term 'organizational behaviour' was first used by Fritz Roethlisberger in the late 1950s because it suggested a wider scope than human relations. The term 'behavioural sciences' was first used to describe a Ford Foundation research at Harvard in 1950, and in 1957 the Human Relations Group at Harvard (previously the Mayo Group) became the organizational Behavior Group. organizational behaviour was recognized as a subject at Harvard in 1962, with Roethlisberger as the first area head (Roethlisberger, 1977). The first appointments to chairs in organizational behaviour went to Professor Derek Pugh at London Business School in 1970 and to Professor David Wier at Glasgow Business School in Scotland in 1974.

In parallel with the development of the behavioural school the Tavistock Institute of Human Relations, London, undertook a series of studies which developed a distinctive research approach in that they proposed that a unique feature of business organizations is that they are both social and technical systems. The socio-technical approach emphasized that the needs of both the technical and social aspects should be served by organizations. Interestingly, the Tavistock Group related their work to the construction industry in an important early study of communications in the construction industry (Higgins and Jessop, 1965; Tavistock Institute, 1966). This view contributed to combining many of the previous approaches, some of which considered only technical needs whilst others considered only social needs. As mentioned earlier, Bennis (1959) labelled the former 'organizations without people' but also labelled the latter 'people without organizations'. Scott (1992) notes that the goal should be one of 'joint optimization' of the needs of both the technical and social systems since the two systems follow 'laws' and their relationship represents a 'coupling of dissimilars' (Emery, 1959).

From this base, OB has strengthened and broadened as a serious field of study. However, it still remains descriptive as it 'attempts to describe behaviour rather than to prescribe how behaviour can be changed in consistent and predictable ways' (Moorhead and Griffin, 2001). They believe that 'organizational behaviour is descriptive for several reasons: the immaturity of the field; the complexities inherent in studying human behaviour; and the lack of valid, reliable and accepted definitions and measures'. They add, 'Whether the field will ever be able to make definitive predictions and prescriptions is still an open question.' Nevertheless, the field of OB has much to offer by asking practising managers and those being managed to think about the impact of their behaviour and the behaviour of others on the performance of their organizations.

1.6 Organizational behaviour, common sense and research

Many people see OB as no more than common sense as a result of having experienced living and working with people for most of their lives. But this base line of a pre-existing common sense understanding of the social world is incomplete as it is drawn from a limited range of experience, and also common sense views expressed by individuals may vary in value as a function of the intelligence of the person propounding them. Common sense by itself is often a poor guide (Fincham and Rhodes, 2005). Advancing from a common sense understanding of human behaviour and society relies on the development of theoretical frameworks through appropriate research methods.

We are all biased as a result of our own experiences and interpret what we see selectively and personally. We develop entrenched attitudes and stereotypes which oversimplify the way in which the world around us really operates. In contrast, the analytical frameworks and theories that behavioural scientists develop provide a sounder basis for understanding human social behaviour. Behavioural scientists claim they can improve on the base line of common sense understandings by using research techniques to collect data more systematically than the layperson could (Fincham and Rhodes, 2005).

Research methods in organization behaviour are drawn from the major contributing disciplines of psychology, social psychology and sociology. Psychology's focus is the individual and the individual's interaction with their environment in seeking to answer why a particular person behaved in a particular way. Social-psychology is concerned with the social group, in particular how social structures can modify the attitudes, perceptions and beliefs of the individuals who belong to them such that customary ways of behaving emerge, for example rituals and style of dress. Sociology is the study of social systems, that is, of people in relation to their social environment or culture, such as families and occupational groups. Sociologists have contributed to research on organizational culture, gender, communications, power and conflict. From these disciplines are drawn research methods based on laboratory experiments, questionnaires and interview surveys and participant observation (Fincham and Rhodes, 2005).

As referred to earlier, the field of organizational behaviour still remains essentially descriptive as it strives to adopt research paradigms. In this it has similarities with research in the 'softer' topics of construction. This has led to interesting debates on the merits of quantitative (positivist/scientific) methods versus qualitative (interpretive methods) in the construction management field (Chau et al., 1998). The paper points out that many of the research issues in construction management are practical problems which involve generalization of experience and formulation of a hypothesis that can generate empirically testable implications. For problems of this nature, testability of hypothesis and reproducibility of results are said to be important, and the quantitative approach of discovering causal relationships is seen as more likely to produce general practical solutions. However, the paper does not deny the value of the interpretative approach which they say may be more suitable for certain types of problem and the years since the paper was published have seen a strong movement towards behavioural topics as subjects for research using qualitative methods. In 1998 they concluded that:

> We believe that no harm will come from an increased concentration on qualitative methods, for they certainly have been neglected in this field in the past, but that to focus exclusively on a narrowly drawn interpretative approach is inappropriate. We see no one method as containing a panacea: we do not want to become 'one club golfers.' Therefore we suggest that the way forward for the field is through methodological pluralism and paradigm diversity.

1.7 The rationale of this book

Construction is much more than a series of technological phenomena. So the more we understand behavioural issues, which are endemic to companies and projects, the better we understand ourselves and others and so we have more chance of contributing to successful projects.

The focus of this book is on individuals within construction organizations, which include all types of professional practices, contractors and projects, but it should not be overlooked that the issues discussed relate equally to the industry's clients. Whilst we may not understand how behavioural issues impact on our client organizations to the extent that we do our own organizations, nevertheless, our understanding of them may also help us to understand our clients a little better.

As stated earlier, OB has assumed an extremely broad spectrum of topics within its purview and not everything within its disciplines can or needs to be covered in a book on OB relevant to construction. Therefore, a selective approach has been taken in order to ensure relevance. As this book is essentially about how people behave within construction organizations, its focus is the individual, not the organization; hence, for example, organization structure is only referred to in so far as it impinges on the behaviour of the individual. The topics covered overlap and interact in such a way that there is no obvious sequence in which to deal with them, but in order to make them as coherent as possible they are dealt with from the 'bottom up'. So we begin with the individual, then progress to groups of individuals working together and on to the broader issues of working in organizations. We begin, therefore, with the individual per se in Chapters 2 to 5 where we examine characteristics, personality, perceptions, emotions, feelings, ethics and motivation within the context of the various organizations which contribute to construction projects. The overarching topics of communications, politics and culture follow in Chapters 6 to 8. Team working is paramount in construction, perhaps more so than in any other commercial activity, hence considerable emphasis is placed on it in the next section, Chapters 9 and 10, dealing with working in teams and decision making. The final Chapter 11 deals with leadership which also incorporates aspects of organizational learning and organizational change.

2 Individual Behaviour

2.1 The constituents

It is perhaps obvious, but nevertheless necessary, to repeat that the behaviour of individuals is vital to the successful performance of business organizations. But this simple statement disguises the complexity of the constituents and interconnectedness which make up individuals' behaviour and hence the quality of an individual's performance and, particularly importantly, an individual's job satisfaction. An individual's ability and attitudes are fundamental to job performance and job satisfaction. Whereas ability can be seen to depend on intellect, training and experience, attitudes are formed by personality, perceptions, emotions, feelings, morals and ethics. Each aspect will be covered in this and other chapters, as will other issues affecting job performance and satisfaction including role perceptions and situational factors.

2.2 Ability

Ability is an individual's capacity to perform a task successfully. People differ in their abilities depending their intelligence and experience but also on their aptitude for the task allotted to them. Aptitudes are the natural talents that help people undertake specific tasks more effectively and can be on a broad scale – for example, as designers – or on a more limited technical scale – for example, the dexterity which allows someone to sew. Hence managers who allocate jobs are interested in the difference in ability and aptitude between employees and in finding a match between an employee and a job position; known as achieving a high ability–job fit. If an individual is working within their comfort zone then they can be expected to successfully perform their task and gain job satisfaction. However, that may not be so if they find the task monotonous or lacking in challenge. As a result of their personal characteristics, some individuals wish to be challenged by the tasks they are given in order to gain job satisfaction. If they are not, they may become frustrated by the limitations of the job.

Organizational Behaviour in Construction, First Edition. Anthony Walker.

Such a situation requires a manager to exercise judgement on whether an individual should be placed in a challenging situation. If successful, everyone will be satisfied; if not, there could be serious consequences for the organization. In the case of an architect given a design task beyond his/her capabilities, but in the belief that this is not so, the consequences can be significant in terms of project progress, cost and client satisfaction. The characteristics of the people involved can be seen to be paramount to a sound decision. How the task is allocated depends on the risk characteristics of the managing architect – will he/she take a chance and present the challenge to a junior architect or is he/she too cautious to do so? If the junior architect is allocated the job will he/she welcome the challenge, will he/she have belief in or doubts about his/her ability and will he/she be capable of providing a successful project? The outcome of such decisions by managers will also determine whether a firm retains its staff. A wrong decision about the characteristics of a staff member may determine whether he/she remains with the firm, which may or may not be in the best interests of the firm.

The general management literature recognizes intelligence as one of the best predictors of performance across all types of jobs. (Ree et al., 2001) However, the range of intelligence of people occupying professional and managerial jobs in construction is likely to be narrower than those in managerial roles generally due to the consistency of the educational standards required, so it is to be expected that the range of performance would also be narrow. Nevertheless, the level of performance by individuals in construction may be affected to a greater degree by the nature of the job/project they are involved in if it does not present the level of challenge which they believe their intelligence/experience requires. This is reflected in the finding that intelligence does not ensure that people are more satisfied in their jobs. In fact, the correlation between intelligence and job satisfaction is about zero. The reason has been put forward that this is because, although intelligent people tend to have more interesting jobs and perform better, they are more critical of their job conditions and expect more (Ganzach, 2003). Members of the construction industry and its professions are likely to recognize this phenomenon.

'Employee competencies' are recognized in the general management literature as the characteristics of people that lead to superior performance. Competencies are seen as not only a person's technical and professional skills but also his or her values and personality traits (McShane and Von Glinow, 2003). This is increasingly so in construction where, traditionally, expectations were essentially about competency in profession-specific skills. These are now generally taken as given and the distinguishing features of an individual are seen to be the ability to work effectively in teams, communication skills, cultural awareness and other abilities necessary for social interaction which has come about through the increasing complexity of the environment of the industry and its clients.

2.3 Attitudes

Attitudes are evaluative statements, or opinions, which people have about events, people and objects. Different people are found to have a wide range of different attitudes towards the same thing. These different attitudes are a result of an individual's characteristics, personality, values, experience, emotions, etc. Most people have attitudes towards an enormous range of things – the Iraq War, the Labour Party, Premiership football, Beethoven – the list is long; but their attitude to their job is probably the most significant after attitude to their family, as it infects nearly every aspect of their life.

Attitudes comprise cognition, affect, and behaviour (Breckler, 1984). The statement of an opinion is the cognitive component of an attitude, for example 'high rise buildings are of an inhuman scale'. The affective component is the critical part of attitude and reflects emotion and feeling, for example 'I do not respect Peter as an architect as he is interested in designing only high rise commercial buildings'. Affect may lead to behavioural outcomes that represent the behavioural component which refers to an intention to behave in a particular manner, for example 'therefore I will not recommend him to a potential client'. Although categorising attitudes is useful for analysis, in reality cognition, affect and behaviour are difficult to separate, particularly cognition and affect.

2.4 Attitude and behaviour

An interesting aspect of attitude is people's need for consistency between their attitudes and their behaviour (Schleicher et al., 2004) to the extent that they will change what they say so that it does not contradict what they do. An example is an architect who has always been opposed to design-build contracts who finds himself/herself as the principle architect on such a project from when he/she turns into a champion for such contracts. People seek to reconcile divergent attitudes and behaviour in order to appear rational and consistent; nevertheless, it remains common to see many examples of people exercising the 'do as I say not as I do' phenomenon.

Research has shown that attitudes significantly predict future behaviour (Ajzen, 2001), which is what common sense tells us, but nevertheless inconsistencies do occur. But it is possible to say something about the likelihood of behaviour matching attitude and the strength of the match in specific circumstances. Robbins and Judge (2008)[1] refer to these features as 'moderating variables' which they describe (with this author's illustrations relating to construction added) as:

- '*Importance.* Important attitudes reflect fundamental values, self interest or identification with individuals or groups that a person values. Attitudes that individuals consider important tend to show a strong relationship to behaviour.' This is illustrated most strongly in construction by the attitudes

[1] From Robbins and Judge, *Essentials of Organizational Behaviour, 9th Edition.* © 2008, Printed and Electronically reproduced by permission of Pearson Education, Inc., Upper Saddle River, New Jersey.

engendered by the professional allegiance of project team members, often referred to as sentience. For example, generally the attitudes of architects, engineers and contractors differ substantially along an aesthetic–rationality spectrum which in turn affects the way in which they behave in project meetings to such an extent that their behaviour is often predictable.

- '*Specificity.* The more specific the attitude and the more specific the behaviour, the stronger the link between the two.' For instance, asking a judge for an architectural design competition specifically about how they intend to vote is more likely to identify their architectural style preference than if they were asked for their favourite architectural style.
- '*Accessibility.* Attitudes that are easily remembered are more likely to predict behaviour than attitudes that are not accessible in memory. Interestingly you're more likely to remember attitudes that are frequently expressed. So the more you talk about your attitude on a subject, the more you are likely to remember it, and the more likely it is to shape your behaviour.' For example, a project team member who speaks a lot about collaboration between clients, consultants and contractors is likely to propose a partnership arrangement or other formal collaborative arrangement.
- '*Social pressures.* Discrepancies between attitudes and behaviour are more likely to occur when social pressures to behave in certain ways hold exceptional power. This tends to characterise behaviour in organizations.' This can occur in project organizations; for example, one member of, say, the quantity surveying group in the project team does not agree with his colleagues but does not split ranks in the face of opposition from other members of the project team with other professional allegiances.
- '*Direct experience.* The attitude–behaviour relationship is likely to be much stronger if an attitude refers to something with which the individual has direct personal experience.' Asking an architect how he would respond to working for a particularly demanding client who he had not worked with before is less likely to predict actual behaviour than asking the same question of an architect who had worked for the client previously.

The corollary to the idea that attitude can predict future behaviour is the idea that behaviour influences attitude. This is known as *self perception theory* (Bem, 1972) and states that individuals infer their attitude from their past behaviour. Robbins and Judge (2008) express it well:

> 'Self-perception theory, therefore, argues that attitudes are used, after the fact, to make sense out of an action that has already occurred rather than as devices that precede or guide action. When people are asked about their attitudes and they do not have strong convictions or feelings, self-perception theory says they tend to create plausible answers.'

While the traditional attitude-behaviour relationship is generally positive, the behaviour-attitude relationship is just as strong. This is particularly true when attitudes are vague and ambiguous. When you have had few experiences regarding an attitude issue or have given little previous thought to it, you will tend to infer your attitudes from your behaviour. However, when your attitudes have been established for a while and are well defined, those attitudes are likely to guide your behaviour.

A simple illustration could be a structural engineer who, asked about his/her attitude to working in an open plan office, may react by saying that as he/she had always worked in an open plan office, he/she must find it acceptable as he/she could always have moved to another firm with traditional offices. So he/she defined his/her attitude to open plan offices by his/her behaviour.

2.5 Job satisfaction

Those who say they have job satisfaction are expressing an attitude towards their job as a result of evaluating the characteristics of their job against their expectations of the job. Low job satisfaction is the way of saying that they are dissatisfied with their job. The term 'employee attitude' is used interchangeably with 'job satisfaction'. Job satisfaction is the major attitude of importance to commercial organizations. Generally, the facets of job satisfaction are seen to be the actual work undertaken, co-workers, supervision, pay and promotion prospects. Generally enjoying the work itself has the strongest correlation to high levels of overall job satisfaction. However, the pattern of the relationship between job satisfaction and pay has similarities to Maslow's hierarchy of human needs (Maslow, 1954) in that job satisfaction, particularly the work itself, only comes into play when an acceptable level of pay has been reached which allows the employees to attain their basic needs. Below that level, pay is the overriding concern but it has been shown that once individuals reach a pay level which allows them a level of comfortable living (in their terms) there is no relationship between pay and job satisfaction (Judge et al., 2005). Of course, this general statement hides the exceptions at the extremes. There will always be those individuals for whom pay and its associated benefits are the only facet to provide satisfaction and those who are immune to anything other than the joy of doing the job.

An individual's characteristics and consequent personality has a significant effect on his/her job satisfaction. At the extreme ends of personality can be seen the type of person who is easily satisfied compared to the type who is never satisfied no matter how well provided for, but most people will lie somewhere between these extremes of positive and negative personalities.

It is to be expected that job satisfaction of members of the construction professions and industry is generally relatively high compared to many other industries (but this may not apply to construction site workers). The rationale for this assumption lies in the challenges which are implicit in developing and constructing projects of all complexities and sizes. It could be claimed that working on larger, more complex and prestigious projects could be the most satisfying but this may not be the case as such projects often include large tracts of relatively monotonous work as opposed to smaller projects where personnel are involved in a variety of smaller tasks with more immediate impact on the progress and completion of the project. Overall, it seems reasonable to assume that the opportunity for an individual to find employment on a project from which it is possible to achieve a reasonable level of job satisfaction is achievable in normal economic conditions.

This rather sweeping statement perhaps needs modifying relative to the various professions and tasks involved. It could be speculated that architects are the profession with the greatest potential to achieve high levels of job satisfaction. Their education, strongly biased towards design, leads them to expectations that they will become significant designers. However, the opportunity to achieve this is very limited. The majority of projects are not very demanding in terms of aesthetics, although they may be demanding in terms of functional design and detailing. Architects may have difficulty facing this reality and so inhibit their chance of reaching high job satisfaction. Much will depend on their personal characteristics which will determine how readily they can come to terms with the real world.

Comparable scenarios can be generated for the other professions and managers in the industry. But in their cases satisfaction is perhaps more easily achieved through the challenges which can arise on any project, no matter what its size or complexity. The monotony of the large project applies to them as do the frustrations which can occur on any sized project but which is more probable on large complex projects. Countering that is the satisfaction that is achieved by solving complex problems which can only arise on large projects. In creating these scenarios the satisfaction to be derived from working on high profile projects should not be overlooked. Again, much depends on the individual as to whether he/she gets a buzz from being involved in such projects but he/she is hardly likely to have applied to work on such projects unless he/she expects to obtain enhanced satisfaction.

Obviously, pay is as significant an issue in construction as in other industries; however, the industry is specific and the professions even more so. As a result, knowledge about salary levels is widespread so the expectations regarding pay are well established, hence other facets of job satisfaction are likely to feature highly in overall job satisfaction. A further aspect of the construction industry which, whilst not unique to the industry, contributes significantly to job satisfaction is that the industry is fundamentally about projects with little emphasis on general management. The sense of achievement in seeing a project completed contributes substantially to job satisfaction. One would expect, therefore, that project specific staff would be likely to achieve greater job satisfaction if they see a project through to completion than those who work intermittently on a number of projects, although the latter does tend to apply to a large proportion of staff in the industry.

This scenario, based on assumed expectations, may not be borne out in practice and, although little research has been carried out on job satisfaction of the professions and managers associated with construction, that which has been done challenges some of these assumptions. However, in a study essentially aimed at looking at construction managers' health (Love and Edwards, 2005) using the full job strain model (JSM) (see, for example, Karasek, 1979; Fox et al., 1993), the model was found to significantly predict employees' job satisfaction as well as their psychological wellbeing. It was also found that job control, seen as the degree of autonomy project managers have during a project, demonstrated a main effect on job satisfaction. Reference was made to the job satisfaction of engineers and foremen in the Thai construction industry (Ruthankoon and Ogunlana, 2003) in their application of Herzberg's Two-Factor Theory of Motivation (for details see Chapter 5 on motivation). They found that many of the facets of satisfaction occurring generally also featured for Thai engineers and foremen, but this was not the focus of their study. However, they did review the previous literature on motivation of construction personnel but this was solely at the tradespersons and foremen level of employee and did not include professionals and was undertaken between 1974 and 1982. A later study (Taylor, 2002) was also concerned only with tradespersons, specifically early school leavers.

A further reference to job satisfaction in construction is made in Love and Edwards (2005) but in the context of being an element in a study of stress in construction project managers (see Chapter 3). More specific to professionals in construction is Sang et al.'s (2009) examination of the job satisfaction of UK architects and the relationships with work–life balance and turnover intentions. They found that between 20 and 40 percent of respondents to their survey were dissatisfied with their pay, practice management, prospects, working hours and opportunities to use their abilities. This represents a very wide range of topics of dissatisfaction. But, significantly, a large proportion of the dissatisfaction of those sampled was related to organizational issues rather than with the inherent work of architects. They point out that these are similar to the findings of Symes et al. (1995), suggesting that in some ways the experience of architects has changed little since the early 1990s. Not far behind dissatisfaction with pay comes dissatisfaction with the way in which their practices are managed and the resulting organizational issues which has profound implications for architects' work–life balance (which was induced to a large extent by long working hours) and their intention to leave their organization, but relatively few were actually searching for a job outside the profession. The authors recognized limitations to their work due to a small sample size and other issues and look forward to further research on the subject. Employees of a large Australian construction company were also found to suffer work–life balance problems. It was also found that male employees in site-based roles reported significantly higher levels of work to family conflict and emotional exhaustion than male employees who worked in the regional or head office and that site-based employees were less satisfied with their pay (Lingard and Frances, 2004). Whilst this study was not designed to deal with all facets of job satisfaction, it does demonstrate the complexity of the issues of job satisfaction in the construction industry.

A study of the job satisfaction of South African quantity surveyors (Bowen et al., 2008) found that most respondents would choose the same career again. Factors influencing job satisfaction included personal satisfaction in doing the work, low levels of supervision, participation in decision-making, challenging and creative work and recognition of achievements; however, some motivators were not always present in the workplace, in particular those related to participation in decision-making and regular feedback on performance. These few examples of the job satisfaction of professionals associated with construction show how complex an apparantly simple concept actually is in practice. Writers and researchers of job satisfaction in construction do not seem to have focused on the context in which job satisfaction arises, yet, for example, the job satisfaction of South African quantity surveyors is likely to be heavily influenced by what alternative opportunities are available in the South African economy and the level of job satisfaction that might be achieved from them. The idea of job satisfaction in construction is further complicated by the notion of 'project participant satisfaction' (Leung et al., 2004) which is seen as an affective state reached by the individual through attainment of certain goals (successes) which give rise to rewards. Their work found that various management mechanisms rather than particular project goals directly affected participation satisfaction. These were cooperation/participation, conflict and commitment which were found to be strongly correlated to direct participant satisfaction. They claim that the most significant finding is that high commitment attenuates the negative effect of a difficult situation (high task and team conflict) on participant satisfaction. How project participant satisfaction is related to job satisfaction is not made clear. If a participant is satisfied with a project outcome does that mean they have high job satisfaction? Can someone with low job satisfaction be satisfied with the outcome of projects on which they worked?

An important aspect of job satisfaction in construction which has not received the attention it deserves is gender; that is, the differences in job satisfaction between women and men. Research that has been undertaken has considered job satisfaction within the context of a stress perspective (Sang et al., 2007; Loosemore and Waters, 2004) and is considered further in Chapter 3.

2.6 Job satisfaction and performance

A review of a large number of studies into the relationship between job satisfaction and the level of job performance indicates that they are positively correlated but some are not convinced (Wilson, 1999). At the organizational level the relationship continues to be supported as it is found that organizations with more satisfied workers tend to be more effective than organizations with a lower proportion of satisfied employees (Judge et al., 2001; Harter et al., 2002).

Whilst job satisfaction and, hence, job performance are strongly influenced by the facets identified earlier, they are also influenced by less direct forces, referred to as situational factors, that constrain or facilitate the employee's performance (McShane and Von Glinow, 2003). Such factors can be internally or externally generated. Internal factors are those within the control of the organization but beyond the control of the employee such as budgets, physical work facilities and aspects of uncertainty. External facets are generated by the external environment of the organization (cf. Walker, 2007) and comprise aspects such as economic conditions, legal requirement and external political forces. Organization leaders need to be conscious of internal facets and should deal with them so that employees have the best chance of achieving their optimum performance. External factors are not so readily dealt with, so all managers can do is anticipate them and mitigate them to the best of their ability.

Internal factors can be illustrated for construction using the uncertainty induced by a prospective project which is not guaranteed to go ahead as a result of the client threatening to terminate the commission due to poor performance by the firm on a different project for the same client. A lowering of job satisfaction can occur if, during the design and development stage, the people working on a project are not kept informed of the status of the project. Working on a new and exciting project can generate high job satisfaction, but if doubts begin to creep in whether the client will actually go ahead then job satisfaction may decline swiftly. If this is accompanied by a reduction in staffing levels for the project due to a lack of commitment then job satisfaction will decrease further as morale declines. This is a difficult situation for managers to deal with. Perhaps the only contribution they can make to sustaining job satisfaction in such circumstances is to be transparent regarding the project situation and explain the need for consequent manpower reduction.

The effect of external environmental factors on job satisfaction is shown by a downturn in economic conditions relevant to the construction industry due to political forces (for example, a change of government to one which does not favour capital investment) which reduces expenditure on construction. Employees may not be able to see where future projects are coming from. As a result, progress slows on the projects in hand and staff become disgruntled but there is little the firms can do and staff face the prospect of being laid off.

Associated with job satisfaction and performance but much less studied is the concept of *job involvement* which is the extent to which employees identify psychologically with their job and consider it important to their self worth. An associated concept is *psychological empowerment* which is employees' 'beliefs in the degree to which they affect their work environments, their competence, the meaningfulness of their jobs and the perceived autonomy of their work' (Robbins and Judge, 2008). Both are seen to contribute to improved job performance and also to improved organizational citizenship, fewer absences and lower turnover of staff.

Similarly, *organizational commitment* is found to contribute positively to job performance, but not to a high degree (Riketta, 2002), and also to reduced absenteeism and reduced staff turnover. Whilst job involvement is identifying with one's job, organizational commitment means identifying with one's employing organization. Three dimensions of organizational commitment have been identified (Meyer et al., 1993):

- *Affective commitment* which is a belief in the organization's values.
- *Continuance commitment* which is a perceived economic value of remaining with an organization compared to leaving, e.g., being well paid.
- *Normative commitment* which is to remain with an organization for moral or ethical reasons.

It is possible to observe elements of these ideas in the construction industry. Peoples' allegiance to their profession remains strong which leads to strong job involvement from which they derive their self-worth. A downside to this is the tendency to place the self-worth of those in other professions below their own which is counter-productive in an industry which relies overwhelmingly on team work to produce successful projects. Organizational commitment can also be seen as a force in construction but not perhaps to the same extent as job involvement. Affective commitment is represented by the respect which employees hold for their organization. Within the professions and industry there are many well established and admired firms. Continuance commitment is said to be a weaker commitment (Dunham et al., 1994) as it is based only on there being no better alternative employer which can provide an individual with better pay and/or working environment. This can manifest itself strongly in the construction industry from time to time due to its tendency to have significantly fluctuating workloads as a result of economic conditions such that better alternative opportunities for employees are severely limited. Normative commitment can be visualised for employees who are closely involved in key aspects of projects and who are reluctant to leave at a critical time but, of course, this will depend on the sense of responsibility held by the individual employee. This commitment could also be envisaged for people employed by specific organizations to which the individual is committed by moral or ethical ties, such as charities and religious organizations.

Whilst the above illustrations are related to firms, comparable situations occur for projects. For example, Dainty et al. (2005) identify the notion of 'project affinity' which is stated to be 'the affinity of members [of project teams] towards the endeavour embodied within the project goals'. They illustrate the notion through a case study of the construction of a cancer research centre. They found that whilst project affinity was 'clearly important to project success, a series of questions around the structure and culture of the project reveal other conditions necessary to foster such a strong allegiance to project outcomes (e.g., the nature of the client, the challenges inherent in the project and the level of trust and empowerment amongst the project team)'. It can be imagined that affective and normative commitment are substantially enhanced in cases where they apply to both the firm and the projects on which an individual is engaged.

The idea of commitment is in danger of becoming obsolete as new attitudes towards loyalty take hold. Increasingly firms are employing staff on contract terms for specific periods. In the past employees often worked for a small number of employers during their working life, not infrequently for only one. As a result, contracted employees are now seeking jobs which increase their marketability in order to provide them with the flexibility to apply for a wider range of job opportunities. This movement will induce employees to believe that their job satisfaction will be determined to a large extent by the skills and opportunities the job gives them to obtain their next job, which hopefully will be a better one. Employers will find that staff will leave more readily as they complete their contracts, particularly if they have been able to have enhanced their skills, as there are likely to be many more opportunities as staff play a 'game of musical chairs'. Alternatively, employers may offer enhanced contracts which may lead to a rising wages bill and uncertainty regarding contract negotiations which could lead to lower productivity.

The extent to which organizations value the efforts of their staff and care about their wellbeing, known as *perceived organizational support*, improves job satisfaction and hence performance. Organizations are seen to be supportive when rewards are seen to be fair, employees have a voice which is listened to and managers are seen as supportive. The professional level of employees found in construction are invariably listened to as their knowledge and opinion are what they were employed to give; managers are usually from the same background so are instinctively supportive and information about reward levels is readily available. As a result, organizational support is not usually a problem.

A new concept is seen to be *employee engagement* which Robbins and Judge (2008) define as individuals' involvement with, satisfaction with and enthusiasm for the work they do. They identify that it has been found that high-average levels of engagement show enhanced performance and comment that as it is so new we do not yet know how it relates to other concepts. But perhaps this is a case of old wine in new bottles, which is a tendency OB has in a number of areas as it attempts to define the OB discipline more clearly.

2.7 Job satisfaction and organizational citizenship behaviour

Organizational citizenship behaviour (OCB) has been recognized for many years (Barnard, 1938) and refers to behaviour of employees which goes beyond their normal duties and contributes to the organization's effectiveness. It includes such activities as helping colleagues without selfish intent, sharing resources when not required to, responding positively to requests which they would not normally be expected to fulfil and avoiding unnecessary conflict.

It is to be expected that employees with high job satisfaction would be the ones to demonstrate OCB but the relationship has been found to be only modest. What has been shown to be the main driving force is that the employer treats employees fairly by preventing injustices and is fair with such things as pay. Whilst these actions contribute greatly to job satisfaction, their identification gives a more specific finding relative to OCB. OCB can be encouraged by involving employees in decisions that affect them, which can lead to trust in the employer and hence enhanced OCB.

In terms of the professions associated with construction, OCB should generally be well developed. The professionalism which is instilled during training should in most cases ensure that OCB is second nature. However, although professional instincts should lead to high levels of OCB, there will be limits which will be determined by the fairness of the employer as discussed above and the characteristics and personality of the employee. In addition, the standard of professionalism found now may be seen as lower than in the past as more commercial attitudes have taken hold in the professional workplace. A specific study of the behaviour of quantity surveyors as organizational citizens by Liu and Fellows (2008) found 'that collectivism is positively related to OCB, especially the individual's collectivist norms. Hence, a work culture promoting a norm of reciprocity may have a significant influence on the level of OCB, which will enhance work group performance.' Using Organ's (1988) five OCB dimensions of sportsmanship, civic virtue, altruism, courtesy and conscientiousness, it was found that citizenship behaviours, underpinned by norms and expressed through their values and beliefs, were manifest in sportsmanship for values and beliefs; and collectivist norms positively affect civic virtue, sportsmanship, courtesy, conscientiousness. The paper recommends that organizations seek to foster collectivist norms that support OCB through reward systems to enhance work group performance.

2.8 Job satisfaction and client satisfaction

OB books tend to focus to a large extent on service industries where regular, short-time span contact occurs between the service provider's front line employee and the customer such as shop assistants and airline personnel. It appears that customer satisfaction and loyalty is positively correlated to employee satisfaction (Robbins and Judge, 2008). The benefits to the employer are naturally very high. Satisfied employees are less likely to leave so become more experienced leading to greater customer satisfaction. Conversely, dissatisfied customers lead to lower job satisfaction.

A direct parallel exists regarding the relationship between construction industry clients and the professional and industry firms. However the nature of the relationship differs. In the construction industry the relationship is over a much longer period as the design develops and construction is undertaken. This requires relationships to be developed which bring long lasting understanding and respect between the parties. If this can be achieved, satisfaction with the process will accrue to the client representatives and their contemporaries in the project team which will contribute significantly to their job satisfaction. A major problem in this respect in construction is continuity of the personnel representing each side in the process. It is extremely rare that they can remain the same over the long duration of a construction project due to reorganizations within companies leading to staff reallocations and turnover in both client and construction organizations. Nevertheless, these ideas have contributed to the emergence of partnering in construction.

Whilst the way in which construction staff treat clients, and vice versa, is important to the satisfaction of both, the dominant determinant of satisfaction in construction is the degree to which the completed project satisfies the client's needs of aesthetic and functional design; budget; and timely completion. No matter how well the client may have been treated during the design and construction process, it will be as nothing if the project does not live up to expectations. This factor means that job satisfaction differs dramatically in construction as, whilst in the general consumer market many products which do not live up to expectations can be discarded or changed, this is not so in construction. In the general consumer market, even if the product is unsatisfactory, the customer may still return to the company who provided it if they were treated well by the front line staff. If a project team in the construction industry provide an unsatisfactory project, the likelihood of the client returning to those firms who provided the project team is remote.

2.9 Job satisfaction, turnover and absenteeism

A large part of an organization's knowledge, sometimes referred to as its intellectual capital, is contained in its employees' heads. The importance of documenting knowledge cannot be underestimated, but nor can recognition that all knowledge cannot be documented and much of the most important remains in the minds of employees. A major concern of organizations is therefore to keep those employees who are most valuable in this respect on the payroll.

This is particularly the case in construction where it acts in three major ways; firstly are the really long service members of staff who carry valuable knowledge about the company as a whole, about work process, corporate values and culture and long standing clients; secondly is knowledge about techniques and technology which have been used in specific circumstances on previous projects; thirdly is knowledge about current projects which includes the reasons why particular decisions had been taken and the attitudes of the various firms and employees working on the project. The latter is particularly disruptive if lost as one individual can harbour a large amount of project specific information as identified by Eskerod and Blichfeldt (2005) in their study of how project managers should handle such situations. Whilst all industries may claim that knowledge held by employees is vital to them and a number of industries may be more dependent than construction, nevertheless construction must rank as one of the most dependant on employee knowledge. A very pointed quote is use by McShane and Glinow (2003)[2]: 'At 5 p.m., 95% of our assets walk out the door, says an executive at SAS Institute, a leading statistics software firm. We

[2] From McShane and Glinow, *Organizational Behaviour, 2nd Edition*. © 2003, McGraw-Hill. Reproduced by permission of the McGraw-Hill Companies.

have to have an environment that makes them want to walk back in the door next morning.' They continue by saying that over half of the 500 executives recently surveyed worldwide identified retaining talented employees as the top people issue in the company.

People seek alternative employment due to a lack of job satisfaction but factors such as market conditions, availability of alternative job opportunities and length of tenure in their present job act to limit their opportunity to move. It has been found that job dissatisfaction has a greater effect on turnover than incentives which attract employees to new jobs so the main problem is if current jobs do not motivate good employees to stay (McShane and Von Glinow, 2003). It has also been found that the level of satisfaction is less important in predicting turnover for high performers. This is because organizations make great efforts to retain them through such things as pay rises, praise, recognition and promotions. So even with lower levels of job satisfaction, high performers are likely to stay with the organization. The opposite applies to poor performers as organizations make no effort to retain them. In view of key employees being vital to the prosperity of firms in construction, these findings stress that professional and construction industry firms need to consider carefully the level of satisfaction and rewards (in the broadest sense) of their key people.

Robbins and Judge (2008) identify that, as to be expected there is negative correlation between job satisfaction and absenteeism, but that it is only moderate to weak. Contributing to this view was a finding that generous sick leave arrangements encourage all employees no matter what their level of satisfaction to take time off. Countering this is a study done at Sears, Roebuck (Smith, 1977) which showed that, in a comparison between their Chicago and New York offices for one day when there was a freak snowstorm in Chicago, absenteeism in New York was just as high for satisfied workers and dissatisfied workers but for Chicago workers with high satisfaction had a much higher attendance than those with lower satisfaction. That is, the higher satisfaction workers were more prepared to battle to work through the snowstorm, which reinforces the view that there is negative correlation between job satisfaction and absenteeism.

2.10 Personality

Psychologists would describe personality in terms such as 'those properties of behaviour which are both enduring and set the individual apart from others. These properties concern the individual's typical ways of coping with life' (Huczynski, 2004). Personality theory deals with behaviour patterns that are consistent over time and unique to the individual. Organizational behaviourists are concerned with how personality shapes our behaviour and how it affects the way in which an individual reacts and interacts with others. Many employers believe that personality is related to how someone performs at work and the success they make of their career, hence personality is often used as part of selection processes.

There is a continuing argument amongst psychologists about whether personality is inherited, hence determined by genetics, or whether it is created by environmental, cultural and social factors. This is known as the nature – nurture debate. These days extreme positions are not generally taken, both sets of factors are seen to be involved in determining an individual's personality, but debate continues about the weight to be given to each and their interdependency.

The analysis of personality categorises people into personality types who possess common behaviour patterns, e.g., intuitive types, thinking types and personality traits; personal characteristics which are relatively constant, influence behaviour and are exhibited in a wide range of ways, e.g., shyness, aggressiveness, moodiness, laziness, agreeableness, trustworthiness. Individuals are placed into personality types but traits are characteristics of individuals. Psychology researchers are concerned with complex issues of how traits combine to form personality types.

The Big Five Model of personality has achieved broad acceptance as a descriptive system of trait clusters (Judge et al., 2002). The extremes illustrated by its five basic categories describe the most significant variations in personality:

Extroversion
Extrovert; outgoing, warm, positive, sociable
Introvert; quiet, reserved, shy

Agreeableness
Adapter; straightforward, cooperative, trusting, sympathetic
Challenger; quarrelsome, antagonistic, cold

Openness
Explorer: creative, curious, open-minded
Preserver; unimaginative, close-minded, disinterested

Conscientious
Focused; self-disciplined, responsible, organized, dutiful, dependable
Flexible; unreliable, irresponsible, disorganized

Emotional stability
Resilient; calm, self-confident, secure, contented.
Reactive; anxious, depressed, insecure, self-conscious.

Huczynski and Buchanan (2007) point out that, from a psychologist's perspective, the Big Five are not personality types but are 'super traits'.

Obviously these categories represent extremes on a spectrum on which an individual's characteristics will lie; that is each individual will lie at a specific point on each spectrum and very few will lie at an extreme point of any category.

The personality of an individual will therefore be a combination of the points at which the person lies for each of the categories. On any kind of refined measure, the diversity of individuals in terms of their personality is enormous. As Moorhead and Griffin (2001) state:

> The similarities can be as important as the differences. After all, none of us are exactly alike. We may be similar but never the same. Thus, it is important to note that although two employees may have the same gender, ethnicity, and even university education, they are different employees who may act differently and react differently to various management styles.

Alongside this idea Fincham and Rhodes (2005),[3] who's numerical analyses examine personality and intelligence, state that:

> ... although we are all unique we are also reasonably consistent; the fundamentals of character and intelligence are formed in early life and remain much the same throughout adult life (Costa and McCrae, 1997) ... Even in the shorter term, social life would be unimaginable if this were not the case – i.e., if our personalities changed dramatically day to day, for example, one day conforming but next being individualistic with little regard for our obligations to others.

Their ideas have enabled them, as psychologists, to statistically analyse personality and intelligence and present a significant commentary on the 'Big Five' whilst also pointing out that 'many researchers believe personality is the outcome of a process too dynamic to be captured by any statistical technique.' They also point out that the best known of these dynamic theorists is Freud and that his work in the role of emotion in the workplace and their relevance in understanding 'less-rational' elements of workplace behaviour is receiving renewed interest.

Huczynski and Buchanan (2007) identify trait clusters which seem to suit a range of occupations among which those related to construction include; interestingly, yet not unexpectedly, Openness (Explorer) was useful for architects, whereas Openness (Preserver) was useful for finance managers (which could be taken as akin to quantity surveyors) and project managers although emotional stability (Resilience) was also seen to be useful for finance managers (QSs) and engineers. However, these views appear to be an over-simplification as personalities do not usually comprise one category but are complex amalgams as shown by finance managers (akin to QSs) being suggested for two categories, which can also be seen to apply to project managers. Specific construction professions require a range of attributes which will rely on a range of characteristics, as no doubt will other occupations other than construction. In construction for example, to be successful,

[3] From Fincham and Rhodes, *Principles of Organizational Behaviour, 4th Edition*. © 2005, Oxford University Press. Reproduced by permission of Oxford University Press.

architects will need a range of characteristics in addition to openness and project managers will be well served by having open characteristics. In this respect the findings regarding the general application of some factors are useful when taken in conjunction with occupational specific factors in matching factors to occupations.

Quoting Robertson (2001), they argue that the relationship between personality and performance is not straightforward. In particular:

- only two of the big five personality factors, conscientiousness and emotional stability, are consistently associated with better performance in most occupations;
- conscientiousness is a better predictor of work performance than emotional stability;
- although openness, agreeableness and extroversion are not universally important, any of the big five personality factors could be significant in certain occupations;
- the correlations that have been found between personality factors and job performance are not strong, as performance is also affected by a range of other factors.

The only big five personality factor that predicts organizational citizenship behaviour strongly and consistently is conscientiousness (Podsakoff et al., 2000).

The resulting sense of all this is that personality as it impacts on the work one does is extremely complex. This is particularly the case in construction as the contributing professions require a wide range personality attributes to deal with both the range of skills demanded by the job and the complexity of relationships which construction generates. One is also left with an over-riding sense that whilst organizational behaviourists understand behaviour they do not have as great an understanding of the complexity of how organizations work and hence how behaviour affects performance as the measurement of performance in complex industries is extremely problematic.

Type A	Type B
Competitive	*Can enjoy leisure*
Need for achievement	*No high need for achievement*
Aggressive	*Easy-going*
Works fast	*Works steadily*
Impatient	*Rarely impatient*
Restless	*Calm*
Tense	*Relaxed*
Under pressure	*Has enough time*

Type A and Type B Personalities are well embedded in the psyche of businessmen and women. They were identified by Friedman and Rosenman (1974) as two extreme modes of behaviour which explained different stress levels in individuals which were linked to health. Type A personalities were three times more likely to suffer heart disease than Type B. Their work translated into the characteristics of each type in work situations:

It can be seen that Type A and B are at the extreme ends of a spectrum in which Type As are driven to succeed and have proactive personalities which leads them to seek to make progress (in their terms) and create change in their environments in spite of constraints and obstacles. They are often admired in their workplace, particularly in the North American culture, as they are associated with ambition, motivation and the acquisition of wealth. Type B is entirely opposite with none of the drive of a Type A. Type As are also more likely to be appointed as they tend to do better in job interviews as they are seen to have desirable traits (maybe the same as the interviewers!). They are more likely to be seen as leaders and to achieve success in their careers. But there is a downside to employees with a Type A personality. They may not be able to judge an issue dispassionately or analyse it in a detached manner. Their traits may increase the tension with those with whom they work to the extent that organizations become dysfunctional as a result of their attitudes. They are more likely to manipulate situations to their own benefit to achieve career success at the expense of organizational effectiveness. The actions of proactive personalities can be for good or bad, depending on the organizational situation but a problem is that, if they are bad, colleagues with Type B personalities are unlikely to do anything about it. It will take another Type A to mount a challenge and that could create even greater chaos.

A consolation is that few people are placed at the extremes of the spectrum so most organizations manage to steer a relatively stable course. It does not appear to be known whether the personality characteristics of Types A and B attach themselves to particular occupations, although it could be suggested that Type A personalities predominate amongst dealers in the financial markets and also perhaps developers. Both types are to be found in construction organizations, although conventional wisdom would hold that Type As are more likely to be found in construction firms and Type B in professional organizations, careful examination is likely to reveal that the types are more evenly distributed across all organizations. The situation in which the personality types will have a large and potentially serious detrimental effect is in the many interdisciplinary teams necessary to successfully design and construct buildings. Dominance by Type A personalities in team situations can distort decisions to the detriment of the effectiveness of the completed project.

Other personality attributes influencing organizational behaviour are given by Robbins and Judge (2008):

Core self-evaluation, which can be positive or negative. People with the positive form like themselves, believe they are valuable, have self confidence and feel that they are in control of their environment. Those with the negative form are, of course, the opposite. People with the negative form require approval from others and are likely to follow the beliefs and behaviour of those they respect. The extent to which people believe they are the masters of their own fate, known as their 'locus of control' is an indicator of core self-evaluation. People who believe they control what happens to them will have positive core self-evaluation and those who believe that it is controlled by outside forces will be negative and will lack confidence (Rotter, 1966). Those with positive core self-evaluation achieve greater job satisfaction as they see more challenges in their jobs. They set more ambitious goals and remain committed to them and believe that positive outcomes result from their own actions. But there is a downside. Overconfidence leading to a self-perceived infallibility can lead to bad decisions, although more often people are more likely to sell themselves short. This attribute can be characterised as optimism which has been studied by Dolfi and Andrews (2007) in respect to project managers in general. They found that 'optimism in project managers is an important attribute as only 7 percent of "optimists" in the quantitative survey rated their work environment as negative while 60 percent of "pessimists" rated their work environment negatively.' They also concluded that 'optimism can be learned, even in an unfavourable work environment.' Once more this idea is represented by the extreme ends of a spectrum and most of us will lie somewhere along it.

Machiavellianism is a term used to describe the idea that the end justifies the means not matter how unethical the means may be. It is suggested that people with Machiavellian characteristics are pragmatic, cold, distant, manipulative and persuasive. Their performance is curtailed by situational factors, particularly where ethical and behavioural standards are present.

Narcissism describes a person with an inflated sense of their own importance and entitlement, accompanied by arrogance. Narcissists tend to be selfish and exploitive and have the attitude that others exist for their benefit. They are not seen by their bosses as effective, do not help others and believe that they are better leaders than their superiors think they are.

Self monitoring is the ability to be adaptable in adjusting behaviour to external situational factors. High self monitors can behave differently in different situations. Such high levels of adaptability are seen to be beneficial in organizational settings, particularly in leadership roles. They conform to what is expected of them, are mobile in their careers, receive more promotions and are more likely to occupy central positions in an organization. Low self monitors are consistent in their behaviour; what you see is what you get.

Risk taking differs considerably between people and in different situations. A person's attitude to risk is a powerful behavioural phenomenon which needs matching to their job as the propensity to take or avoid risk affects how long it will take a person to make a decision and how much information they need before making a decision. Whilst the majority of jobs require people to be risk averse (e.g., engineers who adopt large safety factors in their calculations), for some the ability to accept risk is essential (e.g., a city trader).

Personality and its impact on organizational effectiveness can be seen to be complex and not yet fully understood so remains difficult to use in formal analytical studies. Nevertheless, valuable insights are possible using the ideas put forward by organization behaviouralists.

2.11 Personality assessment

Arising from psychology is a wide range of methods for assessing and measuring aptitude, intelligence and personality, referred to as psychometrics. Many are directed at specific occupations with the intention of identifying the most suitable candidates for employment, promotion, redeployment, training etc. They are generally used to complement less formal and more subjective methods. They have been criticized as unfair and misleading and hence poor predictors of performance, nevertheless they continue to be used by reputable companies.

This section deals solely with personality testing, not the many tests for specialist purposes such as intelligence and aptitude testing. There are two major approaches – nomothetic and idiographic – the former being the one used for most contemporary psychometrics which are based on self-reporting questionnaires whereas the idiographic approach uses open-ended questioning. The nomothetic approach is concerned with the study of traits and how they cluster to form personality types, whereas the idiographic approach is designed to capture an individual's unique characteristics. Nomothetic approaches look for universal laws of behaviour and depend on identifying the main dimensions based on which personalities vary. It assumes that an individual's personality can be measured and compared with that of others. This means that an individual's performance is related to the norm of the group being considered such that those with characteristics which deviate from the norm can be identified. The nomothetic approach has been criticized by those who follow the idiographic method which aims to capture the uniqueness, richness and complexity of the individual. The idiographic approach deepens understanding but does not readily lead to generation of universal laws of behaviour. The two approaches also differ fundamentally in that the nomothetic approach believes that personality is primarily determined by heredity, biology and genetics whereas the idiographic approach believes that it is primarily determined by social and cultural processes. Psychologists continue to debate the merits of the approaches without resolution and suggest that they should be treated as complementary rather than as options.

Nevertheless, modern employee assessment and selection mostly use nomothetic methods, the most widely used of which is probably *The Myers-Brigg Type Indicator (MBTI)* (Quenk, 2000). This instrument relies on a 100-question personality test which leads to rating personal preferences on four scales. As a result individuals are classified:

- *Extroverted(E) versus Introverted(I)* – Extroverted individuals are expansive, sociable and talkative. Introverts are quiet and shy and cautious.

- *Sensing(S) versus Intuitive(N)* – Sensing types are practical, focus on detail and acquire information, preferably factual and quantitative, systematically. Intuitives acquire information non-systematically through unconscious processes and look at the big picture.
- *Thinking(T) versus Feeling(F)* – Thinking types use rational processes to approach problems in a detached manner. Feeling types rely on their personal values and emotions and take into account the feeling of others.
- *Judging(J) versus Perceiving(P)* – Judging types want control and prefer the world to be ordered and structured. Perceiving types are flexible and spontaneous.

Individuals are assigned to one end or other of each category which establishes 16 personality types. Each end category is known by its capital letter except Intuitive which is known by N to distinguish it from Introverted. Examples of types are that ISFJs are seen to have a high sense of duty, ENTJs are seen as natural leaders and ESTJs as organizers and business managers but they only indicate preferences not established modes of behaviour. Trompenaars and Wooliams (2002) make the point that the profiles do not necessarily mean that the individuals are trapped within those categories. Some people may be able to shift categories as the need arises due to environmental conditions. Just because we favour practical approaches (S) does not mean we cannot be imaginative (N) in the right circumstances. MBTI may be popular, particularly with management consultants, but questions have been raised about its validity as a measure of personality. According to Robbins and Judge (2008) the evidence appears to be against MBTI being a valid measure of personality (cf. Gardner and Martinko, 1996; Bess and Harvey, 2002; Capraro and Capraro, 2002) using similar arguments to those raised above that it forces a person to be categorized as either one type or another. They agree with McShane and Von Glinow (2003) who believe that 'Overall, the MBTI seems to improve self-awareness for career development and mutual understanding, but probably should not be used in selecting job applicants.'

MBTI has been used in research into construction related topics. It was used by Carr et al. (2002) in examining the relationship between personality traits and performance of members of the engineering and architectural professions providing project design services. They define project design services in the context of this research in four project service categories; conceptual design, contract documents, construction administration and firm management duties. The research found that those with a preference for Intuition (N) supported by a preference for Perceiving (P) develop 'an optimal pattern of personality for optimal individual performance in the project planning phase [conceptual design] of the designer's service'. The same personality traits as for conceptual design (N and I) were also found to be connected with the greatest improvement in performance within the construction administration phase due to 'the way one is open to alternatives and the way options are thoroughly explored' and the likely need for immediate attention in a crisis. Professionals with a personality favouring Judging (J) outperformed in duties associated with the contract documents phase. It is stated that 'the preparation of detailed contract documents needs professionals who are "decisive, not curious" ... These descriptions are those of a person whose personality preference is for Judging (J).' The fourth project service category is general management of the firm for which there was no significant performance differences between different personality preferences. Also, the Thinking(T)/Feeling(F) dimension did not influence the performance of any of the four categories which the paper states is contrary to predictions. This is certainly surprising in the light of the paper's description of the Thinking(T)/Feeling(F) dimension: 'Some people typically draw conclusions or make judgements objectively, dispassionately, and analytically: others weigh human factors or societal import and make judgement with the personal conviction as to their value' and the recognition that the design and construction of projects, particularly in the construction administration category, are strongly characterized by decisions processes.

MBTI has also been used by Culp and Smith (2001), management consultants, in a study designed to examine how individuals with different personalities approach engineering projects by comparing the profile of engineering team members with the profile of the US population at large. Engineers were found to have a comfortably higher proportion of introverts (62%), thinkers (75%) and judgers (67%) but only slightly higher proportion of sensors (54%) than the general population. As engineering team members have a frequent preference for introversion, building relationships with other team members and with clients was often seen to be a challenge. Combinations of preferences were significant in understanding behaviours. ISTJ was the most preferred type for engineers (nearly 25% of all engineers). Engineers with a TJ preference comprised 53% of all engineers, 2.2 times the percentage of the general population. Engineers with an FP preference comprised only 11% of all engineers. Individuals with a TJ preference look for quick closures of issues using logic and rationalisation whilst those with FP preferences keep decision-making process open and consider the impact of decisions on the people involved. As a result it is recommended that engineering teams take into account the effect of their decisions on others, their interaction with other team members and their clients. It is interesting that neither this nor the previous study made reference to the criticisms which have been levelled at MBTI and their implications for the studies.

Psychometric tests are widely used by large employers for point of entry selection with usage estimated at 75% despite them being controversial but are much lower for the professions and small employers at about 16% (Fincham and Rhodes, 2005). The use of personality testing in the construction industry is centred on their use by recruitment consultancies employed by construction associated firms which offer the facility as part of their service. The take up by construction-related companies directly is not widespread as they rely on the traditional process of application, references and interview. In light of the relatively low predictive validity of personality assessments their reluctance may be well founded as the predictive value is no higher than other less expensive techniques with only a small number achieving more than 50% success (Anderson and Shackleton, 1993). Nevertheless, the trend appears to be that as construction-related firms increase in size, they include personality assessment as part of their recruitment policy. Psychometric testing should be incorporated as only one of a range of indicators when assessing the suitability of candidates particularly as the interview has undergone a remarkable rehabilitation in the research literature as a predictor at least as good as some more expensive methods.

2.12 Values

A person's values describes their standards or beliefs, hence personality and values are not the same. People with similar personalities may have completely different value systems; conversely people with similar values may have dissimilar personalities. That is not to say that a person's values do not influence their personality, one would not expect that a person who values inner harmony to be antagonistic. Nevertheless values and personality are not consistently correlated. On the other hand, when put in a corner, people with similar values may ultimately express a common view irrespective of their personalities but may state their position in entirely different ways as a result of their personalities.

A person's values represent what that person judges to be appropriate standards and beliefs such as self-respect, equality, family, courage, honesty, and the relative importance assigned to them. Thus each of someone's standards and beliefs comprise that which is seen to be important to them and their intensity (how important it is to them).

The Rokeach Value Survey (RVS) (Rokeach, 1973) classified 18 values for each of which is given desirable end-states (terminal values) and the means of achieving the terminal values (instrumental values). Some examples are:

Ambitious (hardworking, aspiring) leads to *A comfortable life* (a prosperous life).
Broad- minded (open-minded) leads to *An exciting life* (a stimulating active life).
Forgiving (willing to pardon others) leads to *Family security* (taking care of loved ones).
Imaginative (daring, creative) leads to *Inner harmony* (freedom from inner conflict).
Polite (courteous, well-mannered) leads to *Social recognition* (respect, admiration).

Whilst it can be expected that people in similar occupations or categories may hold similar values, there can be considerable overlap. The significance for the construction industry is that when different groups have to negotiate, different values can create conflict, for example, if a lack of clarity of the objectives of a project occurs between the client and the project team as well as within the client and project team themselves, then differences in values between and within these groups will compound the lack of unity.

An interesting perspective on values was provided by Robbins and Judge (2008) who integrated several recent analyses of work values (cf. Lancaster and

Stillman, 2002) to show that values change over generations. Their findings, whilst acknowledging limitations, show some interesting changes in values since the end of the Second World War to 2008. There has been a shift progressively over this period from employees being conservative, conforming and loyal to their organization immediately after the war, to loyalty to career and dislike of authority, to loyalty to relationships, concern for work/life balance and dislike of rules and finally to loyalty to themselves and relationships, self reliance and focus on financial success. This has been accompanied by a greater affinity to team-working which has been a characteristic of the way in which the management of projects has developed over the same period. As can be seen these changes have been accompanied by a reducing respect for authority which in construction has seen many changes in the way in which projects are organized, typified by a reduction in the authority of the architect which was paramount on projects in the 50s and 60s so that the architect is now seen more as a professional contributor (albeit a most important one) on many projects rather than the figurehead.

People's values form an important part of their attitudes and are likely to have a significant effect on their job satisfaction. People's attitudes to their job often depends on whether the job provides the things they value. Different people may value any of life/work balance, financial security, respect, freedom, friendship, etc., to different degrees. Given sufficient employment opportunities people's values will generally determine the kind of job they aspire to.

The term 'organizational values' is often seen to reflect the reference in chapter one to the term organizational behaviour being shorthand for how people behave within an organizational context; as stated previously there is actually no such study as behaviour of an organization, only a study of the behaviour of individuals in organizations. Similarly it can be claimed that there is no such thing as organizational values only a collection of the values held by the individuals who make up the organization. What are claimed to be organizational values are likely to be values generated by senior management which take little or no account of employees' values. Interestingly, Zhang et al. (2008) undertook a case study of a UK construction company in which they sought to establish organizational values through a bottom-up approach by establishing employees' value profiles and allowing staff representatives to develop their own organizational values statement independent of senior management. They claim that the approach presented offers practical guidance on how to reveal employees' personal values and hence formulate collective organizational values statements that can facilitate the alignment of individual and organizational values and hence be an effective way for shared values to emerge, evolve and enter into the corporate conscience.

2.13 Personality, values and the construction professions

The practical use of OB is to better understand how people behave at work. Hopefully this will result in the right people being assigned to the tasks appropriate to them. This begins by the selection of an appropriate career and appointment of people to the right company. Robbins and Judge (2008) state: 'Because managers today are less interested in an applicants ability to perform a *specific* job than with the *flexibility* to meet challenging situations and commitment to the organization, in recent years managers have become interested in determining how well an employee's personality *and* values match the *organization*.' This may be the case in general business organizations but it is not necessarily the case in construction. Though these characteristics are generally valuable for most occupations, in construction the ability to meet the technical and creative demands of the various professional skills is paramount. Given that, then flexibility and matching values and personality do come into play. This will also be the case for all highly skilled professions, such as the medical and legal professions.

Nevertheless, Holland's Personality – Job Fit Theory (Holland, 1997) is interesting in considering the issue of matching people's personality types to professional occupations in construction. Holland proposes six personality types; realistic; investigative; social; conventional; enterprising; artistic. On the face of it, there is not much difficulty in allocating the construction professions to the personality types. Engineers (of all shades) fit the 'investigative' category, so do construction managers, quantity surveyors would find the 'conventional' category recognizable, architects the 'artistic' one and project managers can be conceived as belonging in the 'enterprising' category, particularly the clients' project managers. But it is not as simple as that. To take such an approach would be to stereotype. Tempting though this is, it would not a true representation of the many personality types which successfully occupy the wide range of skill-related jobs which make up the construction industry. Even so, whilst the professional institutions take care in their recruitment literature not to stereotype their members, they do refer (albeit in passing) to attributes which are desirable in the case of their profession, for example the architects refer to the ability to sketch (or to acquire the ability) and the structural engineers refer to the need to be practical in order to find an engineering solution that fits with the architect's vision for a structure. Significantly, what they all stress is the ability to work in a team.

The reality is that the construction industry professions require their members to undertake a wide range of activities of which their stereotyped image plays only part, albeit usually a significant one. For example, although architects are seen to require an artistic personality, how much of their working lives is spent being 'artistic'? Much of an architect's career is spent on technical and administrative functions for which a predominantly artistic personality is not entirely appropriate. Similarly although engineers are essentially technologists, great imagination has been needed to conceive and design the world's great structures. Construction managers need to be inventive as well as practical and effective as do quantity surveyors if they are to contribute to innovative designs within financial constraints.

Whilst stereotyping is a natural phenomena by which over-simple judgements are made and may not be seen as significant, stereotyping does need treating with suspicion as it can be a two edged sword. On the one hand people may seek out and exaggerate characteristics in others to confirm their own stereotyped view of them. On the other, people may play up to the stereotyped view others can be expected to have of them as that is how they wish to be perceived. Each case can be a way of making a person feel superior. In the case of the stereotyped person, playing up to an image they think complementary and, in the case of the stereotyper, playing up the stereotype when they believe it denigrates the stereotyped (Fineman et al., 2005).

Categorizing clients is fraught with difficulty. Clients are not unitary but will be represented by a number of people each of whom will have their own personality. Whether they will gel into a team with a consistent approach or will demonstrate internal conflict will depend on how well the client's team has been chosen and generally is out of the hands of the project team. If the person or unit within the client organization which selects the client's team is familiar with personality aspects of OB then perhaps a cohesive group can be formed, otherwise it will be left to chance and the construction team will be left to make the best of it.

If this scenario is considered against The Big Five Model it is clear that the extremes used to illustrate the five personality types are useful in a simplistic way but create as many questions as answers. For example if the personality of engineers is considered against the 'openness' category it can be argued that they need to be at both ends of the spectrum and that does not mean in the middle. On the one hand they need to be creative, curious and open minded to invent novel solutions and also unimaginative and close-minded to undertake the conventional analysis need to justify a novel solution. Similar situations often occur for all the professions in construction. The need for individuals to operate at both ends of the 'agreeableness' category is also necessary in construction. For example, a quantity surveyor will need to be an 'adaptor', that is straightforward, trusting etc when dealing with a client but perhaps a 'challenger', that is quarrelsome, antagonistic etc. when in negotiation with an intransigent subcontractor. On the other hand the categories of 'extroversion, conscientious and emotional stability' are overarching characteristics which are not related to specific tasks.

Similarly, Type A and B personalities comprise over-arching characteristics not directly related to specific professional activities but which can greatly affect the impact an individual professional's contribution has on the project as a whole. A Type A personality is likely to try to impose his will on his colleagues which may, in a project context, be to the detriment or to the benefit of the project as a whole and in a company context, to the company. The problem being that Type A personalities' proposals may not be considered objectively because they either do not draft them objectively or because they do not present them persuasively. In drafting them they may have their own agenda which may not fully meet the needs of the project. In presenting them to the client or project team members their over-powering personality may create an adverse reaction in those they are seeking to persuade. If a project team or firm comprises a number of Type A personalities there is a high likelihood of conflict but fortunately people are normally positioned on the spectrum between A and B which should dampen the extent of any conflict.

If we consider construction against the classification of the Myers-Brigg Type Indicator we see similar complications. The professions, in carrying out their professional work, would be required to be both 'sensing and intuitive', both 'thinking and feeling' and both 'judging and perceiving' on different occasions. This is extremely demanding and not within the capabilities of most people. Each member will be naturally located at a point on each spectrum and not readily able to adapt to a different position. So a balance of characteristics between project team members is needed but these issues are not usually considered when selecting team members. 'Extroverted and introverted' is not included here as it is not task specific.

What this means is that the issues of personality - job fit in construction are not simple. Robbins and Judge (2008) states that 'satisfaction is highest and turnover lowest when personality and occupation are in agreement. Social individuals should be in social jobs, conventional people in conventional jobs, and so forth.' But in construction, jobs are not one-dimensional as assumed here but multi-faceted so a simple correlation to a single personality type is unrealistic. This is particularly the case for project managers who have the broadest remit requiring the widest range of understanding of the skills used in producing projects and so need the widest range of understanding of the personalities of the contributors who they are required to coordinate. Whist the predictive power of personality-job fit ideas may not be totally relevant in construction, the ideas of personality which underpin them are valuable in assessing the different and complex individuals which inhabit construction projects and their associated companies to understand better their interactions with their colleagues.

The other fit of importance is the person–organization fit. The OB literature tends to emphasise that it is more important that an employee's personality fits with the organization's culture rather than with the characteristics of any specific job as employees are required to be able to readily change tasks and teams. This requirement implies that employees are generalists rather than specialists. Again, this is not the case for construction where professional skills predominate. The specificity of the work of the professions in construction means that attempting to match personalities to organizational culture will be constrained as they will already be matched to their profession to a large extent. However, it is of great benefit for employees to feel comfortable with their employing company so matching peoples' values to organizations' cultures, if this is possible, is likely to be productive. It has been found that a fit of employees' values with the culture of their organizations predicts job satisfaction, commitment and low turnover (Verquer et al., 2003).

Values strongly affect a person's attitudes and perceptions and whilst not directly affecting behaviour will determine whether a person fits into an organization's culture. This is likely to be the case in construction. For example an architect who values imagination, intellectual activity, broadmindedness and independence will probably be unhappy working for a development company and a quantity surveyor who values ambition, capability, courage and forgiveness and independence would probably be happy moving to a project management company. So a company in the construction industry may be well advised to concentrate on candidates' values when recruiting staff.

A rare example of research on values in the construction industry is a study by Kadefors (2005) who examined fairness in inter-organizational project relations. She believes that there is a strong preference for fairness in human interactions so that people who experience unfairness tend to react with anger, resentment and loss of motivation. Consequently the opportunism present when a contractor is appointed by competitive tender on a lowest price basis with uncertainty in aspects post-contract, it is hard to develop shared perceptions of fairness. This she contends leads to conflicting fairness values between contractor and client and that 'fixed price construction contract procurement seems to be a case where fairness concerns, rather than facilitate cooperative interaction, tend to produce conflicts and hinder an efficient negotiation outcome'. However, it should be pointed out that many in the construction industry recognize that if properly prepared, controlled and executed this type of contract has been shown to lead to successful outcomes for both client and contractor and so that their fairness values reflect this experience. Hence, an individual's attitude to and perception to the value of fairness in competitive lowest bid construction contracts will help to determine whether a person fits into an organization which undertakes mainly this type of contract.

2.14 Perception

Scientific enquiry and knowledge is generally accepted as being objective (although not always). However when it comes to the world around us each of us perceives it in a different way. We interpret what we see in our own way and our behaviour is based on our perception of what is real, not on reality itself. Other people behave in ways we find difficult to understand. It would be of great benefit to be able to see things in the way that others do, particularly in business organization settings. To do so we need to be able to discover how another perceives the context and their place in it. If we can discover that we will find their behaviour more readily understandable. We need to know why we perceived things differently in the first place.

Psychologists link sensation and perception together in what they call 'top down' and 'bottom up' processing in which the former is the way in which raw data is received and screened to filter out redundant and less relevant information so that we focus on that which we consider important and the latter is the process by which we order and interpret this information to make sense of our environment. This process results in individual perceptions so that we do not respond to the world as it really is; each of us has our own view of it. Huczynski and Buchanan (2007) sum up these ideas as:

> The 'real world' as a concept is not a useful starting point for developing an understanding of human behaviour in general or organisational behaviour in particular. We behave in, and respond to the world as we perceive it. We each live in our own perceptual world.
>
> Successful interpersonal relationships depend on some overlap between our perceptual worlds, or we would never be able to understand each other. Our perceptual worlds, however, are in a detailed analysis unique, which makes life interesting, can also gives us problems.'

Our sensory apparatus receives enormous amounts of information some of which comes from within the body such as pain and tiredness which are classified as internal factors and other information from the environment within which we live such as political events, economic conditions and the people with whom we interact, known as external factors. We sift and organize and interpret this information to make sense of it in a process of *selective attention* and *perceptual organization*. The factors which affect selective attention are classified as external and internal factors. *External factors* comprise *stimulus factors* e.g., large rather than small, bright rather than dark, loud rather than quiet, moving rather than stationary (to which our attention is drawn to the first named of each pair) and *context factors* which are the situation in which something occurs which gives meaning to the occurrence. *Internal factors* comprise *learning, personality and motivation* which contribute to expectations which impact on and in part determine our selective attention. Learning is our past experience which leads us to pay attention to specific stimuli, personality is our personality traits which lead us to pay attention to some things but not others and motivation refers to those stimuli we find motivating.

In developing our perceptions we categorise such things as issues, events, people, houses and all manner of things. We create these categories as we grow; we are not born knowing them. Our perceptions do not constitute reality and are known as social constructs. We select information that fits our expectations and play down information which does not. This categorisation process and the meaning and pattern which develops is known as perceptual organization and our readiness to respond to certain stimuli is known as an individual's perceptual set and our unique view of what is out there is known as our perceptual world (Huczynski and Buchanan, 2007).

In order to understand an individual's behaviour we need to know what constitutes their perceptual world and the influences, both informational and cultural, which have created it. To change an individual's behaviour requires a change in their perceptions which is extremely difficult to achieve due to a lack of mutual understanding. As we in turn also find it difficult to develop an understanding of our own perceptual world, communication blockages occur which, in turn, can reinforce both of our perceptual worlds and so changing an individual's behaviour becomes increasingly difficult.

Our perceptions about peoples' actions are strongly influenced by whether we believe they were under their personal control, i.e., internal. *Attribution theory* helps to show whether the cause of an individual's behaviour is either *internally* or *externally* caused. Internal behaviours are under the control of the individual, external behaviours result from outside causes which force the person into the behaviour; a common example is of an employee being late for work which could be caused by oversleeping due to excesses the evening before (internal attribution) or due to their bus breaking down (external attribution). Whether we decide that someone's behaviour is externally or internally caused is largely determined by distinctiveness, consensus or consistency (Robbins and Judge, 2008). *Distinctiveness* is whether an individual displays different behaviour in different situations. For example, if an employee avoids meetings with clients by arranging substitutes to cover for him/her and also does not turn up for other formal meetings such as site meetings, the employee's behaviour will be judged to be internal. If the employee misses a meeting with the client to the surprise of other members of the meeting, the employee's behaviour will be judged to be externally attributed. *Consensus* is when everyone facing a situation behaves in the same way, for example all quantity surveyors in a practice complete their interim valuations forms properly. If one employee did not comply on a regular basis, the cause would be seen as internal. If, alternatively, everyone regularly made the same mistake it would be seen as an external cause. *Consistency* is when a person responds the same way to the same situation over time. For example a project team member who frequently seems to antagonise the team chairperson. Consistent behaviour is seen to be attributed to internal causes.

Our understanding of our social world is based on our attempts at *causal analysis* which allocates causes to either external or internal causes. As developing a construction project is a team activity we are continually observing our colleagues in their work and making judgements of their success or failure as well as our own. Certain patterns have emerged. When we seek to explain the behaviour of other members of our company or project team we tend to attribute it to their personality which we exaggerate whilst overlooking the effect of external factors. For example, concerning a project team member (not from your company) who is on edge, not concentrating and hence not contributing to project team discussions, you may decide the cause is because his personality does not enable him to force his way into the discussion (an internal cause) when in fact he is distracted having heard prior to the meeting that a client has been declared bankrupt and will not pay a large outstanding account (an external cause). Conversely, we tend to explain a person's success or promotion in terms of luck (being in the right place at the right time) or their contacts (both external causes) rather than their knowledge and ability (internal). However, when it comes to ourselves, in accounting for our success we tend to point to our capabilities and when explaining our failures we project blame onto external causes that are beyond our control. This is known as *projection*. Projection also refers to the tendency to attribute one's own characteristics to other people which distorts perceptions made about other people and, hence, to see people as more similar than they really are.

The Halo Effect is an application of the idea of selective attention to our perception of people and is the habit of deciding on first meeting a person the kind of person they are based on a single striking characteristic, such as our perception of their intelligence, appearance, or posture. We are selective with regard to the information to which we pay attention and they do the same to us. So first impressions do count. We also tend to give more favourable judgements to people who have characteristics similar to our own. Subsequently we are inhibited from correcting our first impressions as more reliable information becomes available. This is dangerous ground in business, as when did a single characteristic noticed on a first meeting determine a person's ability to design a building or bridge, manage a project or control construction costs? The halo effect can be a serious negative force in construction projects as teams form and re-form and new people are introduced frequently, particularly as they are often from different companies and are expected to collaborate on complex high value projects. The opportunity to modify first impressions can be very limited unless positive efforts are made to arrange less formal meetings than high pressure project meetings. The problem is most acute on international projects in which culture intervenes but also occurs on national projects with members drawn nationwide.

Stereotyping is the way in which we group together people who we expect to share similar characteristics. It is very prevalent in people outside the construction industry when perceiving the characteristics on those in the industry. Perhaps surprisingly, it is also the case with people within the industry perceiving those within the industry. Stereotyping is an error at the perceptual organization stage in the process of perception (Huczynski and Buchanan, 2007) so we create simple mental images of groups and their characteristics. The word accountant or nurse immediately conjures up mental images to which we attribute personality traits. Such images are reinforced by the media, particularly novels and films (McShane and Von Glinow, 2003).

Construction is beset by stereotypes. It is seen on the one hand as a macho industry at the site and construction firm level, both for the construction worker and the construction manager, very much inhabited by rough diamonds who are prepared to use brute strength to produce the project. Such a view is commonly held outside the industry irrespective of the massively improved technology and sophistication of the industry. Thankfully these views are not now generally held within the industry but there still remains to some extent a mistrust of contractors. An interesting account on how a change in stereotyping of builders may occur is given by Moore (2001). He believes that an opportunity to rehabilitate the image of builders has occurred due to a generation of children having an expectation of builders behaviour framed in terms of positive characteristics as a result of watching the children's TV programme 'Bob the Builder'. He believes that as a result children have formed and maintain a positive stereotype of builders but that the industry does not appear to have taken advantage of this opportunity. Architects are stereotyped as sensitive, sophisticated, artistic souls who have vision and can satisfy the client's needs with elegantly design solutions. This exaggeration is perhaps not too far from the mark amongst the general public and authors of pulp fiction novels but is not the reality of everyday construction activity. Nevertheless, there is a residual perception in the industry that architects are often not firmly entrenched in the industry which can make them seem impractical, unreliable and difficult to pin down. Architects themselves tend to feel misunderstood. None of these images are true reflections; they are over-generalizations. The reason for the need to generalize is that the variety of individual characteristics in the groups being described is too complex to explain as the range of perceptions which can be held about architects is as wide as the number of people having perceptions. This is due to the breadth of architects' functions, spanning design, administration, project management and financial matters (although few architects will undertake all these on any one project). These activities (except for design) are not reflected in the design-related stereotype image of architects which tends to arise from their artistic- and design-related training and from the work of a few world famous design architects. Most architectural students enter their course expecting to emerge as fledgling top designers soon to be in great demand. This is the image picked up by the general public and tends to reflect the way in which architects wish to be seen.

Structural engineers are stereotyped as serious, pragmatic, single-minded, mathematically orientated achievers. This gives them gravitas which they exploit in creating their image. There is a constancy of their image both outside the construction industry and within it. Particularly within the industry they are often perceived to be cautious and lacking in imagination. They are not seen to be as complex as architects.

Services engineers are probably hardly recognized outside the industry. Within it they are perceived to have some of the qualities of structural engineers but without the same reputation for reliability. Civil engineers when lead consultant on infrastructure projects are perceived both within the industry and outside it as steadfast individuals with the characteristics of structural engineers but additionally with a broader ability across project types and importantly with managerial competences arising from their organizational abilities. Quantity surveyors are perceived by designers as cautious, conservative, reliable control freaks who have the potential to inhibit architectural adventure. By contractors they are perceived as argumentative and nit-picking as they stick to the letter of the contract. They have no image of note outside the industry except, perhaps, as falling into the same category as accountants.

Whilst the stereotype descriptions are greatly exaggerated to make the point, they do reflect perceptions held both within and without the construction industry. Huczynski and Buchanan (2007) state 'Stereotypes are overgeneralisations, and are bound to be radically inaccurate on occasion. But they can be convenient. By adopting a stereotype perspective, we may be able to shortcut our evaluation process, and make quicker and more reliable predictions of behaviour'. This view may be acceptable in non-business settings but stereotyping can also be dangerous. None of the stereotypes given above can describe adequately an individual person. The stereotype is overlain by the personality of the individual and the experience obtained in work. So, for example, an architect who has worked mostly on detail design and project administration may demonstrate none of the characteristics of one who has made a reputation in design. Their differences will also be reinforced by their different personalities. So if other professional colleagues deal with an individual from a perspective of their stereotype, their relationship is unlikely to be fruitful. As relationships are so intense and the stakes so high on many construction projects and the integration of the professional contribution so vital, it is important that members of project teams overcome the temptation to stereotype their project team members, or for that matter, judge members on first meeting (halo effect). Awareness of the way in which the phenomena of perception can influence their view of their colleagues in their firm and particularly their project teams in terms of the extent to which they will heed advice given by their colleagues is significant to success in a project's outcome.

An area explored by Singh (2002) in connection with engineers involved in construction is hemisphericity which is the role and dominance of right brain/left brain, each of which has varying roles and characteristics leading to different thinking orientations. The field of hemisphericity states that left-brained people are generally analytical, scientific, methodical, linear, timely, verbal and logical whilst right-brained people are essentially holistic, visual, intuitive, psychic, instantaneous and artistic. His study of engineers found that construction engineers were predominantly left-brained and design engineers were predominantly right-brained. He concluded that hemisphericity explained many differences in types of behaviour between construction and design engineers:

> 'For instance, the desire for change amongst construction engineers here may be explained by their left-brain orientation; the perceived lack of appropriate vision of senior construction engineers may be explained by their predominant left-brain orientation; the errors in design drawings may be explained by the right-brain orientation of design engineers; the ineffective coordination of costs, claims, and schedules by construction engineers may be explained by the left-brained orientation [whilst on the face of it counter-intuitive, this is explained as due to the left hemispheric orientation having the propensity to pay less attention to tasks requiring integrative and holistic thinking]; the discord between design and construction engineers at a group level can perhaps be explained by their reverse orientations. At the same time, the apparent lack of conflict between the individual engineers in design and construction branches may be explained by their hemisphericities as individuals of opposite hemisphericities make good friends and partners. If the hemisphericities were more evenly distributed at 50 – 50, many of the adverse behaviours that are observed may not be existent … Among other conclusions, left-brain construction engineers were unable to value self-actualization principles – a largely right-brain activity. Left-brained [construction] engineers enjoy their work thoroughly – largely a left-brained work – while right-brained design engineers did not enjoy their work as much.'

It is possible to imagine these conclusions extrapolated to a wider construction field than solely engineers. Differences in predominant brain orientation are likely to be found between architects, planners, quantity surveyors and other professions involved in building and engineering works which can explain some of the differences in perceptions encountered. They will manifest themselves predominantly in project teams which would appear to need a balance of right and left- brained members if they are to produce a balanced outcome, or, for example, a preponderance of left-brained members if cost and construction engineering is of paramount importance.

2.15 Creativity

Moorhead and Griffin (2001) identify creativity as an important component of individual behaviour, which has great resonance for the construction industry. They raise the questions: What makes a person creative? How do people become creative? How does the creative process work? They comment that psychologists have not yet completely answered these questions but identify the common attributes of creative individuals as: background experience, personal traits and cognitive abilities.

Background experience of being raised in environments which nurtured creativity is cited as being evident in many creative individuals but there is evidence of famously creative people coming from opposite backgrounds and with very limited education.

The personality traits linked to creativity have been found to be 'openness; an attraction to complexity; high levels of energy, independence, and autonomy; strong self confidence; and a strong belief that one is, if fact, creative'.

Cognitive abilities are generally seen as necessary for creativity but not all intelligent people are necessarily creative. Creativity benefits from the ability to think both divergently and convergently, so seeing both differences and similarities between situations, phenomena and events.

Creativity is a behavioural outcome which in construction is most clearly seen in the work of architects but is also necessary, albeit in a different form, in other occupations in construction. Engineers need to be creative in their solutions to architects' visions, quantity surveyors in their cost advice and procurement strategies and project managers in exercising their organizational skills.

2.16 The significance of individual behaviour for construction

Attitudes, personality, values and perception are fundamental to understanding ethics, motivation, communication, politics and all the other topics covered from hereon even though the former themselves are not yet fully understood. But before proceeding it is worth examining the extent to which they are recognized as fundamental to understanding OB in construction related firms and construction projects. An interesting pattern emerges in which relationships figure large in the construction literature but the factors which underpin relationships are acknowledged but not generally developed in depth.

Nicolini (2002) comes up with the notion of 'project chemistry' as a way of deepening our understanding of the relational dimensions and factors that may lead to successful project performance. He draws upon Akintoye et al. (2000) who believe that new initiatives in construction 'highlight the importance of social, human and cultural factors in the management of construction organizations and projects.' Many other publications start from the same platform. Soetanto and Proverbs (2002) refer to promotion of 'harmonious relationships' between contractors and clients who are likely to have different personality profiles and perceptions. Similar recognition of the importance of harmonious relationships arises frequently in connections with partnering, for example, Bresnen and Marshall (2000a) recognize 'that collaborative approaches did not necessarily remove conflict at source' and 'the difficulties of accurately judging likely future behaviour in the context of a selection process should not be underestimated' and 'people and relationships were considered to be the heart of the problem'. In a further example (Wong et al., 2000) confirmed that 'integrity' and 'demonstrating concern' were fundamental to trust in three public sector infrastructure project management agencies in Singapore. Whilst not focusing specifically on individual behaviour these publications do recognize the significance of psychological factors. Also, whilst individual characteristics were grouped into organizational characteristics, Wong proposed further refinement to cater for a finer level of analysis of personal characteristics.

An interesting paper by Ankrah and Langford (2005) is a comparative study of the organizational cultures of architects and contractors. Although the researchers were essentially concerned with organizational cultures, it found that 'conflicts are likely to occur within the project coalition at the interface level where human interaction elements occur and this could detract from achieving objectives'. Awareness of these differences, however, improves the chances of achieving the right balance when constructing the team and this could lead to the development of synergy and good 'project chemistry' with positive consequences for overall project performance'. This paper sets the context for understanding the behaviour of project participants without venturing directly into the field of OB. The phenomena of only working at a high level of abstraction is also illustrated by Barrett and Stanley (1999) who states that twenty examples of failure in the project briefing process were due to 'human nature' which they see as critical and requiring considering in more detail, which would require analysis of the behaviour of individuals working on the brief.

Relationships in construction obviously take place between people who bring along all their complex individuality to the relationship. The relationships which are formed are founded in individual behaviour. Relationships are the symptoms; individual behaviour is the source. So whether the many relationships which are demanded by the complex organizational arrangement required by construction projects actually work effectively is largely down to the compatibility of the individuals locked into the organizational arrangements in terms of their attitudes, personalities, perspectives and resulting behaviour. Whilst not consciously undertaken (although maybe informally or unconsciously), there may be occasions where it is possible to match members of project teams according to the compatibility of their characteristics in order to enhance the quality of the output of teams. This may be particularly so in cases where firms have worked together over many years and have a good knowledge of their personnel and their performance in project teams. Whilst matching individuals for compatibility in these terms may, generally, be impractical, nevertheless recognition and understanding of these possible causes of acrimony and ineffectiveness can help significantly in ameliorating the worst effects of incompatible characteristics. It may even be possible in some cases to counsel incompatible staff who are expected to work together effectively or to move staff to other assignments in order to overcome difficult relationships.

In research terms, whilst the difficulties and complexities of researching at the level of the individual in a construction setting are enormous, it is nevertheless important for researchers to recognize that it is at this level that the relative success or failure of a firm or a project may lie. The level of integration through team-working demanded by construction places a high premium on collaboration which is essentially about the attitudes, characteristics and perception which each individual brings to their joint work. Currently this is recognized implicitly rather than explicitly by researchers and practitioners but more detailed analysis could lead to better understanding of organizational issues associated with success or failure of project teams and construction related firms.

3 Emotions, Feelings and Stress

3.1 Rationality

Traditionally, management thinking has been based on the idea that business organizations are rational and dispassionate. This strongly held view arose out of the early 'scientific management' school of management which perceived workers as the human equivalent of machines. Whilst this approach to management is now seen as historic (but perhaps still practised to a larger extent than the OB literature is inclined to admit) and contemporary organizational behaviour perspectives have taken over, it is only in recent years that the field of OB has explicitly recognized the part played by emotions in organizational life (Ashkanasy et al., 2002). Even though the field of OB may be incorporating the effect of emotions into organization theory, in practice many organizations cling to the myth of rationality (Putnam and Mumby, 1993), yet 'emotions permeate organizational life' (McShane and Von Glinow, 2003).

That is not to say that emotions are now seen as having free reign in organizations; it is still very much the case that they are generally suppressed in most business situations but are now recognized as bubbling away below the surface and likely to emerge at any time. People cannot ditch their emotions; they bring them to work every day. Organizations need to account for the process and interaction of emotions in the workplace as feelings connect us with our realities and provide internal feedback on how we are doing, what we want and what we might do next (Wilson, 1999). Work occupies the largest part of most peoples' lives so we are bound to react emotionally to events which occur at work whether we show our emotions or not. We come to work in a particular mood that may or may not change during the day and which determines the way we interact with colleagues and clients, how we perform in meetings and other interpersonal situations. Even when suppressed, our moods in such situations can have a significant effect on the synergy required to make progress on the development of construction projects which is invariably achieved through team efforts.

Organizational Behaviour in Construction, First Edition. Anthony Walker.

3.2 Emotions and moods

Affect is a term that encompasses both *emotions* and *moods* which are:

Emotions

- Caused by something specific and directed towards someone or something
- A specific experience such as anger, fear, sadness, happiness, love, surprise
- Usually generate facial expressions
- Create a readiness to take action
- Last for short periods (seconds or minutes).

Moods

- Cause is usually not directed at anything in particular
- Usually last for hours or days
- Can be positive (good moods) or negative (bad moods)
- Do not necessarily generate distinct facial expressions
- Cognitive in nature.

Emotions can lead to moods if you lose focus on what triggered the emotion, leaving you in, for example, a sad mood or happy mood depending on the event and the emotion. An illustration is the joy (emotion) of getting a promotion leading to you being in a good mood for days. The converse also applies, for example, being in a bad mood could lead to an outburst at a colleague over something which would normally generate a mild reaction.

A term used in conjunction with emotions is feelings. Fineman et al. (2005) say that the terms are often used interchangeably but that it is useful to describe them differently. They see feelings as 'essentially private, "internal", experiences which often have both psychological and physical manifestations, such as the stomach churning and sense of apprehension before a job interview. They provide an essential, personal readout on how we are doing, how we are relating to the world as we try to deal with it. We also have feelings about feelings, such as being angry at feeling upset, or anxious about our infatuation with someone. Emotions are the outward presentation of our feelings through learned social codes'.

People vary in the intensity of their emotional responses which can be the result of their personality or their job requirements. In terms of personality, one of the trait clusters of the Big Five Model of personality was 'emotional stability' which contrasted a calm vs. an unstable personality; similarly, the 'agreeableness' category contrasted a cooperative against a quarrelsome personality. There are three aspects to this link: personality may predispose some people to certain moods and emotions such as a greater propensity to feel guilt or anger, or to remain calm and relaxed no matter what the provocation; some people may experience certain emotions and moods more frequently than others; some people may experience any emotion more intensely than others. There does not appear to have been a body of research on these links. McShane and Von Glinow (2003) make reference to the links between emotions and personality. They describe positive affectivity, which is the inclination to be in positive emotional states, and link it to the characteristic extroversion. They contrast it to the opposite, negative affectivity, which is a tendency to be in negative emotional states. Most evidence suggests that the influence of personality on emotions is relatively weak and although it does influence emotions and attitudes in the workplace, its effect is not as strong as situational factors.

A complementary aspect is discussed by Moorhead and Griffin (2001) who point out that it was once believed that a person's emotions and moods could vary from day to day but that it is now believed that there is a tendency for a person's emotions and moods to have stability and some predictability even though short-term fluctuations may occur. This reflects positive affinity through the recognition that some people have the tendency to be optimistic and upbeat relative to others and generally view things positively. At the other end of the spectrum is negative affectivity, which represents people who are generally pessimistic and negative in outlook. Very few of us will be at one extreme or the other but will lie somewhere between.

Members of construction project teams, professional and production firms associated with construction and members of client bodies comprise people from a wide range of personalities and emotional/mood profiles. Most of the time they will be required to suppress their feelings (which is examined later) and this review helps us to understand the differences in pressure experienced by the different members of project teams, firms and clients as a result of their personalities/emotional profiles as they attempt to suppress their feelings in the interests of collaboration. This leads us to the emotional response required as a result of a person's job requirements. Examples are given of surgeons and air traffic controllers who are required to be calm and controlled in all circumstances and evangelist preachers and lawyers who need to alter their emotional intensity according to the situation. Construction professionals will generally hope to keep their emotions under control as they pursue logical/scientific solutions to problems in team situations; however, there may be times when an outburst of emotion does move the project forward when a logjam of conflicting opinions emerges. According to Robbins and Judge (2008), it has been shown that emotions are critical to rational thinking because our emotions provide important information about how we understand the world around us. They also comment that more research has been undertaken on the sources of moods than on emotions which, in summary and in general, include:

- People are in their worst moods at the beginning of the week and their best later in the week and at weekends. During the day moods tend to improve, then decline in the evening.
- Weather has little effect on moods.

- For most people social activities increase positive mood. People in positive moods seek out social interaction and social interactions cause people to be in a good mood.
- Sleep quality affects mood negatively. Poor sleep quality impairs decision making and makes it difficult to control emotions.
- Exercise enhances peoples' positive mood, particularly among those who are depressed. The effect is consistent but not very strong
- Negative emotions seem to occur less as people get older. Periods of highly positive moods last longer for older individuals and bad moods fade more quickly.

3.3 Disguising emotions

It is generally thought that well run business organizations expect to be free of emotional reactions by staff. In construction this is particularly the case for professional firms and many client bodies. But expecting freedom from emotional display does not neutralise emotions, it simply means that staff are expected to suppress them and this is what happens to a large extent in construction organizations. Much of the literature on this issue cites service industries where the issue is even more severe – not only do staff have to suppress their emotions but frequently have to show false emotions as in 'the customer is always right'.

This pretence is known as *emotional labour* and is very stressful. The disparity is known as *emotional dissonance*. Robbins and Judge (2008) point out that 'it can take a heavy toll on employees. Left untreated, bottled up feelings of frustration, anger and resentment can eventually lead to emotional exhaustion and burnout'. Emotional labour occurs in two ways. One is 'surface acting' which is the type described above and which is pretending to feel what we do not feel. The other is 'deep acting' which is deceiving oneself as well as others (Wilson, 1999). Wilson also points out that jobs requiring important amounts of emotional labour are dominated by women and she draws on significant literature regarding flight attendants. However, it is more likely to be rather more evenly distributed between men and women in construction as both men and women are more often placed in similar situations, although women have the additional pressure of being in a world dominated by men.

Keeping calm, giving thoughtful, measured responses and controlling one's emotions is ingrained in the construction industry through both the expectation and training of the professions and also by the macho image of the industry itself. The high incidence of the need for team working requires many diverse personalities to cooperate in developing a project and if they all gave free reign to their emotions chaos would no doubt ensue. Nevertheless, great strain is associated with holding emotions in check and arguing rationally with one's fellow professionals, particularly when they are impinging inappropriately on one's professional area, leading to stress. The demand for emotional labour is extremely high in meetings with clients who can be extremely frustrating in the demands they place on project teams and leaders of professional firms and no doubt clients believe the reverse is also the case.

However, this is not to say that, upon returning to the office after a frustrating project or client meeting, a professional will not let off steam to a close colleague. Such an emotional outlet can be very important in relieving stress. This could only be done in the presence of a close colleague to ensure it would go no further in order to avoid affecting one's image within the firm, and would need to be on an issue with which the colleague could sympathise. For issues within the firm, such as a failed promotion or a disagreement with a colleague, an emotional display would be seen as inappropriate and a professional demeanour would be necessary. An emotional outlet would have to await a sympathetic ear outside the firm, at home or even alone when a good tantrum could suffice!

An interesting aspect in this connection is *emotional contagion*, which is the automatic and unconscious tendency to mimic and synchronise our nonverbal behaviour with those of other people (McShane and Von Glinow, 2003). This is an act of empathising with the feelings of others by, for example, appearing happy when someone describes a happy event, or showing disappointment when they are disappointed. Emotional contagion bonds people together and creates social solidarity which can be valuable in team building both within a firm and within a project, for example, when a client is being particularly difficult.

3.4 Emotional stereotyping and emotional cultures

Emotional stereotyping occurs in most organizations, the commonest being stereotyping according to sex. Wilson (1999) encapsulates the issue: 'Popular myth says that women are too neurotic to be able to cope with public positions of responsibility and are more suited to "instinctive" roles like caring for the young or infirm, or in supportive, helping, administrative roles.' In most organizations men are generally expected to show no emotion and to appear to behave as rational, logical, decisive thinkers detached from their feelings, even though many may not subscribe to such a profile. Women, on the other hand, are stereotyped as more emotional than men and likely to show their emotions much more readily. In male-dominated organizations women are frequently not trusted to act according to male expectations if they are put in senior management positions. So whilst both sexes are expected to disguise their feelings to reflect organizational expectations, women are usually stereotyped as emotional creatures not suitable for top management positions because they do not subscribe to the profile expected of top managers as defined by men. Although the position may have improved in recent years, it remains convenient for men to believe that they are more temperamentally suited to jobs which happen to have the most power, pay and prestige in our society (Fincham and Rhodes, 2005).

It has been said that some women in senior positions have responded by trying to act like men are expected to act and Margaret Thatcher is often cited as an example. But the emotional stereotypes of men and of women are just that – stereotypes which are at opposite ends of a spectrum. Men and women have male and female sides in varying proportions. Some men are emotional creatures just as some women are not; although it is generally believed that overall women are the more emotional, such profiling can be seen as a device for maintaining male dominance. In the current climate it seems that both sexes are required to disguise their emotions but we have seen that emotions can be valuable in decision making so the conventional view of organizations as emotion free may not be helpful to their effectiveness.

Organizations have informal boundaries of what is acceptable and unacceptable where showing emotion is concerned. This is their emotional culture which can break down into a number of sub-cultures in different parts of the organization. Readily expressing emotion may be entirely acceptable and encouraged in creative situations, e.g., software development and in high-pressure situations, such as on the floor of the Stock Exchange. Paradoxically, men dominate the latter, yet the perceived wisdom is that it is women not men who are more emotionally inclined. Conversely, the majority of organizational cultures expect emotions to be suppressed, e.g., accountancy, engineering and medicine. Again paradoxically, the medical professions include many women as doctors, nurses, physiotherapists, etc. who are very able at suppressing their emotions. The issue of stereotyping and emotional sub-cultures is far from clear and seems to be based more on protecting privileged positions than on emotional suitability. Emotional culture is just one specific element of culture generally which is covered in Chapter 8.

Practically nothing has appeared in the literature about the emotional cultures of construction organizations. Nevertheless, it can probably be accepted that design professionals (architects and interior designers) are more likely to accept open expressions of their emotions in organizational settings than are the more traditionally staid engineering, surveying and construction organizations. Having said this, contractors may be driven to expressing strong emotions in certain adversarial situations as also may clients if they are not receiving the service they expect (although this view again smacks of stereotyping).

Similarly, there is little written about the emotional implications for women in the construction industry. What has been written is predominantly about the number of women in construction which shows that women are significantly under-represented in the workforce, both professionally and in skilled construction and related trades. Of those that are employed, about a half is in administrative and secretarial positions which reflects the stereotyping of women generally (Briscoe, 2005). The papers that do touch on emotional issues have the main theme of making a plea for more women in construction. De Graft-Johnson et al. (2005) point to the stereotyping of women as a result of the macho culture of the industry, which includes inappropriate attitudes, treatment and sexist behaviour, resulting in restrictions in their opportunities to develop technical skills and their lack of opportunity to work on site and be involved in negotiations and dialogue with contractors. Whittock (2002) touches more directly on the emotional issues to be dealt with by women in construction. She found that women simply had to adapt to the male environment, for example '... in attempting to gain acceptance from men, it was invariably left to token women to make the "first move"'. Citing a female partner of a construction firm attending a course in a situation where all her co-attendees were men, she found that 'Accommodating to men's feelings in this way, putting them at their ease, sometimes at the expense of their own feelings, was common to women in both the training and employment environments'. So women had to suppress their emotions and behave like men. Interestingly, De Graft-Johnson et al. (2005) identified what they called the 'queen bee' syndrome which is 'characterised by women who have achieved relatively important positions failing to support or in some cases actively obstructing other women from moving up the career ladder'. This behaviour included bullying and other hostile behaviour, all of which is likely to produce an emotional response (or its suppression).

3.5 Emotional Intelligence

Supporters of the idea of emotional intelligence (EI) claim that people who are in touch with their own emotions will be more able to readily sense the emotions of others which will enable them to interact in social situations and so be more effective in their work. EI comprises:

- Self-awareness of one's emotions
- Self-management of one's emotions
- Self-motivation in the face of setbacks
- Empathy with others' emotions
- Social skills in managing the emotions of others.

So the perceived wisdom is that people who possess these attributes are likely to succeed in a business setting as a result of their ability to effectively manage the emotions of others with whom they interact, as long as they also have the requisite level of cognitive intelligence. It has even been suggested that EI is more important than technical skills or rational intelligence (Goleman, 1995) but this is hardly likely to be the case in construction with its high demands for professional competence. If anything, it can do no more than reinforce the contribution of an individual professional and not necessarily to the overall benefit of a project if the advice induces bias in the final outcome by the persuasion of others through their EI. This point reflects much of the criticisms levelled at the idea of EI.

Robbins and Judge (2008) provide the pros and cons. They point to the intuitive appeal of EI in which we recognize that people who are good in social situations are popular, that there is some evidence that high levels of EI mean that a person will perform well in their job, and that there is some biologically-based evidence in support of EI. Against this they say that EI can be seen as too vague a concept, hence the difficulty of agreeing a definition. The question is raised whether EI is really a form of intelligence and asks whether it can be measured. There has been criticism of the exaggerated claims made for the novelty and power of EI; it is claimed that EI is actually old wine in new bottles (Woodruffe, 2001) as, if you control for personality and cognitive intelligence, there is nothing left for EI. The argument that EI is derived from personality traits such as extroversion, agreeableness, emotional stability and openness to experience (McShane and Von Glinow, 2003), and not additional to them, is particularly strong. Woodruffe (2001) concludes that 'emotional intelligence is neither a new nor a useful concept'. Fincham and Rhodes (2005) say that EI has been popularised by management consultants as an instant fix solution to complex organizational problems and has been the subject of training programmes offered by those consultants to aid the development of peoples' EI. They doubt whether this can be realised and whether a person's EI can be changed. They also point out that EI does not always mean being sympathetic to others' emotions; there are situations such as in the military when it would be emotionally intelligent to be tough and ruthless. Such situations are also likely to occur at critical times in business organizations, particularly construction when, for example, on a project of national importance, it is vital that a task on the critical path is completed on time and even more so when undertaking emergency construction work following a national disaster. The idea has been proposed that not only individuals but also groups have EI and this idea is referred to in Chapter 9.

Some work on EI related to construction has been undertaken. In a study in the USA, Butler and Chinowsky (2005) found a relationship between EI and transformational leadership behaviours (see Chapter 11) in construction executives, which is perhaps unsurprising as the nature of both components can be said to come from the same root, nor did the paper refer to the mainstream management doubts about the acceptance of the ideas of both EI and transformational leadership. In a similar vein but related to project managers, Clarke (2010) examined the relationship of EI to key project manager competences and to transformational leadership using a sample of project mangers from a variety of industries. They make the point that emotional associated abilities in project management was recognized over 30 years ago which reflects the earlier point that it can be seen as a reinvention of earlier ideas. But they continue by saying that project managers with high EI should be better equipped because for projects trust and commitment need to be established quickly; knowledge exchange would be supported; ambiguity and change are present; and conflict is likely. However, EI was found only to be associated with project management competences of teamwork and managing conflict and that personality differences seem to be far greater predictors for these two competences. Although the third competence of attentiveness (building strong relationships) showed no significant relationship with EI, it is said that this could be due to EI being mediated by the variable empathy.

Surprisingly, no significant relationship was found between the fourth competence – communication. The study also showed a relationship between EI and transformational leadership in two of the transformational leadership dimensions. A further study, this time carried out in Thailand (Sunindijo et al., 2007), also examined the relationship between EI and leadership by interviewing project managers and engineers to determine the benefits to project management. They found that EI affected the leadership behaviour of the project leader such that project managers and engineers with higher EI used open communication and proactive behaviour. As in the previous papers, this paper did not raise issues about the acceptance of EI and these doubts are reflected in the outcome which may have been achieved without the need to incorporate EI.

3.6 Stress

Stress is now recognized as a widespread phenomenon in society, arising not only from work but also from the pace and pressure of life in general, not least from family life. At one time it was seen as the preserve of the executive and managerial classes but is now recognized as affecting people in all walks of life. In the case of lower occupational levels it arises, in particular, from factors such as repetitive work, shift working, and fatigue. Wilson (1999) points out that stress is increasingly being recognized and treated but nevertheless it is still frequently seen by organizations as an everyday problem that the individual is expected to deal with (Kunda, 1982) and that being able to handle oneself in stressful situations is seen as the mark of a professional. This is the case in the construction professions and industry, particularly in the light of its essentially macho profile. Stress does not arise solely from employment but can equally arise from the pressures of life in general and from specific events such as moving house, getting married and conflict with friends.

Stress can have serious medical consequences and has been linked with heart disease, strokes, headaches, backache and stomach and intestinal disorders. As well as the trauma of personal distress, there is a huge cost to the economy. Whilst we are concerned here with stress in relation to behaviour in organizations, it is not possible to separate the consequences of stress arising from the workplace from the sources of stress from wherever they arise. Nevertheless, we will limit ourselves to how stress arises in the workplace but will deal with the consequences from wherever it arises.

Fincham and Rhodes (2005) present a valuable discourse on the meaning of stress, about which there is considerable academic debate, and point out that there are three distinct 'stress literatures' which exist in parallel but generally ignore each other: the social and psychological effects of stressors (the focus in this chapter); an ergonomics literature which looks at the impact of stressors on skilled performance, e.g., man–machine interfaces; and a neuroscience approach. Pinning down what we mean by stress for the purpose of this chapter is not easy but Moorhead and Griffin (2001) provide a useful account:

> Stress has been defined in many ways, but most definitions say that stress is caused by a stimulus, that the stimulus can be either physical or psychological, and that the individual responds to the stimulus in some way. Here, then we define stress as a person's adaptive response to a stimulus that places excessive psychological or physical demands on him or her.
>
> Given the underlying complexities of this definition, we need to examine its components carefully. First is the notion of adaptation … people may adapt to stressful circumstances in any of several ways. Second is the role of the stimulus. This stimulus, generally called a stressor, is anything that induces stress. Third, stressors can be either psychological or physical. Finally the demands the stressor place on the individual must be excessive for stress to result. Of course, what is excessive for one person may be perfectly tolerable for another. The point is simply that a person must perceive the demands as excessive or stress will not result.

By far the major understanding of stress is negative, that is, associated with work overload, insecurity and the increasing pace of life, which is referred to as distress. But there is positive stress, known as eustress, which is stress in moderation, sufficient to activate and motivate people to achieve things and successfully meet challenges. As McShane and Von Glinow (2003) say, 'we need some stress to survive'.

A pioneer in stress research, Hans Selye, discovered the biological and physiological pathways of the stress response (Selye, 1936). He devised the *general adaptation syndrome* which shows that everyone has a normal level of resistance to stressful events. Some people can tolerate stress to a level far higher than others but everyone has a threshold at which stress begins to affect them. The first stage of the syndrome is the stage of alarm in which an organizm recognizes that its environment is threatening. In people, this is shown through the tensing of muscles and increasing rates of respiration, heartbeat and blood pressure. It cannot remain in this state for long. If the threat is too powerful the organizm dies. If it can survive, the organizm enters the stage of resistance in which it gathers resources to resist the threat. People activate biochemical, psychological and behavioural mechanisms which, for example, lead to higher levels of adrenaline. If the demand for resources to resist the threat continues for too long these adaptive resources are worn out and the stage of exhaustion is reached. At this stage people become exhausted and have stress-induced illnesses such as headaches, heart and stomach problems and psychological damage. Fortunately, in most work situations tensions are resolved before the worst effects of stress occur. Coping strategies may be employed where the worker seeks aid to remove or reduce the source of the stress, such as recruiting a co-worker to assist and share the pressure, or they may be able to withdraw from the stressful situation, either permanently or temporarily, so that they can recharge their batteries. To try to avoid stressful situations it is necessary to manage work-related stress by understanding and dealing with potentially stress-inducing events in their early stages.

3.7 Causes of stress

3.7.1 Work stressors

We now recognize stress as widespread in society in a way that previously we did not.

The pace of work and of living appears to have increased over the years generating, it is claimed, increasingly stressful situations or perhaps it could be said that people have become less resistant to stress; for example, compare today with the period before, during and after World War II. Nevertheless, it is important to recognize the causes of stress to better avoid them and develop approaches to stress management. Various surveys seem to show that construction has some of the most stressful jobs. In 1996 the *Wall Street Journal* placed architects as the 25th most stressful job in the USA, in 1997 construction was placed as the 12th most stressful in the UK by Professor Cary Cooper of UMIST on Channel 4's website and in 2008 engineering (generally) was ranked 3rd by tiscali's website. Against this, in 2008 in the USA, *Health Magazine* was showing civil engineering as the 7th least stressful job. So information is far from complete and not entirely consistent but, nevertheless, construction features in some guise, which is not surprising given the complexity of projects, the complexity of the organizations needed to deliver them and the emphasis on deadlines.

At the simplest but nevertheless important level is the *physical work environment* where stressors can include issues such as excessive noise, poor lighting and poor air quality. For construction staff and workers on site these are very real stressors, which seriously affect the capacity to concentrate. They can to an extent be managed to minimise discomfort, with the possible exception of noise pollution. Interestingly – but often ignored by designers and employers – are the noisy open plan offices which can create significantly high stress levels (Evans and Johnson, 2000).

Task demands are stressors arising from a person's job. Some jobs are obviously more stressful than others, e.g., surgeons, air traffic controllers and firefighters. It may be that people with a high threshold to stress are attracted to such jobs, but even for someone in a job which matches reasonably well with their stress threshold, task stress can arise from other sources. Overload is a prime example and occurs in two modes. One is simply too much to do and too little time in which to do it. Deadlines are in themselves stressful and if they lead to sub-standard work are even more stressful. In construction this can lead to reworking which means deadlines are missed, leading to more stress, leading to other deadlines being missed and so the vicious circle continues. A coping strategy used by many is working long hours but long hours can exacerbate the problem by inducing more stress from other causes such as work/family conflict. The second mode of overload is a person's ability to do the job or their belief that they are not up to the job. Either of these is stressful and can lead to reassignment of the task in the first case with a consequent 'loss of face' which may increase stress levels even further. The second case will lead to the person being 'on edge' most of the time unless they are given a confidence boost to help to confirm their ability. As mentioned early, low level stress is valuable for generating energy and motivation, but can easily escalate in the circumstances described here.

Task demands have a large influence on the stress to which construction professionals are subjected, and a major problem is the actual workload. This arises because they are interdependent in carrying out their work so the management of their workload is not in their own hands but depends on information being delivered by others. Hence workload can peak when information which is late being delivered eventually arrives, producing an extremely tight deadline. This situation frequently happens to many of the professional contributors simultaneously and increasingly near to a primary or key decision (Walker, 2007) which cannot be taken rationally without the completion of the individual tasks which lead up to it. Such situations inevitably result in escalating levels of stress. A prime example of a situation where this often happens is producing estimates prior to presentation to the funding body for a project. Estimates are built up from a series of contributions from the project team members, for example, form and function of the project, outline specification and programme, all of which feed into the build-up of the cost plan. If any of the decisions on these matters is delayed, deadlines are missed or become very tight making it difficult to finalise the estimate. Finalisation is thereby rushed increasing the likelihood of mistakes and inducing high stress levels in the project team, particularly in those responsible for signing-off the estimate. Stress due to a person's inability to do his/her job or his/her confidence in their ability is strongly influenced by their capacity to keep abreast of the rapid advances in the construction industry, both technological and organizational. These include developments in construction technology, such as in aspects of structural forms and in new finishing materials and techniques and by having to adapt to work in unfamiliar organizational forms such as partnering and prime contracting. Putting together the task stressors of overload and ability can create great stress.

Role stressors include *role ambiguity* and *role conflict. Role ambiguity* exists when employees are uncertain about what is expected of them in their job in terms of duties, performance and the level of their authority. In the general OB literature ambiguity is seen to arise from two sources. One is shortage of information which employees need to do their job and is closely aligned with the overload task stressor discussed above, and the other is created by the organizational structure. In construction, stress due to information deficiency may not be too severe as professional contributors have ways of dealing with it. An example would be a quantity surveyor preparing contract documentation and being short of information on aspects of construction with the deadline for submission of the documentation looming. What should he/she do? Delay submission of the documentation or assume the details he needs and meet the deadline with the risk of creating problems down the line. Being 'between the devil and the deep blue sea' can be stressful but techniques exist to alleviate such problems. However, shortage of information from the client may create more stressful situations. Ambiguity due to the organizational structure is much more prevalent in construction. Matrix organization structures are generally seen as a source of role ambiguity and are the norm in construction. Members of the industry have always worked within such structures to a greater or lesser extent (Walker, 2007; Rowlinson, 2001). The need to answer to at least two authorities, the head of their professional team and also the project manager or managing architect (and often the client), is fundamental to developing construction projects. Such organizational structures are seen to be highly stressful to their members. For construction professionals they are commonplace, which is not to say they are not stressful but many have learned through experience to cope with their demands. *Role conflict* is when employees face conflicting demands and can be either inter-role conflict or intra-role conflict. Inter-role conflict is when roles conflict with each other, for example, an architect working on a heritage project with high aesthetic demands being placed on him/her by the planning authority but with a shortfall in funding from the funding authority within which he/she is expected to work. The expectation that he/she should satisfy both sources as well as the client can be extremely stressful. Intra-role conflict is when contradictory messages are received from different people about how to perform a task. An example is when a managing partner requires one approach to designing a project but the partner directly managing the project wants another and they cannot (or will not) reconcile their differences.

Fincham and Rhodes (2005) identify what they call *boundary-spanning roles* as stress inducing. These are described as roles which involve representing the organization to outside organizations. The diplomatic service is cited as an example at the highest level. In construction, developing relationships with potential clients would be at the highest level but there is a galaxy of similar boundary-spanning activities, such as maintaining existing client relationships in the expectation of new work, and all the complex interface work with statutory authorities and other professional team members, all of which affect the firm's reputation. They expresses the pressure succinctly '... what can make a boundary-spanning role especially difficult is the availability of hard performance data which means the employer can press the individual for results and ignore the process involved in building relationships with customers.'

Organizational stressors arise from non-task-related events within the firm. Such stressors are widespread and begin when a new employee joins a firm. Many find the new, uncertain environment traumatic and stressful until they understand what is expected of them. Any type of organizational change within a company can induce stress in employees, for example mergers and acquisitions are extremely stressful events. Common in recent years has been restructuring the way in which work is organized leading to a reduction in the number employed (downsizing) which is extremely stressful for those who remain (but even more so on those let go) due to a great sense of insecurity. Insecurity is also brought on by an increase in employment by fixed term contract which has also now become more prevalent in the construction industry. Concern for what happens at the end of a contract, particularly in an uncertain economic environment, can induce stress and lead to under-performance. Wilson (1999) quoting Fineman (1995) emphasises that stress is 'an emotional product of the social and political features of work and organizational life'. The implicit norms of social structures in organizations have to be followed and cannot be changed by the individual who may have to camouflage their feelings, creating tensions leading to stress. As she says, 'We learn feeling rules; we have learned how much emotion to display, about how to appear, and appropriate demeanour for the workplace.'

There are a group of stressors which may be called *relationship stressors.* Difficult relationships in the workplace can arise from a number of sources. These include 'simple' personality and behavioural clashes and more complex ones – for example, when a member of a team is promoted to oversee the work of those with whom he/she was previously on a par – and from group pressures to conform to group norms.

The leadership style of an individual's manager may cause stress if it is incompatible with the individual's expectations, for example, an autocratic boss with a participative employee. Bosses are the greatest conduit of stress in organizations as essentially their job is to ensure the job gets done so they have to put pressure on their team members to achieve this. Whilst in most cases placing stress on staff is unavoidable, if leaders can appreciate the stress they are inducing and the possible consequences for individuals and the project they may be able to temper their demands in order to reduce stress on their team. A start in achieving this is for leaders to understand stress, stressors and the consequences in order to help their team develop coping strategies. Notwithstanding the stress the leader exerts, the leaders themselves will probably be under stress from their superiors and so the cycle continues. Office and project politics and power structures can strain relationships and create stress as individuals are coerced to declare for one camp or another, perhaps at some potential risk to their future with the company. Other serious stressors are sexual harassment, workplace bullying and workplace violence. A relationship stressor which is particular to professional relationships is a difference of opinion on a professional matter. In construction, examples may be a difference in design philosophy between architects on a project for which they are jointly responsible and differences between quantity surveyors regarding the way in which a contractor's claim should be handled. Whilst such differences are usually reconciled amicably, the potential for stressful situations to emerge is present.

Uncertainty is the root of much occupational stress. On the face of it, taking a decision removes uncertainty and should remove stress but it does not necessarily work like that. *Decision-making stress* is generated in making a decision as the process intensifies the stress created by the uncertainty surrounding the decision. Even when a decision is made stress will not necessarily cease as some individuals will continue to worry about whether a decision which has been made is the best one even though they may now not be able to change it. Fincham and Rhodes (2005) find that 'decisional stress [is] a particularly potent form of stress which may result in a significant reduction in the quality of an individual's decision-making'. Whilst many decisions relating to construction projects may be seen as procedural, there are a number of primary and key decisions (Walker, 2007) which are critical to project success, such as design concept, structural form and contractual arrangements which if made under pressure and with conflicting opinions within the project team could induce stress in the decision maker and lead to poor judgement.

Love and Edwards (2005) tested the predictive capabilities of full job strain model (JSM) (see, for example, Karasek, 1979; Fox et al., 1993) by investigating the influence of perceived job demands, job control and social support on the psychological wellbeing of a sample of UK construction project managers. The model was found to significantly predict employees' psychological wellbeing in terms of worker health and job satisfaction. However, they point out that construction project managers have to deal with many different parties within their own and other organizations associated with the project, particularly clients, which can induce stress and which they believe may have led to a result which is contrary to previous research; that is, social support from outside the workplace was found to be more significant than work-related support. This, they believe, was due to the autonomy of project managers working against the development of workplace support leading to them feeling that they were blamed for problems by the project team which in turn led to project managers seeking non-work support from their family and friends which could create problems with home and work–life balance.

An interesting stressor was identified by Zohar (1997) who found that *daily hassles* contributed significantly to burnout in nurses and to stress-inducing factors in parachute trainers. It was suggested that daily hassles could be as important a source of stress as role ambiguity and role conflicts. Also, workplace hassles were found to impact more on low-complexity tasks but this may be because more complex tasks offered more coping options (Zohar, 1999). Therefore, whether daily hassles in the construction professions induce stress may be doubtful.

Whilst the focus has been on stressors to which the project team has been subjected, the project team should also be conscious that client representatives with whom they interact are also likely to be subjected to equal or greater stressors from within their own organizations as pressure is brought to bear on them to deliver the project on time and within budget. An understanding of stress, including both work and non-work stressors, is therefore valuable not only for understanding and adapting to colleagues in the construction industry but also for understanding their impact on members of client organizations.

It is suggested by Loosemore and Waters (2004) that the under-representation of women in the construction industry may produce higher levels of stress in women. They found upon investigation that: 'Although there are common sources of stress for both men and women, there are also some differences. In particular, men appear to suffer more stress in relation to risk taking, disciplinary matters, implication of mistakes, redundancy and career progression. In contrast, the factors that that cause most stress in women were opportunities for personal development, rates of pay, keeping up with new ideas, business travel, and the cumulative effect of minor tasks.' They believe that these differences reflect women's traditional and continued subjugation in the construction industry. Subsequently, Sang et al. (2007) compared the occupational health and wellbeing of male and female architects and found that female architects report poorer health and wellbeing than their male counterparts, accompanied by lower overall job satisfaction and higher work–life conflict and turnover intentions. As with the previous paper, they consider that it is probable that these findings relate to women's subordinate position within the construction industry and the architectural profession.

3.7.2 Non-work stressors

We all know that work is not the only source of stress in our lives. Numerous sources exist outside the workplace and compound the stress arising from work and vice versa. Holmes and Rahe (1967) identify 43 life changes that create stress, varying from the death of a spouse, ranked as the greatest stressor with a score of 100, to minor violations of the law which ranked 43rd with a score of 11 and with, for example, a jail term scoring 63, retirement 45 and beginning school 26. But the more frequent causes of non-work-related stress are stressors that arise from the time and attention required by the everyday demands of home and family alongside the ever-present work stressors. The feeling of guilt experienced by employees who hold strong family values can be particularly acute. The family/work conflict is also exacerbated by the requirement to work long hours and to travel extensively on business.

People vary in their abilities to handle such multiple role conflict and it is considered that women, for whom such role conflict is usually particularly acute, have generally developed adaptive strategies which suggests they are very adept at juggling the conflicting demands of work and homecare and childcare (Fineman, 1995). But as McShane and Von Glinow (2003) point out 'Until men increase their contribution to homemaking and business learns to accommodate the new social order, many of these "supermoms" will continue to experience "superstress"'.

Stress at work combines with stress at home to create a combination of stress on an individual that can more readily reach breaking point as behaviour in one domain spills over into the other. The difficulty of separating the domains has increased rapidly with the development of communication technology which has extended the workplace into the home, and vice versa, making it ever more difficult to separate and deal with stressors separately. Similarly 'role behaviour conflict' (McShane and Glinow, 2003) shows the difficulty of separating work and home in cases where people are expected to behave in a certain manner at work which is not compatible with the more compassionate behavioural style required in their personal lives, particular examples could be correctional services officers and military personnel.

3.8 Differences in individual responses to stress

Differences in the characteristics of people mean that different people are affected differently by stressful situations. Each person sees the same situation differently. A common example is the way in which two people may react to the same very tight deadlines for the completion of a task. One may be focused, resolute and see the deadline as a challenge, the other may verge on panic, be unable to concentrate and see the deadline as a threat. One significant difference is the well-established personality trait of *self-esteem*. Those with self-esteem are self-assured; they perceive a stressor as less threatening than those who internalise their anxieties and they have the confidence to believe that they have the abilities and motivation to deal with the situation and, as a result, are less prone to stress. They are able to shrug off difficulties and do not allow setbacks to affect their self-esteem. Similarly, people who are inclined to be in a positive emotional state (positive affectivity) are more able to cope with stressful situations. Conversely, those with negative affectivity tend to be generally less resilient. Fincham and Rhodes (2005) believe that those with negative affectivity identify more stressors and create more stress, not just for themselves but also for others, and that they tend to be located in jobs which are less attractive and more stressful. So, given the same situation, people have different thresholds of resistance to stressors. It is also felt that younger employees experience fewer and less severe stress symptoms than older employees because they have a larger store of energy to cope with high stress levels (McShane and Von Glinow, 2003).

The Type A personality described earlier would be expected to be more susceptible to stress than a Type B and evidence has shown an association between Type A personalities and coronary heart disease. It has been shown that Type A behaviours are elicited in situations in which Type A's feel challenged, under pressure or threatened, leading to differences in blood pressure, heart rate and adrenaline levels between Type As and Bs (Fincham and Rhodes, 2005). But, interestingly, they also refer to a study some years ago of air traffic controllers which found that Type A air traffic controllers had higher rates of stress than Type B but that the largest single chronic illness in the sample, hypertension, was suffered by Type Bs. Air traffic control can be thought of as a Type A job; as such it produced more stress-related illness in Type B controllers (Rose et al., 1978).

Another idea is that of hardiness (Kobosa, 1979), that is, some people are seen to have hardier personalities than others and as such have a greater ability to cope with stress. They are strongly committed to what they do and welcome change as a challenge to provide opportunities for developing their careers and themselves. Fincham and Rhodes (2005) say, 'Their sense of control means they feel they make things happen rather than things happen to them.' However, they also state that 'It appears to represent no more than the 'rebranding' of a well-established personality dimension'. Hardiness, together with the other ideas discussed here, illustrates that the study of differences in the reaction of individuals to stress and what enables some to cope better than others is still in its infancy (Moorhead and Griffin, 2001).

The role of project managers illustrates the implications of stress in construction professionals. Intuitively it seems that a Type A personality is best suited to the role of project manager due predominantly to the drive they will bring to the job, but they may suffer higher rates of stress which can be bad for them, for the project team members and for the project. On the other hand, if a Type B personality was a project manager, the air traffic controllers' experience may be expected; that is, a Type B personality in a Type A job, so there could be more stress-related illness in Type B personalities. But a Type B personality may be better suited to the job as collaboration and integration of the project team is the essence of a project manager's task. Either way, stress is always likely but as most people are somewhere between A and B, stress may be manageable, particularly if the project manager has a hardy personality.

Leung et al. (2006) examine the nature of stress experienced by estimators in the construction industry in Hong Kong and the stress-coping behaviour they exhibit. They view coping as a major component of the overall stress process citing Newton and Keenan (1985) and Folkman et al. (1986). They distinguish coping behaviours in terms of their focus, i.e., problem-focused; emotion-focused. Problem-focused coping focuses on the task situation and problem solving through a number of coping behaviours, e.g., direct action to remove the stressor (direct and control action); seeking advice, assistance or information (instrumental support seeking); planning and organizing to come up with actions to handle problems (preparatory action); and identifying the cause of problems, possible actions and related consequences (problem appraisal). Emotion-focused coping deals with distressful emotions of individuals through, for example, suppressing impulsive acts and then bolstering morale through positive thinking (affective regulations); expression of unpleasant emotions such as smoking and drinking to reduce tension (emotion discharge); obtaining moral support, sympathy or understanding (seeking emotional support); isolation from the stress encountered by ignoring or escaping from it (denial/escape).

Their results reveal the impact of stress-coping behaviour on the estimating performance of senior estimators and project estimators separately but find that both generally apply problem-focused coping behaviours but with project estimators relying to an extent on emotion-focused coping. Senior estimators do not find the estimating process a significant source of stress but that 'preparatory action' and 'instrumental support seeking' are helpful in improving their estimating performance which helps them develop good cooperation between colleagues. Project estimators who had less work experience than their senior colleagues prefer to apply 'direct action' and 'negative emotional discharge' coping behaviours to manage their stress and hence a moderate amount of stress was seen to improve estimating performance.

Earlier, Haynes and Love (2004) examined coping among male Australian project managers employed by contracting organizations (they appear to use the terms construction project manager and site manager interchangeably). They found that 'Site managers who engaged in more problem-focused style of coping, such as active coping were better adjusted than those who engaged in more emotion-focused styles of coping such as cognitive avoidance coping, social coping, accepting responsibility and self controlling coping'. Problem-focused coping found in construction project managers is reflected in the same coping style used by construction estimators, discussed above, and is also reflected in a survey of project managers generally (i.e., not solely construction project managers) by Aitken and Crawford (2007).

3.9 Consequences of stress

The consequences of stress impact directly on the individual and are then transmitted to his private and working life with the potential to cause major problems in both domains. Early symptoms may be behavioural, and manifest in smoking, alcohol and drug abuse, proneness to accidents and appetite disorders. Medical symptoms may appear such as susceptibility to colds, headaches and muscle contractions. Other more serious illnesses such as hypertension, heart disease, strokes and stomach disorders are associated with stress. Psychological consequences include sleeping disorders, unpredictable mood changes and depression.

The effect of these symptoms on an individual's contribution to his employing organization can be extremely disruptive. The employee's performance will suffer as high stress levels contribute to mistakes, and impair his ability to remember and make effective decisions and take follow up action in a timely manner. Rational debate becomes difficult and, at its worse, employees engage in verbal conflict. In addition, absenteeism and sick leave may increase and an employee's attitudes may change so that job satisfaction declines as does morale, motivation and organizational commitment. Such an occurrence is damaging to any organization but is particularly so in construction due to the dependency on team performance. If a member of the team underperforms, the effect on the performance of the team is significant. Each member brings a specific skill to the combined output such that other tasks cannot be performed effectively if a specific skill is not making its contribution. Even if the person under stress can be replaced, he will take with him undocumented knowledge of the project in much the same way as when a member voluntarily leaves the project team (Eskrod and Blichfeldt, 2005). But, in the case of stress, the situation can be less clear cut and therefore more complex as both the person under stress and his colleagues, both within his firm and project team members from other firms, attempt to cover for him or ignore the problem during which time the whole project team performance and his firm's reputation suffer before the situation is openly recognized and help is provided to the individual.

3.9.1 Burnout

Burnout is a general feeling of exhaustion that develops when a person simultaneously experiences too much pressure and has too few sources of satisfaction (Lee and Ashforth, 1996). Fincham and Rhodes (2005) believe that three factors have emerged to describe burnout:

Emotional exhaustion – originally seen as the core of burnout. A person's feeling that their emotional resources are inadequate to cope with emotionally demanding problems with clients and customers at work. This aspect is now seen as generic.

Cynicism – previously known as depersonalisation. Arose from origins in social service settings. Creates feeling of indifference to clients and their problems such that clients are seen as objects rather than as people. The burnout idea now appears to apply more widely than to human services but to work more generally.

Professional efficacy – feelings of inadequacy at work that little had been achieved (again originally in social services) with clients. Generalizing the concept to other jobs has meant that this element now refers more broadly to people being ineffective at work so not achieving their expectations.

Whilst burnout was originally identified in the caring professions such as nursing, burnout is seen as particularly common in large bureaucratic organizations where employees feel trapped in roles and career paths are blocked leading to stagnation, frustration, apathy and eventually burnout. They point out that these feelings of entrapment are more likely to be experienced by older employees. Large organizations may seek to prevent burnout through career plans which could entail sideway moves of personnel and secondments. In small firms, of which the construction industry comprises many, such solutions are not readily available but it may be possible to switch people between projects if care is taken not to disadvantage the project from which employees are removed. In more general terms, Moorhead and Griffin (2001) say that 'people with high aspirations and strong motivation to get things done are prime candidates for burnout under certain conditions. They are especially vulnerable when the organization suppresses or limits their initiative whilst constantly demanding that they serve the organization's own ends' and add that 'Loss of self-confidence and psychological withdrawal follow'.

As a result of the emergence of research into the phenomenon, the construction industry has recognized burnout as an important issue. The literature stresses that the demanding work environment generates many job-related stressors such as high technology, tight schedules and budgets, low profit margins culminating in high risks. Lingard (2003) was the first significant study. It examined the impact of individual and job characteristics on burnout amongst civil engineers in Australia. She found that 'burnout cannot be attributed to a single cause but is a result of a complex interaction of individual characteristics and issues in the work environment' so there is no single quick-fix solution. But she suggests that, as job characteristics figured highly in creating burnout, job redesign may be an effective preventative measure. Lingard and Francis (2006) examined whether a supportive work environment moderated the relationship between work–family conflict and burnout among construction professionals and found that interventions designed to alleviate or prevent employee burnout should focus, at least in part, on the creation of a supportive work environment. They conclude that such interventions should focus on directly reducing emotional exhaustion by enhancing employees' perceptions that the organization is supportive of them and fostering a caring workforce that is willing and able to provide appropriate emotional support to co-workers and subordinates. Also, that interventions should include strategies to increase the amount of practical support provided as it appears to decrease the extent to which employees who experience work–life conflict suffer from emotional exhaustion.

Yip et al. (2008) followed by considering coping strategies specifically in the relationship between role overload and burnout in a sample of professional engineers in Hong Kong. The model was based on: the coping strategies of rational problem solving (I tried to analyse the problem in order to understand it better); resigned distancing (I went along with fate: sometimes I just have bad luck); seeking support/ventilation (I let my feelings out somehow); and passive wishful thinking (I wish the situation would go away or somehow be over with). The findings showed that only rational problem-solving moderated role overload in all three dimensions of burnout: emotional exhaustion, cynicism and reduced professional efficacy. Resigned distancing was effective only on emotional exhaustion and cynicism and passive wishful thinking on none. Yip and Rowlinson (2009a), again using a sample of engineers in Hong Kong, found that the recognized reasons for burnout (emotional exhaustion, cynicism and diminished professional efficacy) were valid, also that engineers working for contracting organizations suffered higher levels of burnout than engineers working within consulting organizations. Burnout was largely due to job conditions, with qualitative overload and lack of promotion prospects the major reasons for consulting engineers and long working hours, role conflict, role ambiguity and lack of job security the major reasons for contractors' engineers. Again job redesign was suggested as a way to reduce burnout depending on the type of engineering organization. Developing this study using a wider group of construction professions, Yip and Rowlinson (2009b) again found the use of job redesign was effective as an intervention strategy in reducing burnout. This significant group of papers has shown that burnout has a detrimental effect on the construction industry in Hong Kong as it is also likely to do so in most countries and also in many other industries.

3.10 Managing work-related stress

Managing stress takes place at two levels: that of the individual and that of the organization. An individual's own initiatives for coping can be harmful rather than useful. The harmful ones have been referred to before as undesirable consequences of stress such as smoking, alcohol and drug abuse and can also include excessive caffeine consumption all of which may increase stress. More positive approaches include analysis of the situation causing stress so that the issues are seen clearly, enabling the stressed person to tackle the problems identified. This may be achieved through discussion with a sympathetic boss which could result in innovation or modification of working practices or even removal of or withdrawal from the stressor. Both permanent removal of a stressor and withdrawal of an individual from a stressor can disrupt organizations. Short-term temporary removal and withdrawal also leads to inefficiencies as other employees have to pick up the pieces and, in the latter case, later withdraw themselves. In construction, this is another manifestation of the loss of long-term continuity of project teams discussed earlier. An alternative coping strategy, often attempted unsuccessfully, is denying or trying to forget the problem. Another approach often found to be effective is to gain support from individuals or groups of co-workers or friends having experience of the situation. They can help to change the stressed person's perception of a situation in a way that improves self-esteem and hence the ability to cope. Also role-related stressors can be minimised by allocating employees to jobs which match their competences and/or by innovating in their roles to improve their fit. In addition, it has been found that an effective way to reduce workplace stressors is to empower employees so that they have more control over their work, but this can have other consequences for the organization (see Chapter 7). If any of these approaches are not possible, Fincham and Rhodes (2005) suggest an emotion-focused strategy to help individuals manage their anxieties. This approach includes avoiding paying attention to stress or, put another way, by showing a 'stiff upper lip' or by resigned acceptance individuals can stoically accept the situation and expect that the worst is likely to occur. To these approaches they also add emotional discharge or 'letting off steam' but add that this approach can increase rather than reduce the problem if it includes unacceptable behaviour.

Ng et al. (2005) looked at the manageability of a range of stressors among construction project participants in Hong Kong. The most difficult stressors to manage were found to be 'bureaucracy' (which is particularly high in Hong Kong), 'lack of opportunity to learn new skills', 'work–family conflict' and 'different view from superiors'. Different patterns of stress manageability were found between clients, consultants and contractors but this theme was not developed particularly. Stress is a phenomena which impacts on the individual so aggregated data of the type on which this paper was based gives only a broad steer and, as the paper says, previous studies have 'mainly concentrated on the effect of stress and its effect on the performance of an individual and the project outcome, very few have touched on stress experienced in the construction industry'. Also, Leung et al. (2005) in a paper on stress on estimators in Hong Kong state, 'Each estimator has distinctive personal characteristics and working experiences, and individuals could therefore have different levels of resistance to stress.'

So far the focus has been on alleviating the causes of stress but techniques to build an individual's resistance to stress may be valuable, and to this end some organizations provide stress management programmes. Such programmes will often include courses to help employees understand stress: what it is, what causes it, how to recognize it and how to cope with it. Then there are courses covering specific aspects of stress management such as, for example, time workload management, which focus on how one manages one's time and distributes workload more appropriately. They also include opportunities for exercise, relaxation (both mental and physical), medical examination to detect stress-related symptoms such as hypertension, health promotion programmes, personal development programmes to improve self-esteem, career and promotion strategies, and task and role definition and design. Counselling may be provided through employee assistance programmes (EAPs) for dependency problems and health issues.

Programmes can be bought in from consultants and tailored to suit specific companies but larger companies may develop their own in-house programmes. However, many such programmes are not necessarily productive. Fincham and Rhodes (2005) state: 'Despite the claims of the stress management industry, research results are inconsistent. Some interventions produce marginal or negligible effects.' Similarly Wilson (1999) believes that 'Evidence of the success of such schemes is generally confusing and imprecise . . .'. However, medical monitoring is obviously valuable and it would seem that physical exercise leading to increased fitness is a successful method of reducing and coping with stress. But encouraging staff to exercise is one thing, getting them to actually do it is another. Even the provision of exercise facilities, which is expensive in terms of space and equipment, may not succeed in enticing employees into the gym.

4 Morality, Ethics and Trust

4.1 Introduction

Morality, ethics and trust arise from an individual's values and attitudes such as honesty, fair play and respect for others. They are the manifestation of an individual's values and attitudes in the actual actions taken by the individual. Such personal ethics underpin the business ethics of organizations as a whole and define the level of ethical behaviour in organizations. Naturally the impact of an individual's ethical standards on an organization as a whole is likely to be a function of his/her position, that is, the ethics of senior members will have a proportionately greater impact on the organization's business ethics. Thus, when reference is made to business ethics it is necessary to recognize that they are underpinned by personal ethics and are not something detached from and independent of the members of an organization.

Good ethical behaviour has proved difficult to define, not least because society's values have shifted over time and the rate of change is accelerating. Individual morality is one person's distinction between what is right or wrong by which they judge their own actions and the actions of others in life as a whole, including the way they make such judgements in the business world. However, the picture is made fuzzy as many people may not live by the same standards in business as they do in other aspects of their life. The meaning of ethics generally (as opposed to professional ethics specifically) overlaps with the meaning of morality and is said by Fineman et al. (2005) to be concerned with the rules and principles which shape moral (good/bad) actions. McShane and Von Glinow (2003) say that ethics refer to the study of moral principles or values that determine whether actions are right or wrong and outcomes are good or bad and that we rely on our ethical values to determine the right thing to do.

McShane and Von Glinow (2003) identify three ethical perspectives condensed from a range of views of ethics generally devised by philosophers and scholars – utilitarianism, individual rights, and distributive judgement – and believe that all three should be applied to put ethical issues to the test:

Organizational Behaviour in Construction, First Edition. Anthony Walker.

- *Utilitarianism*: Looks for the greatest good for the largest number of people. However, it focuses on the end result and not on the means by which the end is achieved. This can result in moral dilemmas between methods and results and can also lead to oppression of minorities.
- *Individual rights*: Believes that everyone has the freedom to act in certain ways associated with, for example, speech, physical security, travel. Conflicts can occur, particularly between the rights of an employee and an employer, for example, in respect to privacy.
- *Distributive justice*: Believes that inequality is acceptable if it has resulted from equal access to all positions in society and that the inequalities favour the worst-off in society. The latter proposes that inequality is acceptable if the better-off accept risks which, by doing so, benefit the less well-off. However, they add that, 'The problem with this principle is that society cannot seem to agree on what activities provide the greatest benefit to the least well-off.'

It can be seen from these attempts to identify ethical principles that variations in values within societies, and particularly between societies, throw up imponderables and moral dilemmas which make business ethics a minefield of conflicting views and unsatisfactory outcomes. It should be stated at this point that when the term 'business ethics' is referred to in the text, 'business' refers to both private and public organizations. The pressure to uphold the highest standards of ethical behaviour generally falls on those in the public sector, as public servants are more directly responsible to the public, particularly in regard to public finances. Small indiscretions which may be accepted within and between private companies may not be so easily forgiven for public servants. At the highest level, the ethical behaviour of those elected to public office should be unimpeachable. Sadly this is often not the case, as demonstrated by the behaviour of some Members of Parliament and some members of the House of Lords in the UK.

It seems from reports in the press that the standard of morality and ethics in business generally are increasingly questionable as people in business behave in ways which do not subscribe to the ethical principles outlined above and are seen to be unacceptable to the public at large, although whether others in the same situation would behave any differently cannot be determined. The scope of behaviour seen to be unethical is extremely wide ranging from, for example, company directors awarding themselves huge pay rises and bonuses whilst at the same time laying off staff, and charging customers for work which was not necessary, to fiddling expense claims and taking a packet of paper clips for personal use. An important distinction within the range of unethical behaviour is whether such behaviour is illegal, or legal but nevertheless unethical. The former has, of course, formal sanctions to act as a deterrent; the latter does not but can be equally damaging or even more damaging to an organization as it is corrosive and corrupts the organization's underlying ethic.

Companies and individuals in the construction industry are as subject to the same issues and problems of morality and business ethics as all other organizations (cf. Walker et al., 2008). But, in addition, members of professional bodies associated with construction (engineers, architects and surveyors) have to abide by the rules of conduct of their professional bodies as a condition of being allowed to practise. Such codes of conduct are essentially designed to protect their clients and the public against unethical behaviour on the part of members of the profession. Sanctions for breaching the rules can, at their extreme, bar a member from practising, with a range of lesser sanctions for lesser breaches of the rules. The defining nature of a profession is self-regulation through rules of conduct with sanctions, and the construction professions share such a regime with other professions such as medicine, law and accountancy. The term 'profession' is much more widely used now than in the past but many other so-called 'professions' do not pass the test of self-regulation through rules of conduct with sanctions. Infringement of expectations of ethical behaviour by members of non-professional business organizations does not carry such sanctions so the professions are more stringently monitored, although increases in the number of regulatory bodies monitoring aspects of business generally are moving in a similar direction but essentially with their focus on the organization rather than the individual. A major informal sanction applying to both professional and non-professional businesses and individuals engaging in unethical behaviour is bad publicity should their behaviour become public and find its way into the media. Although this is likely to affect only prominent organizations and individuals, even local circulation of misbehaviour can be very damaging. Such condemnation is frequently hypocritical as those condemning may also behave in a similar manner; such can be the extent of unethical behaviour in business.

This chapter also includes the topic of trust. Trust requires that there is mutual respect between the parties based on the judgement of each party regarding the behaviour of the other party. As trust only arises when there is no monitoring or control, each party must be satisfied with the ethical standards of the other for trust to be developed and sustained. If the level of ethical behaviour expected fails to be achieved or sustained, trust will break down. This link between morality, ethics and trust is vital for cooperation in today's increasingly interdependent and integrated global economy in which organizations and individuals are required by the complexity of the business environment to work together. This is no more so than in construction as projects increase in complexity. We therefore arrive at a paradox. If commentators are to be believed, there is increasingly lower morality and lower ethics in business at the same time as there is an increasing need for greater cooperation which, to be productive, requires higher levels of trust.

4.2 The nature and scope of ethical behaviour in business organizations

The problem of defining ethical behaviour reflects the complexity of both its nature and its range. At the least serious level, unethical behaviour in business may be undesirable and create distrust in others but in reality may have little serious effect on those in the business environment. Such actions could include, for example, denigrating competitors and colleagues and taking unauthorised time off work. Such behaviour can in many cases rebound on the perpetrator as their reputation may be seriously damaged if they are found out. At a higher level may come such activities as falsifying expenses claims and feeding information of company's activities and contract details to competitors. At an even higher level come such unethical behaviour as poaching competitors' senior staff on the condition that they bring their clients' accounts with them, bribery and fraud.

The range is enormous and is a mixture of the relatively innocuous to the serious and from the unethical but legal to the illegal. The illegal is, of course, proscribed by law and is the height of unethical behaviour in business of which there have been many high profile examples. At a less significant level there is a whole raft of immoral behaviour which is illegal but rarely pursued in the courts, such as petty pilfering and fiddling expenses. An aspect of ethics which is rarely recognized is that of abrogating responsibility, of which Pontius Pilot's behaviour is an early example, and which has continued to be prevalent, particularly in the higher reaches of the public services as blame is passed from one public officer to another.

From this complexity McShane and Von Glinow (2003) have highlighted two issues which assist in analysing ethical issues:

Moral intensity – the degree to which an issue demands the application of ethical principles. The higher the moral intensity, the more that ethical principles should provide guidance for resolving the issue. Intensity is seen as being manifest in a number of ways: the seriousness of the consequences which the moral issue produces; the view of the issue held by society; how close to the issue the decision maker feels and is able to influence the issue (Frey, 2000).

Ethical sensitivity – the extent to which a person recognizes an ethical issue's importance (Sparks and Hunt, 1998). Ethically sensitive people are not necessarily more ethical but are more likely to recognize whether an issue requires ethical consideration and are able to estimate the moral intensity of the issue. Sensitivity can be specialty related, for example, architects could be expected to be more ethically sensitive to aesthetic issues such as modifying listed buildings.

Fineham et al. (2005) believe that the principle of not harming someone through one's actions, directly or indirectly, lies at the heart of moral concepts. They see this as distinguishing morality (through feelings of guilt) from embarrassment. If one's actions directly harm someone, physically or in other serious ways, such as getting them into serious trouble leading to feelings of guilt then the issue relates directly to morality and ethics but if an action transgresses an expectation of another person causing embarrassment, but does not harm another, it is not felt to be a moral issue. They also draw attention to moral dilemmas – situations in which moral principles are at odds with each other and there is no clear right or wrong, a situation which occurs often in business. Moral dilemmas lead to choices of the lesser of two evils. In construction, such a dilemma could be whether a professional project team member supports the architect or the client where they have a difference of opinion but in which one can sympathise with both points of view. In the family setting, it could be whether to support one's partner or child over an issue on which they disagree.

These issues raise the question of where the values and standards that determine our sense of morality come from. We start at a very early age to learn from our parents, and others with whom we are closely associated, what is right and what is wrong. As we grow, the culture within which we live, arising from society in general, influences our judgement, as do our peers, so we eventually create our distinctive code of ethical behaviour as a result of the environment in which we have been reared. Thus the boundaries of the ethical code of our society may be discerned. We also know that in societies a very large group of people lies at its centre; however, the extremes are very wide apart, from the pious at one end to the criminally inclined at the other. But even in the central mass there are variations, particularly in relation to how easily a person's ethical standards may be compromised under the influence of others, such as employers, family members and friends.

Robbins and Judge (2008) ask whether there has been a decline in ethical standards in business. They suggest that as employees' orientation towards their employers has changed since the 1970s, from loyalty to their employers to loyalty to their careers and now to themselves, as discussed in Chapter 2, such self-centred values could be consistent with a decline in ethical standards. It can also be argued that in the Western world during the same period there has been a dilution of the forces which stood for high ethical behaviour such as the church and schools. Additionally, increased transparency has allowed the media to expose unethical behaviour in politicians, church representatives and others who could be expected to set standards to the extent that people have taken the attitudes: 'If they can behave like that so can I'; 'It doesn't matter if you are not found out'; and 'I don't care; it is simply what I do to get on here'. But maybe in the past standards of conduct were no better than now but, due to a lack of transparency, breaches were not exposed to the same extent by the media and other agencies. Bowen et al. (2007) citing Hood (2003) believe that corporate ethical performance has been monitored with increasing interest and that, citing Carlson and Perrewe (1995), 'interest is generated by numerous factors, such as diminishing confidence in ethical corporate practices and a greater emphasis on quality of life' to which can be added the increasing number of high profile failures in ethical performance.

Companies in the construction industry are subject to these issues of ethical behaviour to the same extent as all other industries; some may say even more so as the opportunities to benefit from unethical behaviour are very readily available. Predominantly for this reason, when professional institutes associated with the construction industry were formed in the 19th century they required their members to abide by ethical rules of conduct designed to establish their credentials, protect themselves, their clients and the general public from the unscrupulous, and to which were attached sanctions which can ultimately result in disbarment from practice. It can be claimed, therefore, that the professions are subject to a higher level of ethical behaviour than other companies and individuals as, although they are subject to the law, many unethical behaviours which are not illegal in business generally are the subject of rules of conduct and the companies and individuals are thereby able to be sanctioned by their professional bodies.

The perspective taken of ethical behaviour has been on its presence in business and professional organizations emanating from the actions of individuals, in accordance with the focus of this book as a whole. However, it is appropriate to draw attention to a broader view of ethics which has particular significance for the wider built environment. This broader view is encapsulated by Farmer and Radford (2010) in their editorial to a special issue of *Building Research and Information* devoted to the moral dimensions of built environment practices in which they ask, 'Should designers, for example, retain and strengthen established professional ethics or be looking towards new approaches, ones that may open the site of moral deliberation in the built environment to new agendas?' One method of review referred to is by sub-field consisting of the sectors: architecture; urban design; and construction. Considering architecture the broader issues are seen to include sustainability, 'environmental ethics, architects' responsibilities to non-human fauna and flora, the wellbeing of ecosystems and the world as a whole.' In urban design and planning they point to three ethical dilemmas: between freedom and autonomy and government regulation; overall society, represented by issues between economic efficiency and social justice; and conflicts in the environment, for instance between environmental protection and development. And in construction and contracting they highlight the conflict between acting ethically for the client and acting ethically for a wider public.

Quite how such issues relate to individuals' ethics in their traditional meaning rather than movements of concerned individuals is not examined. They are challenging issues but room to pursue them is not available in this book which takes a narrower perspective.

4.3 Organizational pressures

An organization's culture can be a powerful force in determining the ethical behaviour of its workforce. The view of organizational culture adopted here is the managerial (or functional) position, which is normative and prescriptive and also controversial compared to the social science perspective as discussed in Chapter 8, but provides an appropriate setting in which to place an account of ethics in organizations. How an organization expects its employees to behave, as represented by the values of its leaders, percolates through the whole organization, is enshrined in the folklore of the organization and soon transmitted to new recruits. The way in which pressure is brought to bear on employees to adopt particular ethical behaviour is more likely to be covert than explicit; for example, the mis-selling of financial products in the UK during the 1990s. However, employees have to face more blatant unethical behaviour by companies: 70% of bars and restaurants in the UK only sell measures of wine larger than the standard 125 ml in contradiction of their trade association's voluntary agreement as part of a national campaign to stop binge drinking. As Fineman et al. (2005) point out, 'Leadership is a moral activity and is never value-neutral.' Leaders set the ethical standards of their organizations by their own behaviour which becomes the norm for the organization and enshrined in company folklore, so leaders need to emphasise a deep sense of social and moral responsibility if their organizations wish to establish and maintain a high ethical reputation. 'Such organizations may attract public attention as journalists, politicians and academics test the strength of such claims, especially when they clash with harsh financial realities' (Fineman et al., 2005) But if leaders do not set an ethical example, their organizations may receive even more rigorous attention for any misbehaviour.

Employees will soon realise that to 'get on' they need to subscribe to the organization's ethics (good or bad). Employees who find their organization's ethics difficult to accept will be faced with a raft of ethical dilemmas. If they cannot cope with them they will be well advised to leave. Conversely, employees who work for an ethical company for which they have respect will take great comfort from their work. McShane and Von Glinow (2003) describe what they term 'corporate cults' in which organizations establish such a strong culture that they risk taking over their employees' lives and their individualism. The ability of such companies to influence employees' ethics is enormous and requires such companies to accept responsibility for developing cultures which are consistent with society's ethical values. Although there have been a number of high profile companies which have been shown to have deeply embedded unethical practices, the vast majority do not. Most may from time to time cross the line from good to bad and many employees learn to live with such situations and resolve ethical dilemmas as best they can. But it cannot be overlooked that unethical employees can find themselves in ethical organizations. One would expect them to be found out sooner rather than later but nevertheless they may do serious damage to a company's reputation before this happens and the company has to accept responsibility for their actions.

The pressures that the expectations of an organization place on individuals' ethical values may arise from a wide range of sources, many of which they may not have been able to identify before joining the organization. Issues such as institutionalised racism, evidence of bribery and corruption and membership of cartels are of such seriousness that the employee is faced with whistle-blowing, turning a blind eye or leaving. Other less serious but nevertheless important issues, such as manipulation of employees, institutionalised fiddling, pilfering and overcharging, present employees with tests of loyalty that provide challenges to their ideas of right and wrong in which their ethical values may be questioned by those who will be affected. Fineman et al. (2005) give an account of institutionalise fiddling in building repair, restaurant and clothing businesses. In the case of the building repair workers fiddling was justified in terms of the workers being in lowly paid jobs which fiddling subsidised into a 'wage' the workers could live on. They continue, 'Fiddling is taken for granted in some big corporations, by fiddlers and fiddled alike. [Fiddling] is built into the overall running costs, and is tacitly accepted as part of the informal organization of work.'

On the other hand, employees may hide behind organizational norms, which may be formal, written organization rules or unwritten enshrined expectations. Employees may rationalize their actions on the basis of the norms to justify their behaviour in an attempt to overcome their feeling of guilt at what, deep down, they feel is unethical behaviour. Associated with such a position is the excuse for immoral behaviour that 'I was only obeying orders'. Adopting the moral position of refusing to obey an order may well be tested against the sanction for not obeying which in a business setting could mean being fired. McShane and Von Glinow (2003) identify that 95% of Fortune 500 companies in the USA and 57% of the 500 largest companies in the UK have codes of ethics but say that there is considerable debate about whether they improve ethical conduct (Schwartz, 2001). They also state that one step beyond having codes of ethics is to have employees receive ethics training, require employees to sign ethics codes and provide facilities for employees to report unethical behaviour. Whilst the focus here has been on unethical rather than ethical behaviour in business, it should be recognized that the vast majority of companies and individuals do not indulge in any remotely serious unethical behaviour, therefore any serious breaches are widely publicised in the media which may give the wrong idea that serious unethical behaviour in the business world is widespread.

Corporate social responsibility is a term which can be seen to be an aspect of business ethics. It refers to an organization's obligation to contribute to the social environment within which it functions (Moorhead and Griffin, 2001); that is, an obligation towards others who are affected by its actions. Increasing focus on corporate social responsibility has arisen to a large extent due to environmental awareness and the need for organizations to eliminate any detrimental effect they have on the natural environment. Oil and chemical spills, deforestation, fly tipping and many similar activities have seriously damaged companies' reputations and made them aware of their corporate social responsibilities. But corporate social responsibility extends much wider than this and includes organizations supporting their more general environment including, for example, providing employment for the underprivileged and supporting community initiatives.

In her paper on the development of corporate social responsibility in the Australian construction industry, Petrovic-Lazarevic (2008) defines corporate social responsibility as 'a set of principles established by an organization to meet societal expectations of appropriate behaviour and achieve best practice through social benefits and sustainable competitive advantage'. However, she points out that there are many diverse opinions of what it actually constitutes. In relation to construction companies she sees (drawing on Yadong, 2007) corporate social responsibility as 'including the following activities: a moral obligation to be a good citizen; sustainability; reputation; relationship with employees and unions; relationship with suppliers and community representatives; and commitment to reporting on CSR'. She also sees corporate social responsibility relevant to the construction industry as incorporating 'a corporate governance structure that supports socially responsible business; organizational culture that clarifies the corporate social responsibility trends of a company; healthy working; environment measures either recommended by government or initiated by a company itself; and proper communication with all stakeholders of organizational values and ethics'. Following interviews with representatives of 17 large Australian construction companies she found that they generally espoused the moral obligation to be 'good citizens' but that further improvement of occupational health and safety measures was needed as were better relationships with the local community and attention to sustainability. An earlier paper by Glass and Simmonds (2007), whilst not considering the full scope of corporate social responsibilities, examined 'considerate construction' (Considerate Constructors Scheme, 1997), particularly in the area of community engagement practices used by contractors. The major impacts caused by contractors on site were identified as: noise; dust; traffic congestion; parking; water pollution; health and safety; dirt; and security. A case study of four projects being built by the same contractor that routinely registers projects with the Considerate Contructors Scheme was adopted. The research found evidence of a range of effective practices but with scope for improvement, particularly in community relations. It was also pointed out that 'there is a lack of literature on the extent to which contractors have mastered the art of managing community relations'. Interestingly, better examples of good performance were found in partnered projects.

Leavitt (2005) takes a practical view of a manager's position when faced with a moral dilemma in which his/her personal integrity is at stake and he/she must either do his duty to him/herself and his/her personal standards or compromise his/her standards to protect status or income. For the manager to maintain his respect in such difficult situations requires great strength of character. Failure to do so carries many costs, not least of which is a sense that his integrity has been irreparably undermined. He continues that experienced managers with high ethical standards use their understanding of themselves and their organizations to make their way through such ethical minefields and develop the authoritative presence which enables them to cope with such dilemmas, maintaining both integrity and position. Others, who get tired of dealing with too many such dilemmas, call it a day and leave and some just cave in and do whatever the hierarchy requires. Those leaving opt for a different way of life, those who give in finish up being deadwood and/or become preoccupied with all they have left, their formal authority. He warns against 'such dangerous people who toady up and dictate down'.

Competing complex forces within organizations can create situations in which organizations carry out activities which are unethical, illegal or both. Knowledge of such situations can become known to members of the organization or to members of external organizations who find it impossible to ignore them due to their standards of morality and feel it necessary to 'blow the whistle' on what is happening. There can be many diverse situations which generate such a reaction, increasingly these days they involve damage to the environment but illegal administration of pension funds and other financial irregularities such as 'under the counter payments' often feature. External pressure groups such as Greenpeace and Friends of the Earth together with the trade unions and human rights groups are active as whistleblowers but they can act without the direct personal threats which accrue to whistleblowers from within the offending organization. The sense of moral outrage felt by internal whistleblowers must be enormous for them to risk the ire of their organization and the actions which can be taken against them. Persecution and dismissal without references with little hope of future employment in the same industry may be expected as the company closes ranks as others, with the same sense of morality as the whistleblower, usually fail to appear to support the original whistleblower. In the face of an internal whistleblower, a company is likely to become defensive and do all it can to discredit the whistleblower. There are, therefore, few internal whistleblowers. Pressure groups, on the other hand, have the resources to counter the companies, so internal whistleblowers may seek their support. Whilst it may be extremely difficult to survive in the competitive world in which organizations operate without transgressing ethical standards at some level, particularly in the light of ethics being difficult to define in absolute terms, generally organizations manage not to exceed acceptable standards of behaviour to the extent that employees feel the need to 'blow the whistle' but employees may leave rather than do this.

4.4 Specific issues of ethics in business

Focus on ethics in business arises when decisions which involve ethical dilemmas have to be made. Writers on OB have identified a number of situations in which the context of a decision creates distinctive issues. A major area is that of attempting to influence employees' organizational behaviour which encompasses employee monitoring, organization development (OD) and organization behaviour modification. McShane and Von Glinow (2003) refer to the first two of these areas. Monitoring of employee behaviour and performance is common in most large organizations and has been enhanced at an enormous rate with the increasing sophistication of the technology available for this purpose such as remote cameras, telecommunications, satellite tracking systems and the internet. The ethical issue is whether such activity is an invasion of privacy which shows a lack of trust and undermines employment relationships. Associated with monitoring is organizational development (OD) which is about managing planned change in organizations to increase effectiveness. Again the ethical issues are concerned with the individual's privacy rights. They identify that OD processes require employees to provide personal information which they may not wish to give and which may be identified to a particular employee. Also, that OD activities potentially increase management's power by inducing compliance and conformity in organizational members. Another concern is that the process requires employees to question their own competence or personal relationships and so undermine their own self-belief. Moorhead and Griffin (2001) refer to the other issue, organizational behaviour modification (OB mod), which aims to improve motivation by increasing the frequency of desirable behaviours by linking them to positive consequences and decreasing undesirable behaviours by linking them to undesirable consequences and they report on the approach's success and failure in improving motivation. At its simplest level bonuses can be used to encourage desirable behaviours. Ethical issues regarding manipulation of employees and limitations of freedom of individual choice have caused concern. OB mod's purpose is to shape the behaviour of others and may have no regard for what is best for the individual. Managers may push employees towards a limited array of options which suit the organization but which are not in the best interests of the individual. The construction literature appears to contain only one reference to OB mod (Bresnen and Marshall, 2000b) which refers to the use of financial incentives in partnering as a form of OB mod. They also refer to it as the 'carrot and stick' approach intended to shape behaviour through the selective application of rewards and punishments that encourage certain actions and discourage others. Their analysis of case studies demonstrates a number of important limitations in the use of incentives in partnering amongst which are that it 'emphasised the importance of understanding that participants evaluation of rewards, expectation of performance and perceptions of equity are highly subjective and may differ' and that 'it has been demonstrated that motivation and commitment may be the result of intrinsic as well as extrinsic rewards'.

The ethical perspective 'distributive justice' referred to earlier – which believes that inequality is acceptable if it has resulted from equal access to all positions in society, that the inequalities favour the worst-off in society and that inequality is acceptable if the better off accept risks which, by doing so, benefit the less well-off in society – has particular connotations for ethics in business. McShane and Von Glinow (2003) believe that this means that everyone in an organization should have access to the senior positions (as well as other valued positions in society). They say, however, that the problem with this principle is that society cannot seem to agree on what activities provide the greatest benefit to the least well-off. Nor does it say what constitutes 'risky'; is it physical risk, financial risk, risk of being fired? This principle brings into question the ethics of the extremely high pay and bonuses awarded to senior executives of companies (e.g., bankers as a prelude to the financial crisis 2008) either by themselves, their colleagues or by remuneration committees of boards of directors staffed by non-executive directors who are directors of other companies who in turn are awarded high pay and bonuses by their remuneration committees often staffed by those to whom they have awarded large pay deals and so on around the merry-go-round.

They also point to ethical problems manifest in stereotyping as it assumes all in the group are the same and creates prejudice of which racial discrimination is perhaps the most prominent. But categorising people for whatever reason means that people who cannot overcome the prejudice associated with their stereotype are treated unfairly, hence unethically. They also point out that stereotyping is also partly responsible for sexual harassment as it is mainly caused by the harasser's abuse of power; that is, harassment is more likely to happen to people who are stereotyped by the perpetrator as subservient and powerless.

Importantly, OB researchers have recognized that ethics come close to home when conducting OB research which involves direct action with the people whose behaviour is being studied, for example, the privacy of participants must be protected, involvement must be voluntary and withdrawal allowed at any time, participants must be informed of any risks (e.g., psychological) before partaking and results must be reported accurately and objectively.

The globalization of business has meant that ethical perceptions and standards have assumed great importance so it is important to recognize that ethical standards are different in different countries. Understanding the position of other countries on what is ethical, what is unethical but not illegal and what is illegal are of paramount importance. It is not reasonable to assume, which tends to be the case whichever country one comes from, that other countries have ethical standards lower than one's own and time and patience need to be devoted to understanding other countries' ways of life. What is seen as acceptable in one country may be unacceptable in another. However, it appears that ethical principles are not dissimilar between cultures but application may vary (Buller et al., 1997). Whilst such issues have been brought to the fore by globalization of business, they have been familiar to diplomatic services for aeons. For example, in China 'guanxi' is an elaborate relationship network originating in the mists of time which in the West is sometimes thought to be a form of corruption by some but has been shown not to be so (So and Walker, 2006). In commercial terms, a further example could be of severe problems arising when negotiating construction and other contracts for overseas work such as when three companies each from different countries compete for a contract in a fourth country. The dilemmas which arise are often insuperable if the ethics involved conflict with those acceptable in a company's home country and requires the company to relinquish its opportunity to compete. It is in such circumstances that the legal status of its actions in its home country can become paramount. If such situations are mishandled major scandals can emerge. But development of this topic is beyond the scope of this book.

4.5 Professional ethics in construction

Individuals working in the construction industry who are members of professional institutions (architects, engineers, surveyors and builders) are faced with the issues of business ethics and personal ethics in the same way as everyone else. They will respond on the basis of their values and attitudes which determine their morality and personal ethics and underpin the business ethics of their organization. But, as a condition of continuing membership of the professional institution to which they belong, they are also subject to the ethical rules of conduct of their institution which demand a higher level of ethics relating to the activities of their profession than found in society in general. A business ethics survey commissioned by the Independent Commission against Corruption in Hong Kong showed that the ethical conduct of a company or profession affects consumers' decisions to buy goods or services from it (Hong Kong Ethics Development Centre (HKEDC), 1996). Fan et al. (2001) comment that it is quite clear that in Hong Kong the general public has greater expectations of the professions than the non-professions and believes that they can uphold the consumers' interests. Membership of a professional institution is expected to influence the personal ethics of members in a positive way but this may not be so in some individual cases, resulting in some members breaching the rules of conduct for which sanctions are applied by the institution.

Most of the professional institutions in the UK associated with construction were established in the mid to late 19th century and subsequently gained royal patronage which gave them great respectability. They have spread in a similar form to many Commonwealth countries with numerous other countries having associations for each profession but which may differ in orientation and objectives from those originating in the UK. The UK institutions were established by each of the specialist contributors to the construction process with the prime objectives of ensuring high standards of education and skills in their members through examinations on the one hand and high ethical standards on the other in order to protect the public and their clients from the unscrupulous. The integrity induced by these standards may explain why UK professionals are generally held in high regard internationally.

Applebaum and Lawton (1990) define a profession as a group of people organized to apply a body of specialized knowledge in the interests of society. There is no doubt that construction professionals have a great duty of care to the public as well as to their clients as their level of competence can have serious repercussions on the public, physically, aesthetically and economically. Acting with integrity in this position is often perceived as 'treating the public and their clients as they would expect to be treated themselves'. However, applying this principle can frequently lead to conflict when the interests of the public and the interests of their clients do not coincide. Professional ethics have been said to be a system of behavioural norms (Bayles, 1988). They are enshrined in rules of ethical conduct that are promulgated by the professional institutions as a statement of what the public and clients can expect from members, and so lay a moral responsibility on all members practising a particular profession. Any failure by a member to meet ethical standards reflects on the profession as a whole so pressure on members to conform to ethical standards is high. Rules of ethical conduct are supported by sanctions which at their most severe can result in expulsion from the profession. Increased and increasing public awareness of environmental issues and the impact of construction projects on society in many other ways have put pressure on professional institutions to be responsive to such issues, which have in turn caused them to place increasing responsibility on their members, often creating conflicts with the objectives of their clients. Codes of conduct also protect members by providing a code which, if followed, is seen to be acceptable as the norm for their professional performance and, hence, against which their performance can be judged. As rules of conduct impose common standards of behaviour on all members, a level playing field is created which brings a measure of fairness to competition for all members. Some countries, for example, South Africa, go so far as to require by law that each professional body has a code of conduct for its members (Bowen et al., 2007).

In a survey of architects, contractors, project managers and construction managers in a major Australian conurbation regarding professional ethics (Skitmore and Vee, 2003), 90 per cent belonged to professional institutions which have ethical codes of conduct and 45 per cent worked for organizations which also have their own codes. A similar survey in South Africa (Bowen et al., 2007) showed that 98.3 per cent of the respondents surveyed belonged to professional institutions which have codes of conduct. It is to be expected that similar coverage will occur in the UK and other countries drawing their traditions from the UK. With such coverage the lack of an existing ethical infrastructure can be ruled out as a justification for unethical behaviour (Skitmore and Vee, 2003). However, severe competition in the industry presents challenges to both contractors and professionals in maintaining ethical standards. Contracting has always been a competitive business and competition between professional firms in the UK increased rapidly following the deregulation of their fee scales as a result of Monopolies Commission reports in the 1970s. Whilst professionals and professional firms continue to be required by their institutions to abide by their codes of conduct following deregulation, it is felt by many that ethical standards declined with the advent of increased competition. Contracting companies are generally not constrained by ethics to the same extent as professional firms even though the people employed by them will be constrained by the rules of any professional body to which they belong, with the potential for ethical dilemmas. That is not to say that professional organizations and their employees will not face ethical dilemmas and behave unethically, as there is much scope for this in the milieu of construction and development, but that contracting companies are particularly vulnerable to ethical challenges which may lead to unethical behaviour. Fellows (2006) makes the point that ethical problems may occur incrementally through small, negligible, unethical acts that accumulate into an ethical problem. He says, 'Thus, although it may be easy and tempting to dismiss a marginally unethical (or immoral) action as being insignificant, at least in its own consequences, that is a dangerous perspective due not only to accumulation but also to its possible impact on the person's perspective on what is of ethical significance in the future – a possible progressive change or erosion of ethical standards.'

In Skitmore and Vee's (2003) survey, 84 per cent considered good ethical practice to be an important organizational goal and 93 per cent considered that business ethics should be governed by personal ethics. Eighty-four per cent believed that a balance of client requirements and the impact on the public should be maintained. No respondents were aware of any cases of employers forcing their employees to initiate, or participate in, unethical conduct. They go on to say: 'Despite all this, all the respondents had witnessed or experienced some degree of unethical conduct, in the form of unfair conduct (81 per cent), negligence (67 per cent), conflict of interest (48 per cent), collusive tendering (35 per cent), confidentiality and propriety breach (32 per cent), bribery (26 per cent) and violation of environmental ethics (20 per cent).' A similar survey by Bowen et al. (2007) had a broader group of respondents, including additionally quantity surveyors and consulting engineers. The results were consistent between the two surveys with Bowen's finding that 78 per cent of respondents experienced some breach of ethics, with 32 per cent finding conflict of interest, 24 per cent finding confidentiality and propriety breach, and 22 per cent environmental damage. Interestingly, quantity surveyors were aware of significantly greater numbers of violations of professional responsibilities than the other professions. This was seen to be because quantity surveyors were more closely involved with all participants in projects, spanning clients through to contractors. In addition, this survey found that between 30 per cent (engineers) and 54 per cent (contractors) had witnessed or experienced breaches in professional obligations to the public, notwithstanding that 82 per cent saw a balance between professional obligations to the client and general public as being very important. An interesting result emerged from a survey by Fan et al. (2001) in which they found that 'respondents who are older, have more work experience and hold senior membership [of professional bodies] believe the interests of the general public to be more important in decision-making. However, the attitudes of young QSs are very different. The overall responses tend to rank employer, self and client as more important, whilst the interest of the general public is in a relatively low position'. A study by Ho and Ng (2003) examined the relationship of the background and training of quantity surveyors in Hong Kong to their ethical concepts, which built on Fan et al. (2001) and found that 'ethical perceptions of professional quantity surveyors are found to exist among professional quantity surveyors of different ages, membership levels and work experience. In predicting ethical perceptions, it is confirmed that the more experienced and the higher the education level of quantity surveyors, the more optimistic they are concerning recent declines in ethical standards. This group of quantity surveyors is more willing to sacrifice self-interest when facing ethical dilemmas'. These outcomes have a hint of the perspectives on values referred to by Robbins and Judge (2008) when comparing generations discussed earlier (in this chapter and in Chapter 2). These results also illustrate the paradoxical nature of professional ethics in practice in construction. There is evidence of respondents' belief that ethical behaviour is of major importance, contrasting with evidence of a wide range of unethical behaviour in practice. It is recognized that the surveys are of opinions and are not of validated evidence of actions which could reflect a difference between what is said and what is actually done. Nor did the surveys include the opinions of clients or the public or the reality of the issues affecting them.

The complexity of the construction industry and its projects and the interdependency of the participants present many opportunities for unethical and illegal behaviour. However, the vigilance of professional bodies and their members have to a large extent been able to curb such behaviour through their codes of ethical conduct. Nevertheless, in such a large and complex milieu, unethical and illegal practices continue. Distinguishing between actions which are illegal or actionable at law (and so by definition unethical) and those which are solely unethical can be difficult at times. Corrupt actions, such as bribery and fraud, are obviously illegal; others are not illegal but are unethical in terms of ethical rules of conduct and may be liable to sanction by a member's professional body, for example revealing a contractor's bid rates to others, an engineer working for the client and the contractor on the same project, misuse of a client's funds. Many breaches may be illegal but no action taken; instead they may be pursued through the transgressor's professional ethics mechanisms. A wide range of detailed illegal/unethical actions are identified in Skitmore's and Bowen's surveys, examples of which are:

- Collusive tendering, including price fixing, cover pricing and hidden fees and commissions
- Client or consultants revealing costs to competitors or favoured contractors
- Contractors removing trees marked for keeping
- Waste dumping in unauthorized locations
- Professionals siding with the clients in situations of dispute, where their fees are paid by the client
- Use of other designers' drawings without permission
- Employees of clients and consultant firms having personal interests in a project
- Awarding contracts to former employees and friends
- Unauthorised clearing of vegetation
- Charging clients for work not done, and claiming costs not incurred or overstated.

The giving of gifts and corporate entertainment is an interesting area in ethical behaviour. They can be viewed as a 'thank you' for good past relationships or an inducement to curry favour; as a mechanism for oiling the wheels of business; or to generate an obligation to reciprocate in some way in the future. If small in value they may be seen as innocent but when do they become to be seen as more than that? Certainly corporate entertainment can become substantial – an invitation to a major Premier League football match with chauffeur driven transport, meals, etc.; similarly, a day at the races or at the British Open Golf requires a large financial outlay. The business ethics of some companies may not allow staff to accept such offers but the scale of sponsorship of such events seems to argue against that generally being the case, except perhaps in the public sector where rules of conduct of public servants may not allow such entertainment to be accepted. The Royal Institute of British Architect's (RIBA) code of conduct for its members makes specific reference to this issue and states that [accepting] '(such as promotional gifts and corporate hospitality) is not prohibited so long as the value to the recipient is not such that it exerts an improper influence over them', showing that the RIBA is conscious of the issue.

Skitmore and Vee's (2003) findings indicate that contractors are the most unethical of the groups surveyed by them. They were rated the most unethical on all areas except for negligence where architects rated higher. This view reflects a general public perception of builders, often gleaned from small jobbing builders rather than major contracting companies. But contractors do exist in a business environment which differs from those of professional firms. The reputation for integrity of professional firms is determined by the ethical way in which they conduct their businesses and, although they frequently bid for work, their reputation as well as their price is often taken into account by the client. In addition, professional firms carrying a professional designation have to meet the criteria of their professional body; also the large majority of their employees will be professionally qualified and so subscribe to their professional code of ethical behaviour. Nevertheless, professional firms are under increasing commercial pressure which can lead to behaviour which does not reach the perception of the highest standards of the past. Contractors, on the other hand, are in an entirely commercial business environment and although they may have membership of trade organizations these tend not to have the force of professional bodies. Therefore they subscribe to the organization's specific business ethics for which there is a wide range amongst contracting companies in contrast to a much narrower band of ethical behaviour amongst professional firms. Complicating the situation is that many contractors' employees will be professionally qualified and required to abide by the ethical code of conduct of their profession. Thus, they are more likely to face ethical dilemmas when their professional code conflicts with the ethics of the contracting company. Although business ethics are said to be strongly influenced by the personal ethics of employees, and contractors are often staffed by professionals, nevertheless business ethics of contracting companies are likely to be determined by the personal ethics of senior management which may not be constrained by professional bodies.

The perception of the need for construction professionals to take cognisance of both the public interest and their client's interest frequently faces them with ethical dilemmas, as favouring one can disadvantage the other. Whilst they may try extremely hard to balance conflicting demands it is felt by commentators that they see their obligations to their client outweighing their social responsibility to others such as the public (Skitmore and Vee (2003) quoting Johnson (1991), but the latter reference refers to engineers in the USA where cultural differences may be present). Ambivalence towards this issue is shown by Bowen et al. (2007) in their statements that 'At the most fundamental level, the client must be able to trust the professional to operate in the client's best interests, and it is the professional's responsibility to do so' contrasted with 'The importance of balancing professional obligations to clients and the general public is emphasised by a majority of the respondents' and '... professionals who are supposed to demonstrate professionalism, morality and social responsibility'. The tension in these situations is created by clients who place pressure on professionals to act entirely in their interests whatever the circumstances. In Skitmore and Vee's survey the incidence of clients who act unethically was second highest only to contractors, causing them to conclude that 'What is new, however, is the emergence of clients and government bodies as contenders in the unethical stakes'. But this outcome is hardly surprising if one considers the vast range of private client types served by the construction industry which is enhanced by the increasing commercialisation of public sector organizations. Generally, private client organizations do not subscribe to codes of professional ethics but subscribe to business ethics which, although it is claimed that they are generated from personal ethics of employees, are in general not likely to include, in writing or in deed, the stringent requirement to take into consideration social responsibility issues to the extent expected by professional ethics. However, in some cases businesses may bind themselves to formal corporate ethics codes. Thus a clash between the ethics of clients and those of their professionals is to be expected from time to time, with the tendency for the clients' values to predominate, although their professional advisors will no doubt seek to persuade their clients to their views. If such issues cannot be resolved in favour of the professional advisors what are the professionals to do? Lose a client and income or protect their reputation? Maybe it is a matter of degree but where to draw the line? Although public sector clients are becoming more commercialized, they are likely to have more stringent codes of business ethics than the private sector. However, this is not to say that their members exhibit higher ethical standards in practice.

Whilst the focus here has been on the direct clients of the construction industry, the broader perspective of the ethics of stakeholders is also of significance. In this respect, Moodley et al. (2008) have proposed a stakeholder ethical responsibility matrix (SERM) to assist firms in the construction industry in identifying and understanding their potential ethical responsibilities to a wide and diverse range of interested and influential parties at a corporate and project level and orientating themselves to the dilemma of conflicting stakeholder and ethical demands. They used a modification of linear charting and linear responsibility analysis (Walker, 2007) to identify the relationships between the stakeholder, ethical responsibilities facing the firm and the interaction that exists between them from the viewpoint of ethics, social contracts and corporate responsibility. The vertical axis of the matrix lists the ethical and related issues; the horizontal axis lists the stakeholders. Once the horizontal and vertical elements have been established, the matrix is developed by inserting symbols in each box of the matrix which define the relationship between the elements such as the importance of the responsibility to the stakeholder, cross-cultural influences, ethical risk, ethical support, and ethical consultation. The broad categories of ethical issues on the vertical axis, each of which is broken down into detailed sub-categories are:

Accountability	Conduct	Labour & human rights	Health & safety
Environment	Community	Product	Supply chain

The stakeholders listed on the horizontal axis are:

Shareholder	Staff	Company executive	Financiers
Unions	National government		Local government
Regulators	Major client 1	Major client (environmental)	Alliance partners
Tier 1 supply chain	Tier 2 supply chain (overseas)		NGOs
General public	Community groups		Trade associations

The authors believe that 'the SERM offers a way forward for managing the ethical dimension of construction industry decision making'.

There is no doubt that many of the issues arising in the field of professional ethics are far from being black or white. There is much that is grey in which professionals' integrity may be the only guarantee of good practice. Whilst professionals' integrity is formed by their values and attitudes, it is also strongly influenced by what is expected of them as professionals as determined by the culture of their profession as a whole. Essentially, it is to maintain the level of collective integrity of a profession that professional bodies create codes of professional conduct.

4.5.1 *Codes of conduct*

In the codes of conduct of the UK professional institutions associated with construction the emphasis is on integrity, but they document their codes differently in terms of style and context. The Royal Institute of British Architects (RIBA) introduced a principle-based code in 2005 which lays down clear and simple rules for integrity, competence and relationships which avoid ambiguity. However, they are complemented by relatively more complex guidance notes and are extended to deal with profession-specific issues such as registration, competition to win appointments and continuing professional development. The rules for individuals and firms are contained within the one code. Similarly, the Royal Institution of Chartered Surveyors (RICS) introduced principle-based rules in 2007 but separately for members and for firms. The principles of the member's code cover integrity, competence, service, lifelong learning and solvency and the principles of the code for firms cover integrity, competence, service, training, complaints handling, client's money, professional indemnity insurance, advertising, solvency, incapacity or death of sole practitioner and use of designations. Whilst the RICS cover a wider range in their principles, the RIBA cover the same ground but with more in guidance notes which are more detailed and deal with architecturally-specific issues. The rules of conduct of the Chartered Institute of Building (CIOB) cover essentially the same ground but are articulated in detail rather than being presented as principles.

The structure of the codes of conduct in engineering is rather different from the RIBA and RICS as there is a more complex array of institutions involved. Individual members are governed by their professional institutions, for example, the Institute of Civil Engineers (ICE), the Institute of Structural Engineers (I Struct E) and the Chartered Institute of Building Services Engineers (CIBSE). Individuals also need to be registered by the Engineering Council and are associated with the Royal Academy of Engineering (RAE). The ICE has endorsed the RAE's 'Statement of Ethical Principles' which, in common with the RIBA and RICS, are clear concise principles under headings dealing with 'accuracy and rigour', 'honesty and integrity', 'respect for life, law and the public good' and 'responsible leadership: listening and informing' but has in addition its own code of conduct which has extensive guidance notes. The ICE rules do not incorporate engineering firms separately; engineering firms can become members of the Association of Consulting Engineers (ACE) which publishes its own code of business ethics dealing with client relations, professional standards, independence, reviewing or taking over work, insurance and fairness to others.

The codes of conduct of professional institutions are a special case of imposing standards on individuals and businesses, particularly in relation to the construction industry, but more generally corporate ethics codes are to be found in many types of businesses. They go under a variety of names but essentially aim at ensuring employees adopt high ethical standards in order to create confidence in their companies by society at large. In construction they may run alongside the requirements of professional institutions but construction companies are most likely to have corporate ethical codes due to the nature of their activities in which the potential for ethical dilemmas is likely to exist. What constitutes a corporate ethical code has not been established, nor has it been shown that they are effective in affecting company employees' ethical behaviour (Ho, 2010). Nevertheless, she cites the Hong Kong government's review of the construction industry which, among other things, led to the Hong Kong Housing Authority requiring all general contractors on its registered list to develop a corporate code of ethics/conduct (with the help of the Hong Kong Ethics Development Centre). If they did not comply by the given date they would not be invited to submit tenders. Necessary conditions for codes to be beneficial are an appropriate organizational culture and effective communication.

4.6 Trust

Only rarely are ethics linked to trust in the literature but ethical behaviour is the foundation of trust. A person's values such as integrity and honesty are the cornerstones which create trust in others and which encourage others to trust them. Howell and Costley (2006) illustrate the relationship through the unethical behaviour of high-level leaders in large organizations who have benefited themselves and inflicted harm on others, including followers, customers, and investors who trusted them.

Trust has long been recognized as a significant feature in business but it did not emerge as a focus for organizational research on any scale until the 1990s. Why it became of interest to researchers then is unclear but Westwood and Clegg (2003) suggest that it could be due simply to neglect or to the emergence of loosely coupled organizational structures with extensive networking and a move from explicit to implicit controls and coordination in organizations which rely more on trust. Kramer and Cook (2004) added to this that such organizations were 'embedded in high-velocity environments in which rapid change necessitated swift assessment and action' which could not rely on traditional organizational theories. Kramer (2003) comments that 'trust has rightly moved from bit player to centre stage in contemporary organizational theory and research' and that 'recent theory and research have sharpened our understanding of the complexity of trust within organizations and enhanced our appreciation of the myriad and often subtle benefits such trust confers'.

It is perhaps somewhat surprising to those in construction to find organizational theorists' recent recognition of trust as an important field for research as those in construction have long found trust – or, more accurately, mistrust – significant. Nevertheless, official reports on the construction industry just before and after the turn of the century (Walker, 2007) focused on reducing the adversarial nature of the industry and increasing trust, which reflected the resurgence of trust as a mainstream organizational theory topic about the same time. Focus on trust is now evident in construction in the increase in conciliation and other more benign dispute resolution processes and particularly as a major element in the use of relational contracting as a procurement method.

Kramer (2003) and Sievers (2003) are most concerned by the emergence of 'popular' management books promoting trust as 'a panacea for all types of organizational ills' as such shallow treatments disguise the complexity and paradoxes inherent in the concept of trust. Sievers is particularly concerned about the decline of trust within organizations and believes that the current interest shown in trust is 'an indication of an underlying, even subconscious, anxiety about the decline of relationship values, meaning and trust' and, in support of this view, identifies a whole range of organizational breaches of trust and continues that 'we do not secure a better future with simplistic quick fixes'.

The ambiguity, complexity and paradoxes of trust have been recognized for many years (Gambetta, 1988) and continue to perplex (cf. Murnighan et al.,

2004; Kadefors, 2004). Gambetta's (1988) classical edited review included his significant early definition: 'trust (or symmetrically distrust) is a particular level of the subjective probability with which an agent assesses that another agent or group of agents will perform a particular action, both *before* he can monitor such action (or independently of his capacity ever to be able to monitor it) *and* in a context in which it affects *his own* action … When we say we trust someone or that someone is trustworthy, we implicitly mean that the probability that he will perform an action that is beneficial or at least not detrimental to us is high enough for us to consider engaging in some form of cooperation with him. Correspondingly, when we say that someone is untrustworthy, we imply that that probability is low enough for us to refrain from doing so.' Whilst Gambetta considered, in his edited review, that there was a degree of convergence of the definitions of trust, Costa (2003) believes that there is still a lack of a consensus, although Rousseau et al.'s (1998) definition is probably the most frequently cited:

> Trust is a psychological state comprising the intention to accept vulnerability based on positive expectations of the intentions or behaviour of another.

And follows that of Johnson-George and Swap (1982) who defined trust as:

> The willingness of a party to be vulnerable to the actions of another party based on the expectation that the other will perform a particular action important to the truster, irrespective of the ability to monitor or control that other party.

Thus, trust is personal and reciprocal. The evolutionary nature of our understanding of trust is demonstrated by the transition from Gambetta's (1988) definition to the current focus on the willingness to be vulnerable (Rousseau et al., 1998), with much detailed debate in between (cf. Costa, 2003).

The innovative work of Gambetta and Hamill (2005) continues to pursue the theoretical complexity of trust. They state that, 'In general, we say that a person trusts someone to do X if she acts on the expectation that he will do X when both know that two conditions obtain; if he fails to do X, she would have done better to act otherwise, and her acting in the way she does gives him a selfish reason not to do X'. They also state that 'the trust we have for someone in one context does not necessarily extend to other contexts'. In their theory's development they state that 'the assumptions about people's motivations that are characteristic of the rational choice approach [see below] to trust play no role in the theory we develop here'. They believe that their approach which incorporates moral principles and social norms as well as self-interest in trustworthiness is more realistic. Their ideas also deal, for example, with the information on which we base trust, trusting such information, the difficulty of communicating trust and fraudulent signs intending to suggest trustworthiness.

4.6.1 Types of trust

Classification of different types of trust has had similar problems to defining trust, with many authors seemingly having to work out classifications in their own terms. An example related to construction contracting is Wong et al. (2008) who identify a range of classifications by different authors (most of which contain common elements) and then condense them into their own classification of systems-based, cognition-based and affect-based trust, as shown in Table 4.1.

They see cognition-based trust as describing 'a trusting relationship that builds on mutual understanding through fruitful information exchange and acquaintance'. 'System-based trust focuses on formalized and procedural arrangements with no consideration of personal issues.' 'Affect-based trust builds on a sentimental platform. It describes an emotional bond that ties individuals to invest in personal attachment and be thoughtful to each other.' The empirical results of their work to identify a framework for trust in construction contracting carried out in Hong Kong, 'suggests that all three forms are of almost equal importance in trust building'. Another study of trust in construction by Kadfors (2004) summarizes and uses Rousseau et al's. (1998) more recognized classification of three basic forms of trust:

Calculus-based trust (also known as the rational-choice approach): In this view, individuals are regarded as motivated primarily by economic self-interest and such 'trust' is often based on economic incentives for cooperation and contractual sanctions for breach. But it has been found that, in such situations, the presence of a contract between the parties meant that the parties gave credit to the contracts rather than to their counterparts for their cooperation (Murnighan et al., 2004). They therefore found that formal binding contracts between parties interfered with the process of forming trust and so reduced the likelihood that the parties would easily develop trust. These issues do raise the question of whether this perception of trust is really trust in the terms of Rouseau et al's definition.

Relational trust is between individuals who repeatedly interact over time and who must decide if they can trust each other. Thus the perception of the trustworthiness of each party by the other is paramount, which in turn depends on their view of each others' values and ethics, such as integrity and honesty. Reciprocity is required as the truster wants his trust honoured and also want to be trusted. Murnighan et al. (2004) make a significant point in saying that greater risks are taken by the truster earlier in the interaction of two parties and that 'initial trusters who take considerable risks may suffer considerably from their choices, sometimes very quickly'. They continue: 'the paradox here is that because partial trusters are judged harshly, greater risks are necessary to increase the chance that the trusted party will reciprocate and that trust will develop. But the greater risks are serious gambles that can result in considerable net losses for trusting initiators.' The relational concept of trust closely reflects the focus on vulnerability in Rouseau et al's definition and represents

Table 4.1 Classification of trust in previous studies (Wong, K. W., Cheung, S.O., Yiu, T. W. and Pang, H. Y. (2008) A framework for trust in construction contracting. *International Journal of Project Management,* 26, (8), pp. 821–829, Elsevier).

Authors	Definitions and scopes		Dimension of trust
Luhmann (1979)	Personal trust	Personal trust involves an emotional bond between individuals. Its emotional component acts as a protective base of trust when experiencing betrayal or destructive events	Cognition-based and affect-based
	System trust	System trust contains no emotional content. It rests on a presentational base and it is essential for the effective function of money or power exchange	System-based
Lewis and Weigert (1985)	System trust	System trust refers to trust in the functioning of bureaucratic sanctions and safeguards in terms of the legal system	System-based
	Cognitive trust	This trust is a combination of low emotionality and high rationality and is based on some ‘good reasons’ constituting evidence of trustworthiness	Cognition-based
	Emotional trust	A mix of high emotionality and low rationality constitutes emotional trust which involves bonds of friendship and love	Affect-based
McAllister (1995)	Affect-based trust	Affect-based trust refers to the intention to provide extra help and assistance that is outside an individual’s work role without remuneration	Affect-based
	Cognition-based trust	Cognition-based trust develops based on the success of past interaction, the extent of similarity, and organizational context considerations	Cognition-based

Rousseau et al. (1998)	Calculus-based trust	This trust describes the perceptions about one to another regarding some beneficial issues. References, certificates and diplomas are the media promoting calculus-based trust	Cognition-based
	Relational trust	This trust arises from continual interactions between individuals. Emotions and personal attachments are also influential to the trusting relationship	Affect-based
	Institution-based trust	Legal systems, conflict management and cooperation, systems regulating education and professional practice were suggested as tools to shape trust in institutions	System-based
Hartman (2000)	Competence trust	Competence trust is defined to be based on one's perception of the other's capacity to perform	Cognition-based
	Integrity trust	Integrity trust is founded upon one's perception of the other's attitude to act ethically and be motivated not to take one's advantage	Cognition-based
	Intuitive trust	Intuitive trust involves emotion and intuition about one's impression to the other	Affect-based
Kramer (1999)	Dispositional trust	This trust develops based on the build-up of general belief on early trust-related experience	Affect-based
	History-based trust	Individuals develop history-based trust based on previous interactional information and experience	Cognition-based
	Third parties as conduits of trust	Individuals adopt second-hand knowledge in order to assess the other's trustworthiness	Cognition-based
	Category-based trust	This trust develops according to knowledge acquired from one's membership in a social or organizational category	Cognition-based
	Role-based trust	Role-based trust grows based upon knowledge of role relations, rather than specific knowledge about one's capabilities, dispositions, motives and intentions	Cognition-based
	Rule-based trust	Rule-based trust is subject to shared understandings of the system of rules concerning appropriate behavior	System-based

the commonly-held view of trust. It builds up slowly as the parties interact and learn to trust each other and take further steps. This process is no better elaborated in both personal and business settings than in the Chinese relationship of guanxi in which relationships develop into extraordinarily tight bonds of trust without contracts (So and Walker, 2006). Relational trust leading to cooperation is strongly influenced by emotional and intuitive attachment and the behavioural characteristics of the parties, such as the parties showing respect and concern for each other.

Institution-based trust refers to the role of institutions in providing the conditions necessary for trust to arise. They are the legal systems and societal norms relating to conflict management and cooperation and systems relating to education and professional practice. Cultural rules affect our views of trustworthiness of categories of people and organizations, reflecting aspects of stereotyping. Whether this is a form of trust is questionable, it is rather the context of trust.

4.6.2 Antecedents of trust

An example of what causes complexity and confusion in our articulation of trust is Zaghloul's and Hartman's (2000) and Hartman's (2003) identification of 'three bases of trust'. Wong and Cheung (2004) call them 'competence trust, integrity trust, and intuitive trust' but rather than types of trust these are attributes which may assist in establishing relational trust and equate with the commonly accepted antecedents of trust which are, according to Mayer et al. (1995) ability, integrity and benevolence. Ability is competence in the subject of the trusting topic, integrity is the ethical values of those involved and benevolence is a good feeling towards the topic and the people involved. All three antecedents are required for trust to occur. Antecedents of trust do not often feature in the construction literature; a rare example is Kadfors (2005) who examines fairness in inter-organizational project relations which has also been referred to in Chapter 2 in relation to values. Fairness can be seen to be an aspect of both benevolence and integrity but Kadefors does not relate fairness directly to trust or values, although it underpins both, but her paper presents an interesting perspective.

Pinto et al. (2009) use Hartman's model of competence, integrity and intuitive forms of trust as a basis for carrying out an empirical assessment of owner/contractor relationships because the model was purposely developed to address trust within a project setting to a greater degree than other theoretical work. In their work the operationalization of intuitive trust was problematic as it was not statistically differentiated from integrity trust. They state that 'given the manner in which intuitive trust is defined, it is reasonable that there may be some blending of these constructs ...'. They continue that 'the lack of evidence for this third form of trust actually supports alternative two-factor models ...' Their overall conclusion states that 'This study supports ... the importance of trust as an antecedent variable for project performance. Specifically, trust is argued to *enhance* the strength of working relationships, to solidify partnering roles, and to increase the willingness of various project stakeholders to cooperate in non-self-motivated ways. However, the the study also demonstrated that the perception of the value of trust may differ depending upon which project stakeholder is surveyed. For our study, owners appeared to value integrity and competence trust from their partnership with contractors, while contractors themselves rated only integrity trust as a necessary predictor of positive relationships'.

Further clarifications of trust are identified by Murnighan et al. (2004) by distinguishing trust from cooperation, confidence and predictability, all of which they say have a history of being confused with trust. They cite Mayers et al. (1995) who, in distinguishing trust from cooperation, said '... trust is not a necessary condition for co-operation to occur because co-operation does not necessarily put a party at risk.' Luhmann (1988) is cited as differentiating between confidence and trust by also pointing out that trust requires the recognition and acceptance of risk but confidence in someone does not. Vulnerability can be substituted for risk in both distinctions. Mayer et al. (1995) also point out that trust is not predictability as, for example, one cannot say that someone who predictably always ignores the needs of others is to be trusted because the party is predictable.

The circumstance in which trust is to be formed can have a strong influence on whether it is or is not formed and whether it is sustained. Decisions on whether to trust or not are continuously revised as a result of interventions and information supplied from within the context of where it is established. Burt and Knez (1996) show how third party gossip accelerated the tendency towards, or away from, trust being formed in a relationship. Third party contacts attached to one or other party in a trust relationship may seek to manipulate for their own ends the predisposition for the two parties to trust. Zolin et al. (2004) examined the development of trust in a situation in which co-workers were geographically distant from one another (in different cities or countries) compared to collocated workers. They expected to find less trust between team members who were geographically distant but did not do so. But they did find that perceived trustworthiness was lower than in collocated workers. Also that trust was more stable in geographically distributed workers and that trust was more volatile in collocated groups. Collocated team members more readily updated their trust whereas geographically groups relied on initial perceptions of trustworthiness, so first impressions were important.

4.6.3 The rational-choice approach

The rational-choice approach to trust, in which individuals are regarded as motivated primarily by economic self-interest and is often based on economic incentives for cooperation and contractual sanctions for breach, has complicated the concept of trust as defined by Rouseau et al. (1998) as it includes formal contractual mechanisms to enforce such trust. Murnighan et al. (2004) put it succinctly:

> Contracts and trust do substitute for one another. Thus, a lack of trust suggests that the parties might want to create a contract to specify their rights and responsibilities. Alternatively, when trust is strong, the parties may feel no need for the specifics or constraints of a contract. Instead, they may be able to fulfil a mutually beneficial agreement without having to resort to a contract's restrictions.
>
> Most approaches, whether slanted towards formal contracts or toward trust, acknowledge the value of both, even for the same agreement. Thus, even though contracts are important, they cannot possibly address all of the contingencies that might develop in a relationship. This makes it necessary to cultivate trust. At the same time, it may be crucial for the parties not to underestimate the need to 'get it in writing'. Their task then becomes one of creating contracts while simultaneously cultivating trust.

However, this position sends out mixed messages and we have seen that formal contracts can inhibit the propensity to trust. Where the need to trust is between organizations which need to collaborate to achieve some common objectives but also have independent objectives, as in partnering in the construction industry, tensions can arise which are difficult to manage. The major issue here is whether organizations have the capacity to develop trust between each other. Organizations will tend to reflect the quotation above and require contracts and trust. But trust is seen to be personal and so will exist or not between individual members of the contracted organizations, but their actions will be monitored in accordance with the contract which may communicate to the employees that they are not trusted. Similarly, economic rewards for cooperation hint that the target party will not take any action that is not motivated by self interest (Kadfors, 2004).

4.6.4 Trust in construction

The idea that an organization, often a large amorphous entity, can have an existence which is trusting is difficult to accept, although they have been referred to as such in the construction literature (McDermott et al., 2005). It may be common to hear someone say, 'you can trust that firm' but can a firm as a whole trust and be trusted? Who in a firm will accept the responsibility for the firm to be trusted? In the same vein, reference has been made to 'engineering' collaboration and trust in construction partnering through formal mechanisms (Bresnen and Marshall, 2000b), which hints at manipulation rather than the antecedents of creating trust. The use of financial incentives is also identified as reinforcing collaboration in the short term and helping to cement trust in the long term, which again does not reflect the antecedents of trust. The concept of trust in construction partnering does not appear to subscribe to the theory of relational trust expressed in the mainstream trust literature. This is further confirmed by Wong and Cheung (2004) who identified that, in Hong Kong, what they term 'system-based trust' which they say refers to 'legally binding agreements and terms where trust is relied on the formalized system *(sic)* rather than in personal matters' was ranked as the most important trust factor. By formalised system they mean 'law and contracts'. But this is trust in institutions, not individuals or even organizations. This shows the confusion in construction, particularly in partnering, of the concept and application of trust, the failure of which can mean that recourse frequently has to be made to contractual terms. While trust may play a part alongside contracts, individuals in construction (partnering or not) may trust each other but an agreement made between them based on trust may be overturned by their organizations. What effect does this have on trust, can one trust organizations?

As can be seen, the issue of trust impacts significantly on the construction industry, particularly since the move from the traditional dependence of formal contracts between parties due to the advent of partnering, but trust has not been a focus for research, rather it has been taken as a given. But this may be hardly surprising as the understanding of trust in the mainstream trust literature is not yet fully developed. A study by Munns (1995) addressed trust between construction team members in which personnel employed temporarily on projects were seen to view the development of long-term stable relationships as a secondary concern. He concludes that there are two key factors in determining the relationships which will exist on a project: 'these are the level of global trust experienced by all project members and the early actions of the team members in the formation of the team. It has been suggested that the first actions of the team members will be dictated by their level of global trust and it is the shaping of this global trust that needs to be understood.' It was also concluded that the initial opinions of individuals of their co-team members on entering the project are important in shaping its final outcome. However, this paper was written co-incident with the significant development of trust in the mainstream trust literature. A more recent paper on project team relationships (McDermott et al., 2005) made the point that 'relationships which may otherwise be positive may be hamstrung by institutional, organizational or project factors that will force individuals into taking untrusting positions. It is difficult to ask individuals to behave in a trusting way when the procurement framework or organizational culture means that an individual will be exposing himself or herself to excessive personal risk.' Jin and Ling (2005) undertook case studies of two projects in China in which they found that a deepening relationship gives rise to different types of inherent risks such as a partner's self-interest seeking behaviour and opportunistic actions. To counterbalance these risks, they identify a series of practical trust fostering tools which should be used, including careful selection and effective management of partners.

In construction partnering, trust between the parties is depended upon to an even greater degree but receives little theoretical attention as the literature tends not to be rooted to any great extent in mainstream trust research (Walker and Rowlinson, 2008). The difficulty of dealing with trust is illustrated by

Lazar (2000) in which he concludes that, 'trust based relationships are ... critical to maximizing positive economic outcomes from partnering and maybe necessary to keep the owner/contactor relationship from deteriorating – even those based on mixed (co-operative) strategies.' Yet a mixed strategy comprises collaborative and competitive strategies. He also points out that causes other than trust – for example, fear or coercion – may lead to collaboration. The role and contribution of trust to partnering is far from clear as there is scope for opportunism and hence the need for contractual arrangements to protect against it. Others attempting to relate trust to partnering demonstrate the elusiveness of the concept (cf. Wood and McDermott, 1999; Wong and Cheung, 2004). Bresnen and Marshall (2000b) examined the use of financial incentives in partnering and suggest that 'incentives do reinforce commitment and cement trust between organizations in the long term'. Their perception of trust in partnering appears to be between organizations rather than individuals as in 'significant changes and inconsistencies in internal polices and personnel can confound attempts to establish clear and stable expectations, and make any trust that is developed between organizations fragile and difficult to sustain' which points to trust between individuals not organizations and adds to the confusion regarding trust in partnering. Whilst referring to many papers in support of partnering, but pointing out that most are anecdotal, Wood and Ellis (2005) also highlight those that are sceptical. In their study of the attitudes and opinions of commercial managers with experience of partnering with a leading national contractor, their findings 'indicate a broad agreement that both the process and outcome of partnered projects are beneficial', but also 'as might be expected there are some ambiguities in the results of the study but underneath the veneer of partnering, some of the traits that have characterized the construction industry for many years are still apparent and genuine trust seems some way off'.

Definitions of partnering refer to the need for the members of the partnership to have a shared culture based on trust. But rather than the idea that organizations need to trust one another, which is difficult to conceptualise, organizations should seek to achieve cooperation and collaboration through shared cultures which will then provide a platform for individuals in each organization to trust each other. Whilst a shared culture is much more difficult to achieve on a single project, even when it is of long duration, there is a greater likelihood of achieving it over a long period and many projects. So perhaps the focus in partnering should be more on organizational cultures.

4.6.5 Trust and transaction cost economics

Before we leave trust, a brief look at trust in the context of transaction cost economics (TCE) is necessary. Hopefully, this will suffice for the purpose of this book which is about behaviour not economics. The major aspect of TCE having relevance to trust is the assumption of opportunism. Opportunism essentially means 'self-interest seeking with guile' and has profound implications on the choice and design of appropriate governance structures and organizational arrangements (Williamson, 1975). TCE argues that firms will seek to protect themselves against opportunism, often by contractual arrangements. However, theoretically, construction partnering adopts a fundamentally different approach by basing the relationship between the parties on trust as a protection against opportunistic behaviour. In its conceptual form partnering should not require contractual agreements between the parties but in reality they are invariably used but with the expectation that they will not have to be relied upon. The expectation is that partnering will reduce transaction costs through more amicable arrangements between the parties that will reduce the cost of protecting against opportunism and also, as an additional benefit, by reducing production costs through, for example, joint innovation.

In contrast, for traditionally arranged construction projects, firms enter costly contractual arrangements. The costs of such arrangements are part of the transaction costs of producing the building. Williamson (1985) recognizes such a situation and states '... Parties to a bilateral trade create credible commitments, whereby each will have confidence in trading with the other'. Credible commitments are seen as contractual safeguards, assurances and mechanisms. He goes on to recognize the differential hazards of breach that arise under different investment and contracting scenarios and presumes that suppliers are far-sighted and will bid for any contract for which an expected non-loss making result can be anticipated and that buyers will choose the contractual terms that best suits their needs. The parties therefore have a mutual interest in devising an exchange relationship in which both have confidence (Yates, 1998). Partnering is one such exchange relationship but one in which trust forms a significant part of the credible commitment. However, transaction cost economists find what Williamson (1990) describes as 'the use by social scientists of user-friendly terms such as trust' to be dubious as credible commitments but Westwood and Clegg (2003) claim that the virtues of trust are 'societal in terms of developing social capital and a civic society, and organizational in terms of reducing transaction costs and fostering appropriate corporate citizenship behaviours'. Williamson states:

> The growing tendency to use trust to describe probabilistic events from which the expected net gains from co-operation are perceived to be positive seems to me to be inadvisable. Not only does the use of familiar terms (like trust) invite us to draw mistaken parallels between personal and commercial experience, but also user-friendly terms do not encourage us to examine the deep structure of organization. Rather, we need to understand when credible commitments add value and how to create them, when reputation effects work well, when poorly and why. Trust glosses over rather than helps unpack the relevant micro-analytical features and mechanisms.

Yet he does not dismiss trust, but distinguishes between personal and commercial trust and asks for depth in its analysis. Personal trust can be seen to equate to relationship trust and commercial trust to calculus-based trust (rational-choice approach). In support of trust he cites membership of professions that are self regulating as giving trading confidence to transactions (can be seen to equate with institutional trust) and also recognizes that, for the purpose of economic organization, culture serves as a check on opportunism. Nevertheless, he does say that 'transaction cost economics refers to contractual safeguards, or their absence, rather than trust, or its absence'.

Partnering has focused on trust and culture without explicitly recognizing the relationship with transaction costs. Essentially, partnering is designed to reduce transaction costs through greater collaboration for greater efficiency, by not incurring costs associated with withholding information (through cooperation), and by reducing adversarial attitudes and hence the cost of disputes. In order to make such things happen a change in culture within partnering organizations has to be engendered. That is, a change in organizational culture is needed in order to allow relational trust to flourish and make partnering work effectively. But as Williamson argues, trust has a cost – a transaction cost. The cost of trust is the cost of the risk of trusting your partner without credible commitments, e.g., contractual safeguards, to protect against the performance of your partner not being achieved. Such a breach of trust would generate increased transaction costs which could result in the combined production and transaction costs being greater than would have been incurred using different arrangements, e.g., more conventional arrangements. Thus trust and organizational culture are important in understanding and developing partnering, as they are also in cooperation and collaboration generally within the construction industry.

5 Motivation

5.1 Introduction

Early writers on management, such as Taylor (1911), had a simple view of what motivated people at work; it amounted to power and money. This so-called 'scientific approach to management' was superseded by the human relations approach originating from the famous Hawthorn experiments (Mayo, 1949) which, as Scott (1992) states, 'served to call into question the simple motivational assumptions on which prevailing rational models rested'. By asking questions such as 'otherwise how do we explain what drives people to act selflessly?' the human relations approach identified motivation as much more complex, incorporating needs essential for survival at one end of a spectrum to self-actualization at the other (Maslow, 1954) such that a motive is seen as a person's reason for choosing one behaviour from among several choices and is derived from needs (Moorhead and Griffin, 2001). Whilst higher level needs arise in all aspects of life, they are most frequently found in work settings as a result of the strength of an employee's need for, for example, recognition, status, advancement and personal development (rather than other needs in life generally).

A powerful link is said to exist between motivation and job satisfaction. The idea, as stated in Chapter 2, is that those who say they have job satisfaction are expressing an attitude towards their job as a result of evaluating the characteristics of their job against their expectations. As a result, those with job satisfaction would be expected to be motivated to perform effectively. However, Wilson (1999) believes that 'job satisfaction and motivation are difficult subjects to study as there are so many views on offer, and so many of them offer contradictory positions. It is then a subject where arguments need to be weighed and balanced'. She also observes that as early as 1981 Kleinginna and Kleinginna (1981) had found no fewer than 140 definitions of motivation. She chooses Vroom's (1964) definition – a process governing choices made by a person or lower organisms among alternative forms of voluntary activity – but also adds, 'Dewsbury (1978) says that the concept

Organizational Behaviour in Construction, First Edition. Anthony Walker.

of motivation tends to be used as a garbage pail for a variety of factors whose nature is not well understood.'

More recent definitions include 'motivation is the set of forces that cause people to engage in one behaviour rather than some alternative behaviour' (Moorhead and Griffin, 2001) and 'motivation refers to the forces within a person that affects his or her direction, intensity and persistence of voluntary behaviour' (McShane and Von Glinow, 2003). The latter continue by saying that even when people have clear work objectives, the right skills and a supportive work environment, they must have sufficient motivation to achieve work objectives. The elements of motivation are: preference for a particular outcome; the amount of effort needed to achieve it; and the strength of the urge to persist in the face of obstacles. The motivation for managers to understand the motivation of employees is that, if they do so, they can influence employees' behaviour; that is, managers may be able to devise extrinsic and intrinsic rewards that provide motivation which leads to greater job satisfaction and higher performance.

At this point it may be helpful to explain the idea of intrinsic and extrinsic rewards. Intrinsic rewards are internal to the individual, such as feelings of satisfaction and accomplishment. These feelings of wellbeing do not need payment to be rewarding. Extrinsic rewards are valued outcomes controlled by others, such as salary rises and bonuses, promotion and recognition. It has been argued that intrinsic rewards influence motivation more than extrinsic rewards (Lawler, 1973). But this is likely to be the case generally only after the level of comfort provided by extrinsic rewards is considered to be sufficient for our needs.

Even though definitions of motivation may be moving closer to a common understanding and whilst there is much in the literature which demonstrates that theories of motivation show considerable overlap, they are still not fully compatible. This is hardly surprising considering the variables at play in the idea of motivation. Motivation concerns the innermost thought of individuals, each of whom have different personalities and characteristics which generate an extremely wide range of reasons for being motivated. The complexity of motivation is well illustrated by the feeling experienced by most people at some time of not actually being able to explain why they took a particular action, that is, 'I did it but I don't know why'.

The literature on motivation does not focus on motivating professional employees but is heavily slanted toward motivating a manual/semi-skilled/skilled workforce and needs interpreting to sift out that which is useful for motivating educated high-achievers such as professionals associated with construction. Huczynski and Buchanan (2007) draw on the research of Butler and Waldroop (1999) who argue that educated high-achievers are mobile because they can succeed in just about any job. The same can be said of construction professionals within the context of their specialist skills. Butler and Waldroop argue that *job sculpting* is needed in order to retain valued staff, which involves:

- Listening carefully to discover what really challenges, excites and motivates people
- Understanding the individual's life interests
- Designing both the job and the individual's career path to match those interests
- Using assignments as opportunities to sculpt the job for the person
- Reviewing performance regularly to ensure that work and career interests are consistent.

To what extent are such issues relevant to construction professionals? Traditionally, the type of work available for construction professionals in professional practices and public bodies was relatively narrow, being restricted essentially to carrying out professional tasks with increased challenges arising from the complexity of the projects being worked upon. However, whilst this type of challenge remains and becomes even more challenging, the increase in the scope of work undertaken by practices and public bodies has created opportunities for construction professionals to be involved in work far removed from their original fields of interest. Such development of the work base has required highly and broadly educated professionals who are adaptable and motivated to apply their talents to new fields. To retain and develop such well-educated high-achievers in construction and its spin-off fields, employers in both private and public sectors need to take on board the ideas of job sculpting or talented staff will seek challenges elsewhere. Nevertheless, in competitive commercial construction and related environments such ideas can appear idealistic as flexibility in the type of work available and its timing are not usually within the gift of the employer. But even though the various theories and techniques of motivation are mainly directed at a non-professional workforce, they can, with interpretation in the light of the above, aid our understanding of the phenomena applied to a professional workforce. In addition, although they may not give us absolute answers, they can shed light on how to approach the motivation of the myriad of diverse characters making up the professional construction workforce.

Motivational models commonly fall into two main categories: content theories and process theories. Content theories represent a needs-based perspective on motivation; that is, reflecting Maslow, they consider what a person requires or wants and how their needs emerge. They range from what is needed to simply survive (e.g., water, food, shelter) to more psychological needs (e.g., self-fulfilment). Psychological needs have particular relevance to organizational settings. Understanding what a person needs helps to identify what motivates a person. Process theories do not explain how needs emerge but consider the process through which needs are translated into behaviour which includes an individual's relationships with their environment and judgement about rewards, costs and choices. These ideas help to explain why people act in the way they do and so influence their performance and, hence, are also particularly significant in organizational settings.

5.2 Content theories of motivation

Content theories of motivation were amongst the earliest theories and emerged mainly in the 1940s and 1950s. They are so named because they reveal the content of motives and are concerned with identifying needs. Needs are deficiencies that trigger the behaviours necessary to satisfy them; that is, they motivate people to act. Amongst the foremost early accounts was Maslow's Hierarchy of Peoples Needs (1954), shown in Figure 5.1, which probably remains the most best known theory of motivation, particularly amongst practising managers.

His work identifies that:

- The needs hierarchy is based on needs, not wants.
- It operates on an ascending scale. As one set of needs becomes fulfilled the next higher set comes into play but a satisfied need is not a motivator.

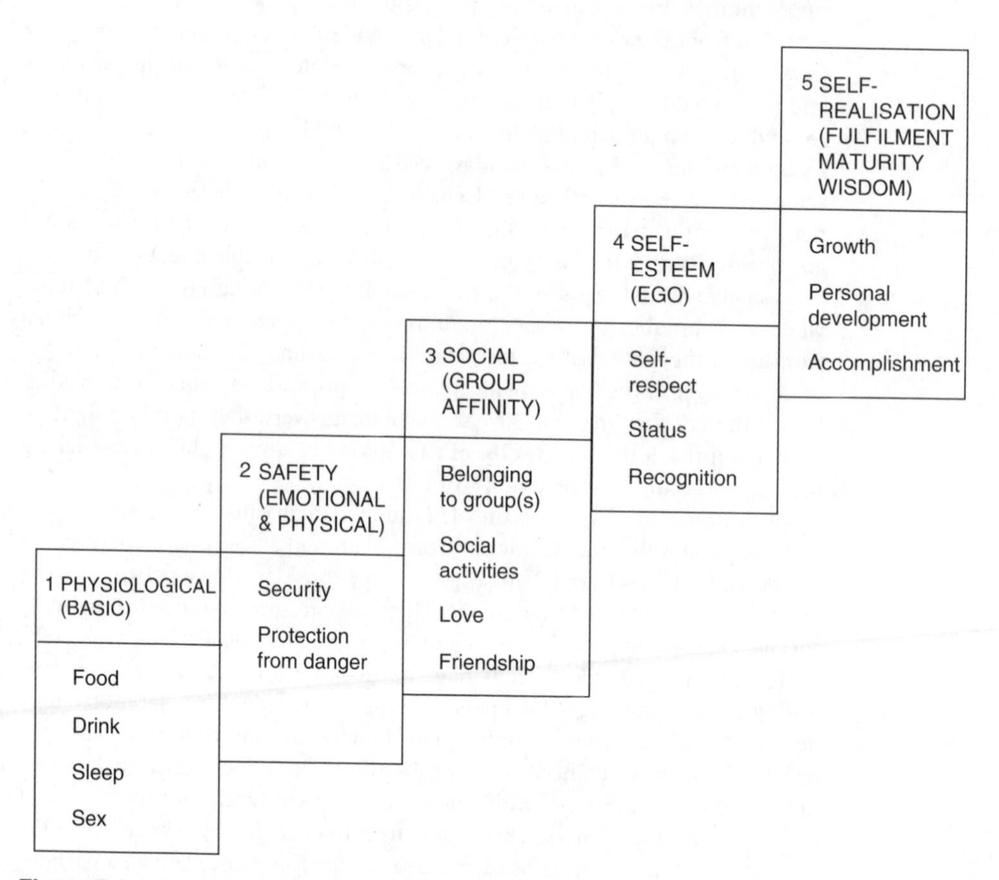

Figure 5.1 Maslow's hierarchy of individual needs.

- We can revert to a lower level. For instance, a person operating at level 4 or 5 will fall to level 2 if a feeling of insecurity takes over. Once the need is met, however, he will return to his former needs area.
- Needs not being met are demonstrated in behaviour. To create an environment in which motivation can take place, managers must therefore understand the level that a person is currently on to be able to recognize behaviour patterns in individuals and work groups. This means developing the ability to 'read' people and situations.
- To avoid apathy, which finally results when needs are unfulfilled, managers must be able to provide motivation at the right time.

The five levels are separated into lower and higher order needs. The lower order needs – physiological and saftey needs – are the very basic needs and are seen to be satisfied externally through the provision of resources. The higher level needs are psychological and are satisfied within a person as perceptions. The latter are therefore particularly important to higher educated professional employees when basic needs have been satisfied.

Maslow's theory has been criticized as lacking empirical support and several studies that sought to validate the theory found no support for it. Other criticisms have been that it is vague and cannot readily predict behaviour and also that it is more of a social philosophy reflecting white middle-class American values (Huczynski and Buchanan, 2007) and has, as a result, been found to be "culture bound". It is seen as too rigid to explain the diverse forces at play in employee motivation. It is also important to recognize that, whilst the ideas of peoples' attitude to work are intellectually satisfying and difficult to disagree with at that level, they are difficult to apply consistently in practice. Eilon (1979) has pointed out that, if everyone is self-actualizing, then the task to be achieved is likely to be neglected when it is routine but vital to success, which no doubt has resonance for many in the construction industry. Nevertheless, Maslow's influence has been important and extensive in showing that behaviour is dependent on a wide range of motives and has provided a significant basis for development of further ideas of motivation.

McGregor (1960) drew on Maslow's ideas in devising his 'Theory X – Theory Y' assumptions about how people behave in organizations. Assumptions underlying Theory X are:

- 'Individuals dislike work and will seek to avoid it.'
- Therefore, 'most people must be coerced, controlled, directed, threatened with punishment to get them to put forth adequate effort toward the achievement of organizational objectives'.
- 'The average human being prefers to be directed, wishes to avoid responsibility, has relatively little ambition, wants security above all.'

By contrast, Theory Y believes that:

- Most individuals do not 'inherently dislike work and that the expenditure of physical and mental effort in work is as natural as play or rest'.

- 'External control and threat of punishment are not the only means for bringing about effort toward organizational objectives.'
- The most significant rewards are those associated with 'the satisfaction of ego and self-actualization needs'.

A Theory X manager instinctively adopts an autocratic leadership style because he assumes, amongst other negative features, that people are lazy and wish to avoid responsibility. A Theory Y manager instinctively adopts a democratic style as he assumes, amongst other positive features, that people are self-motivated and wish to achieve and enjoy responsibility. As with Maslow's ideas, there is no convincing empirical research support for the assumptions.

The next motivation theory of substance was the 'Motivation– Hygiene Theory' proposed by Hertzberg et al. (1959) that focused specifically on attitudes to work. Following from questions asked of people about what they want from their jobs, Hertzberg concluded that the factors that influenced satisfaction were different from those which affected dissatisfaction; that is, satisfaction is not the opposite of dissatisfaction but rather that the opposite of satisfaction is 'no satisfaction' and the opposite of dissatisfaction is 'no dissatisfaction'. The factors leading to satisfaction are known as motivators and are job content factors such as the satisfaction derived from the job itself, and factors such as achievement, advancement, growth, recognition and responsibility and are more under the control of the individual. At the extreme satisfaction level, the outcome is its own intrinsic reward as in the case of musicians, actors and painters but similar intrinsic rewards can be achieved in work generally if perhaps not to the same extent. Factors leading to dissatisfaction are known as 'hygiene factors' and are organizational context factors such as company policy, pay, supervisory style, security and working conditions. They are not motivators but can lead to contentment and provide extrinsic rewards that are valued outcomes controlled by others. Thus this theory implies that the way in which a job is designed can have a great effect on motivation and satisfaction.

However, the motivation–hygiene theory has been criticized widely due to problems with its methodology, the lack of an overall measure of utility, the disregard of situational variables and the assumption of a relationship between satisfaction and productivity but with the research methodology only looking at satisfaction and not at productivity (Robbins and Judge, 2008). Also, subsequent studies have shown that employees are motivated by more than just the nature of their job itself. Some hygiene factors, which are said to lead to dissatisfaction, are actually widely used to motivate people, particularly through pay. Comparably, the image of a 'cool' working environment (a hygiene factor) in the high-tech and other industries is also seen as a significant motivator (McShane and Von Glinow, 2003). Hertzberg's work is widely known to managers as it is intuitively appealing but that does not make it acceptable. But, at the time of its creation in the late 1950s and early 1960s, it put a new interpretation on motivation by focusing on the motivational power of the actual work undertaken by an employee and so provided a new platform for further research and understanding.

As stated by Fincham and Rhodes (2005), 'If anything, the work of Maslow and Hertzberg has become too commonplace. It now suffers to some extent from being taught as if it were 'true' rather than a set of ideas.'

Hertzberg's theory appears to be the only content theory of motivation to be applied in the construction industry (Ruthankoon and Ogunlana, 2003). They tested the theory on Thai engineers and foremen and found that Hertzberg's theory was not entirely applicable in the Thai construction setting. Recognition, advancement, possibility of growth, and supervision contributed to job satisfaction, whilst working conditions, job security, safety on site, and relationships with other organizations contributed to job dissatisfaction. Recognition, work itself, company policy and administration, interpersonal relations, personal life and status contributed to both satisfaction and dissatisfaction.

McClelland's 'Theory of Needs' (McClelland, 1961) differs from the previously described content theories in that, rather than examining the individual's primary or instinctive needs and their relative importance, it is concerned with secondary drives, or needs, for *achievement, power* and *affiliation*. People with a high need for achievement are identified by their need to reach reasonably challenging goals by their own effort. Their major aim is personal achievement rather than the rewards of success. They prefer tasks of intermediate difficulty and moderate risk to provide a reasonable chance of success and perform best when they perceive a 50:50 chance of success. People with a need of high achievement are not gamblers. They prefer to work alone and desire rapid feedback by which they can judge their performance, and recognition of their success. Successful entrepreneurs tend to have a high need for achievement as they prefer to work individually on challenging problems and accept personal responsibility for success or failure. Those with a high need to achieve are unlikely to be good managers or team leaders because they are not good delegators or team players. The need for power as a motivator is the desire to control all aspects of one's working situation, including people and resources. Its primary concern is not performance but the ability to influence events and is concerned with the need for things such as prestige, status and maintaining a leadership position. McShane and Von Glinow (2003) point to the difference between the need for personalized power and socialized power. People with a need for personalized power enjoy power for its own sake and use it to advance their personal position. Those who need socialized power want power in order to help others by improving society or improving organizational effectiveness. Whilst corporate and political leaders have a high need for power to influence others, McClelland argues that they need socializd power, not personalized power, as they need to be concerned about the consequences of their actions on others within a framework of moral and ethical standards.

The need for affiliation as a motivator is the desire to be approved of and accepted by others. People with this need seek friendship, cooperative situations and positive relationships. Their mode of operation is to support others, seek to overcome conflicts and look for compromises. Whilst they fit into coordinating roles in organizations, they have been found to be indecisive and ineffective in allocating scarce resources as they are reluctant to generate conflict. Robbins and Judge (2008) point out that the needs for affiliation and power tend to be closely related to managerial success and that the best managers are high in the need for power and low in the need for affiliation but this idea conflicts with much of the literature on leadership, particularly in relation to team leadership.

McClelland argued that these needs were learned through childhood learning, parental styles and social norms rather than being intuitive and so his ideas contrasted with earlier theories. What McClelland did was to introduce the way in which personality influences motivation. His ideas were forerunners of elements of the 'Big Five Model' of personality, and particularly Type A and Type B personalities.

Following on from the above theories, 'ERG Theory' (Alderfer, 1972) aimed to enhance Maslow's needs hierarchy theory. ERG represents the first letters of three basic needs categories – existence, relatedness and growth. Existence needs correspond to Maslow's physiological and safety needs, relatedness needs are similar to Maslow's social needs, and esteem needs and growth needs are comparable to Maslow's self-realization needs shown in Figure 5.1. Two major differences exist between ERG and Maslow's theories. One is that ERG theory allows for more than one kind of need to be motivating a person at any one time. This, for example, can mean that people may try to satisfy their growth needs even though their relatedness needs are not fully satisfied. The second and particular difference is that, like Maslow, ERG theory includes a satisfaction–progression component, but it also includes a frustration–regression component. Like Maslow, ERG suggests that after satisfying one category of need a person progresses to the next level (satisfaction–progression). The difference lies in ERG suggesting that a person who is frustrated at trying to satisfy a level of need will eventually regress to the previous level (frustration–regression). An example would be where existence and relatedness needs have been satisfied but the fulfillment of growth needs have been frustrated, so relatedness needs become the dominant source of motivation. McShane and Von Glinow (2003) believe that ERG theory seems to explain the dynamics of employee's needs in organizations reasonably well. It provides a less rigid explanation than Maslow and a more accurate explanation of why employees' needs change over time. However, Huczynski and Buchanan (2007) believe that 'despite the differences between these two theories, the practical managerial implications are similar'.

There is significant correlation between the theories discussed under the caption of content theories of motivation as shown chronologically in Figure 5.2. Together they formed the early bedrock of understanding of motivation. All are concerned with hierarchical needs which have their roots in Maslow's work and from which the needs in each theory arise.

McShane and Von Glinow (2003) identify the practical implications derived from content theories as:

- The need for corporate leaders to be sure that rewards fulfill needs
- That different people have different needs at different times
- Financial rewards should not be relied upon too heavily as a source of employee motivation as intrinsic rewards are powerful motivators.

Maslow's hierarchy of individual needs	Hertzberg's motivation-hygiene theory	McClelland's theory of learned needs	Alderfer's ERG theory
Early 1950s	Late 1950s	1960s	1970s
Self-actualization	Motivators (job content factors e.g. satisfaction)	Achievement	Growth
Self-esteem		Power	
Social (Group affinity)	Hygienes (organizational context factors e.g. pay, working conditions)	Affiliation	Relatedness
Safety		(Not concerned with primary needs)	Existence
Physiological needs			

Figure 5.2 Content theories of motivation in chronological order.

However, Moorhead and Griffin (2001) place the current situation in context:

> Unfortunately, despite the many conceptual similarities among needs theorists that have emerged over the years, the theories share an inherent weakness. They do an adequate job of describing the factors that motivate behaviour, but they tell us very little about the actual process of motivation. Even if two people are obviously motivated by interpersonal needs, they may pursue quite different paths to satisfy those needs.

Process theories of motivation aim to shed light on this aspect of motivation. Content theories are concerned with the generality of the factors that influence motivation and are not specific to particular circumstances or particular types or groups of workers. Therefore, they are of general interest to those in construction in the same way as anyone else but they cannot be applied to construction professionals in any way which is specific to the nature of their work or the circumstances in which they work.

5.3 Process theories of motivation

The criticisms of content theories of motivation are about the assumption that they apply to everyone and the implication that we do not have choices about our goals and how we achieve them. Process theories, on the other hand, focus on how we make choices about motives, desired goals and the means by which to pursue them. They recognize that individuals are subject to social influences, motivated by different outcomes and that different cultures encourage different patterns of motivation. The practical objective of process theories is to suggest how and why employees behave in the way they do and so help managers decide on rewards and other outcomes designed to enhance employee motivation and performance. Essentially, process theories seek to describe the process through which needs are translated into behaviour. Three major process theories of motivation are 'Equity Theory', 'Expectancy Theory' and 'Goal-setting Theory'.

Equity Theory is the process theory most easily recognized as the formalization of something most people will have done at some time; that is, compare themselves to others. The theory's most influential proponent was Adams (1963). It begins from the premise that in work people look for what they perceive to be a just or equitable return for their efforts in comparison with others and that if they perceive that the situation in which they find themselves is unfair they are motivated to do something about it.

Adams proposes that people compare the ratio of their rewards or outcomes to their effort and contribution to the same ratio for others with whom they are comparing themselves. Rewards and outcomes are things such as pay, recognition, social relationships and other extrinsic and intrinsic rewards and their effort and contribution consist of such things as time, competence, energy and loyalty. A person's assessment of the values to put into such comparisons is to some extent objective, for example in the case of pay, but mostly assessments are predominantly perceptions, as the person making the comparison does not usually have objective data; not only are most factors not quantifiable but there is no guidance of the relative weighting of the factors involved. The values of each of the perceived outcomes and inputs on both sides of the equation do not have to be equal for equity to be perceived, only that the *ratios* should be perceived to be equal. Employees are most likely to compare themselves to co-workers but may also include colleagues in other organizations, friends, or neighbours and also previous jobs they have held.

A number of interesting ideas arise from this theory. It considers inequity to occur when one gets either more or less than one deserves and that one tends to perceive a modest amount of over-reward as good luck and tolerate it, whilst under-reward is not so easily tolerated. Interestingly, it is suggested that people who are overpaid reduce their perceived inequality by working harder. Also, the more strongly the inequality is felt the stronger the motivation to act. If motivated to act, employees can, for example, reduce their effort and dedication to their work, seek a pay rise or other rewards or prospects, boost their perceptions of the value of their current position, change their assessment of their current contributions, change their perception of the job of the person with whom comparison is being made (this can be seen as a form of self delusion), change the person with whom comparison is being made or simply look for another job. Equity theory can be clearly seen as a psychological phenomenon.

Robbins and Judge (2008) state that 'some of these propositions have been supported but others have not' and also point out that originally equity theory focused on distributive justice, one of the three ethical perspectives referred to in Chapter 4. Distributive justice, interpreted in business terms, is about the perception of each employee of fairness in the amount and allocation of rewards among employees. Fairness is seen to be based on perceptions so what is seen as fair by one person may be seen as unfair by another. The perceived fairness of the process used to establish how much we are paid relative to what we should be paid (procedural justice) is seen to be just as important as how much we are paid compared to what we think we should be paid (distributive justice). There are seen to be two key elements of procedural justice: the opportunity to present to decision makers one's point of view about desired outcomes; and that clear reasons for the outcome are given by management. The importance of this idea is that if people do not get what they want they need to know why. A further development is the idea of interactional justice which is an individual's perception of the respect, dignity and concern with which he/she is treated. Organizational justice comprises the three perceptions of justice: distributive justice, procedural justice and interactional justice. They summarize:

> To promote fairness in the workplace, managers should consider openlysharing information on how allocation decisions are made, following consistent and unbiased procedures, and engaging in similar practices to increase the perception of procedural justice. By having an increased perception of procedural fairness, employees are likely to view their bosses and the organization positively even if they are dissatisfied with pay, promotions and other personal outcomes.

Fineman et al. (2005) put an interesting slant on fairness by describing how the perception of unfairness can eat away at a person's contentment at work and give a different view of motive of the sort: 'I did it because I felt I was treated unfairly.' If such feelings fester they can lead to extreme reactions such as petty larceny, malingering and the like or even cataclysmic results rather than the more normal reactions such as accepting or rationalizing such feelings.

Whilst much of content theories of motivation have appeal to life in general and form a useful background for thinking about motivation at work, process theories are more specific to employment. Of the process theories, equity theory will be recognizable to members of the construction industry. Robbins and Judge (2008) remark that 'upper level employees, those in professional ranks, tend to have better information about people in other organizations', as do those with higher educational levels. Construction professionals have the advantage of being members of professional bodies and, as such, many are likely to attend professional meetings where close contact with others in similar employment in similar organizations is made. This together with long standing friendships arising from university days means that general information is available about salary levels and conditions of service at competing firms and in public sector organizations. Whilst pay secrecy is the norm, speculation on salaries amongst colleagues is usually reasonably accurate. This results in such topics being ready areas of conversations amongst colleagues from which feelings of equity and inequity emerge. However, although construction professionals will have been through the process of comparison on many occasions as they monitor the employment market, they are likely to take into account factors other than those cited in equity theory which appear to focus on the employment situation at levels lower than employees in professional groups. Whilst, status and recognition are cited as factors in perceptions of equity, many professionals in the construction industry will aspire to the highest levels of growth, personal development and accomplishment associated with working on prestigious projects with demanding professional challenges and in employment by the most respected organizations with the aim of reaching the top levels of such companies. As benefits to employees tend to standardize due to high information availability, so professionals are motivated towards employment which provides higher job satisfaction through the nature of the projects and challenges on which they are called to work and significantly also through the challenges inherent in starting their own firm should they decide to go down that route. Thus comparisons may be predominantly internal to their employing organization and not with external organizations unless they can provide comparable challenges. Though this may appear to be an idealized view of employment in professional construction organizations, it holds true for many, not only in construction but also for other major professions.

Expectancy Theory is a congnitive theory. Cognitive theories in psychology assume that we are purposive and aware of our goals and actions. It began with the work of Tolman (1932) but Vroom (1964) developed the application of the theory to employment. It is seen to be more encompassing than equity theory (Moorhead and Griffin, 2001) and is probably the foremost of the process theories of motivation. They state that it attempts to determine how individuals choose among alternative behaviours and that it shows that motivation depends on how much we want something and how likely we think we are to get it.

Expectancy theory is based on three concepts: expectancy (E), instrumentality (I) and valence (V). Expectancy, also known as 'Effort to Performance Expectancy', is a person's perception of the probability that effort will lead to sucessful performance. Instrumentality, the next link in the chain, known as 'Performance to Outcome Expectancy', is a person's perception of the probability that performance will lead to certain outcomes. Several 'Performance to Outome' expectancies might logically be possible so each possible outcome will have its own expectancy probability. The valence of an outcome is the relative attractiveness of an outcome to a person. Valances can be negative or positive. Positive valances may be promotion and recognition and negative valences may be stress and fatigue. The net value of valences can be negative if the negative valences are greater than the positive. Expectancy theory is represented by the expectancy formula:

$$F = E \times I \times V$$

where F (Force) is the force of a person's motivation to work hard.

As F is the product of the three terms, for motivation to occur all three terms have to be positive. If any one is zero then the product is zero and the result is no motivation. What the formula means is that an employee must believe that exerting effort will lead to high levels of performance and that their performance will result in valued outcomes and that the sum of all the valences for the potential outcomes will be positive. The expectency formula shows that the theory represents a mathematical process to determine someone's motivation. People weigh up the consequences of their behaviour in this fashon, albeit perhaps not in such an explicit manner.

These ideas lead to the question of how, as a result, can employees' motivation be increased? Expectancies are influenced by an individual's self-efficacy (see Chapter 2), which effectively is a can-do attitude towards a challenge together with the competencies to undertake the task. Companies can encourage employees to act in this way by providing training and support. Instrumentality is improved by an employer's formal measurement of employee performance, but most importantly by ensuring that rewards, in whatever kind is valued by the employee, do fully recognize performance. McShane and Von Glinow (2003) remark that many organizations have difficulty putting this straightforward idea into practice. Valences are only influencial when they are valued by employees. Outcomes have positive valences when they directly or indirectly satisfy needs and they have negative valences when they inhibit the fulfillment of needs. As employees have different needs at different times (echoes of content theories?) so firms should seek to offer rewards, in whatever form, to suit the specific needs of individuals and reduce negative valences.

Based on Huczynski and Buchanan (2007), expectancy theory can be summarized as:

- States that behaviour results from a conscious decision-making process based on expectations, measured as subjective probabilities, that the individual has about the results of different behaviours leading to performance and to rewards
- Helps to explain individual differences in motivation and behaviour, unlike Maslow's universal content theory of motivation
- Provides a basis for measuring the strength or force of an indivdual's motivation to behave in particular ways
- Assumes that behaviour is rational, and that we are conscious of our motives. As we take into account the probable outcomes of our behaviour and place values on these outcomes, expectancy theory attempts to predict individual behaviour.

McShane and Von Glinow (2003) believe that expectancy theory offers one of the best models available for predicting work effort and motivation and that all three components have received some research support even though it is a difficult model to test. They also believe that expectancy theory applies to team and organizational outcomes not just individual performance and, whilst it has been said to be culture specific, that this is not the case.

Whilst not applying expectancy theory to directly studying the motivation of individuals in the construction professions, Liu and Walker (1998) used and adapted the theory in attempting to develop a way to evaluate construction project outcomes. They recognized that the evaluation of project outcomes had been the subject of unresolved debate for many years and that previous views had tried to find a simple solution to a complex problem. They relied on the work of Vroom (1964) and Porter and Lawler (1968) (see below) to develop a behaviour-performance-outcome cycle analogous to the expectancy theory model in which the outcome was the measure of success of the completed project. They acknowledged that the expectancy theory perspective was of the individual but for construction project organizations individuals work in teams and that an individual's perceptions of outcomes were influenced by being part of a team. The model does not deal directly with motivation but subsumes it within the general concept of behaviour which includes what act (professional task) the individual is to perform and how many resources are going to be committed to performing it.

'The Porter-Lawler Model' (Porter and Lawler, 1968) extends expectancy theory to incorportate other aspects of motivation which were job satisfaction, perceptions of intrinsic and extrinsic rewards, abilities, traits, role perceptions and implicitly equity theory. Feedback loops are included at various points in the model. The major change involving a new point of principle was the view that if rewards are acceptable, high levels of job performance may lead to job satisfaction rather than the conventional view that job satisfaction leads to job performance. Effort results from the perceived value of the potential rewards (valence) and the perceived expectation that those rewards will follow. Vroom's instrumentality then becomes a feedback loop from performance to effort. Expectancy (effort–performance link) is influenced by abilities, traits and role perception to determine actual job performance. Job performance produces two kinds of rewards: intangible rewards (intrinsic), such as a sense of achievement; and tangible (extrinsic) rewards such as pay and promotion. A person then uses social comparison processes such as equity theory to decide if the rewards received are reasonable. If they are, the employee is said to receive job satisfaction. This is a continuing cycle as feedback operates between the perceived equity of the rewards (via job satisfaction) and the value of the rewards (valence).

Huczynski and Buchanan (2007) believe that the model has intuitive appeal and that research appears to support it. But they point out that it has received a number of criticisms, one of which is that it covers a range of inter-related variables and is complex. And so it is, but this is because people are complex beings resulting in complex motives and motivation. They question the assumption that people make decisions in such a logical and detailed manner, and it is unlikely that they do, but the model serves to expose the underlying forces which impinge on motivation. They also highlight the problems faced by researchers in testing such a theory due to the need to take into account the many variables involved and the lack of valid research instruments.

Goal-setting Theory is a process theory of motivation created by Locke (1968) which proposes that people are better motivated as a result of having clear goals. Goals provide a useful framework for managing motivation through employees setting goals for themselves or having them allocated and working towards them. This is a much more purposeful way of working than just being told to 'do one's best' or 'try as hard as you can.' On the face of it, this is a straightforward idea but the challenge is to develop a thorough understanding the process by which people set goals that are appropriate and achievable and then accomplish them.

Goal setting theory has four main propositions which are supported by research:

1 Specific goals define clearly and precisely what is to be achieved. Goals should be specific, measurable, attainable, realistic and time-related.
2 Challenging goals encourage employees to aim higher than they would normally be expected to but should not be above the level of their capability otherwise such aims would be demotivating. Difficult goals focus attention, make us work harder, persist in attaining them and assist in creating new ways of achieving results.
3 Participation by employees in setting their goals may increase commitment to the goals and so improve performance. Goals assigned by a manager without consultation should be fully explained and justified if good performance is to be achieved.
4 Feedback of past performance is necessary for effective goal achievement in the future.

Self-efficacy theory has significance for goal setting theory (and for expectancy theory). It concerns the belief that a person has about whether they are capable of achieving a task. For difficult tasks people who have low self-esteem are likely to reduce their effort or give up altogether but those with high self-efficacy try harder to succeed. Also, people with high self-efficacy accept negative feedback and increase their efforts to improve, whilst those with negative efficacy do the opposite. It is argued that setting a high goal for an employee can lead to employees having higher self-efficacy as their boss' belief in them leads to increased confidence. Also, research has shown that intelligence and personality related to conscientiousness and emotional stability can increase self-efficacy.

A number of other factors which influence the achievement of goals are:

- *Goal commitment* – the extent to which a person is personally interested in reaching the goal. Commitment is most likely to occur when goals are made public.
- *Goal acceptance* – the extent to which people accept goals as their own. Naturally this is more likely to be the case when goals are self-set rather than assigned.

- *Task characteristics* – the nature of the task relating to the goal. Goal-setting theory is considered to work better when tasks are simple rather than complex, familiar rather than novel and independent rather than interdependent. Many aspects of job performance are difficult to measure and do not lend themselve to goal-setting with the result that goal-setting may focus on parts of the task whilst the overall aspects are neglected.

Difficulties can arise from employees selecting easy rather than difficult goals when monetary rewards are linked to performance. By contrast, employees with a need for growth and achievement will set challenging goals irrespective of financial rewards. Nevertheless, goal-setting is one of the most successfully tested motivation theories but mainly in circumstances where short-term goals are unambiguous.

Three aspects of goal-setting theory mitigate against its usefulness in professional construction organizations; that is, it is considered to work better when tasks are simple rather than complex and that it is more suited when tasks are independent and when tasks are familiar rather than novel. Such conditions rarely occur in construction, in fact the opposite is invariably the case. Perhaps the theories' only use would be when training junior employees. In such cases it may be possible to define goals for technical tasks designed to stretch the junior by increasing the difficulty of succeding tasks and defining them as goals. Such a routine would have the benefit of testing junior employees' characters to establish their self-efficacy. Whether goal-setting theory can be applied to teams is an open question. On the face of it, it should be possible but whether it is necessary with a team of experienced and highly-educated specialists who are generally self-motivated is questionable, particularly as the potential need for goals is more likely to be for complex projects which challenge the team but for which definition of simple goals is problematic. Other motivation theories which stress accomplishment and growth are more likely to be relevant. Goals for individuals within teams, such as researching new techniques and information, may be more readily identified but may not be required to be formalized as team members are likely to make their contribution automatically without the need for formal goals. If they did not they would be soon be found out and their position within the team become untenable.

'Management by Objectives' (MBO) operationalizes goal-setting theory throughout an organization. Organizational objectives (goals) are established at the top of the organization and then cascade down, subdivided at the different levels of the organization into simpler specific and more contained goals down to goals for individual employees, the sum of which achieves the overall organizational goals.

It is claimed that many organizations have successfully used the technique (Moorhead and Griffiths, 2003). They say that it helps to implement goal-setting theory systematically and as a result employees are motivated, rewards are clarified, communications are improved, appraisals easier and more clear cut and managers can use the system for control purposes. Against these advantages are weighed the potential lack of full participation by top managers so that employees become cynical. MBO tends to emphasize goals that are quantifiable and verifiable and requires much administration. MBO requires commitment and establishing goals can be time consuming but it is often felt to be effective if organizations tailor it to their particular circumstances.

5.4 Job design and job enrichment

Further perspectives on motivation arise from the job design and job enrichment literature. Job design is the process of allocating tasks to a job and then determining the interdependency of those tasks with other jobs. Job enrichment is the expansion of a person's work to include associated tasks to make the job more challenging which could ultimately lead to empowerment. The earliest management writers focused on job specialization to increase work efficiency (as their idea of job design) but this was always unlikely to improve job performance as it ignored the motivational potential of jobs which has now become the central focus of many job design and enrichment initiatives.

Of these, the 'Job Characteristics Theory' devised by Hackman and Oldham (1974) links the features of jobs, the individual's experience and the outcomes in terms of motivation, job satisfaction and performance. The model also takes into account individual differences in 'growth need strength (GNS)' which is an indicator of a person's attraction to personal development through job enrichment and reflects Maslow's higher levels of needs. Enriched jobs will not lead to positive performances for people whose GNS is low.

The core of the theory is the idea that jobs can be described in terms of five 'core job dimensions' which stimulate three 'critical psychological states'. The five core dimensions are:

1. Skill variety: the extent to which a job requires different skills and abilities.
2. Task identity: the extent to which a job has a begining and an end with a tangible outcome.
3. Task significance: the degree to which a job affects the work of others, both in the immediate organization and the external environment.
4. Autonomy: the degree to which the job provides independence and the use of discretion.
5. Feedback: the extent to which performance is transmitted back to the individual.

The first three dimensions; skill variety and task identity and significance are combined to determine the meaningfulness of the work. Autonomy determines responsibility for the result of the work. Feedback produces knowledge of the result of the work. These three aspects – meaningfulness, responsibility and knowledge – are the 'critical psychological states'.

Work high in core job dimension values produces high critical psychological states. The model proposes that the higher the three psychological states, the greater the employee's motivation, performance and satisfaction with the work and the lower their absenteeism and their likelihood of resigning. The strength of the relationship between the core job dimensions and the work outcome is adjusted by the GNS, by a person's need for self fulfilment.

Both the critical psychological states and the core job dimensions are closely reflected in the work of construction professionals. In the case of the meaningfulness of work, this is usually met by the nature of the project on which the professional is working and is enhanced by working on the project continuously from start to finish (but this is rarely possible). Obviously some projects are more meaningful than others (see Dainty et al., (2005) for the satisfaction of working on a cancer facility) but the vast majority of projects make a contribution to the public good or to industry and commerce and are thereby meaningful to society. And generally, if economic times are good, construction professionals can move to another project if dissatisfied with the one they are working on. Construction professionals are by definition accountable for their work and how they are performing is apparent from the progress of their project. The core job dimensions are all characteristics of the work of construction professionals, except perhaps for autonomy. Whilst they work independently in carrying out the specific task in which they use their discretion, they do so within a team setting. As the model so closely resembles the work of the construction professional, GNS becomes the major determining factor in their motivation

The model also includes ways to operationalize the motivating potential of jobs (Hackman et al., 1975) by:

- Combining tasks to increase the variey of the job by giving employees additional parts of the process to do
- Forming natural work units, for example by creating teams (but this appears to contradict the idea of autonomy)
- Establishing client relations which gives employees responsibility for personal contacts of significance to the company (but this can be impractical in many situations)
- Vertical combining, which means allocating what would normally be the resonsibility of supervisors to their subordinates
- Opening feedback channels which essentially means formalising feedback loops within the organization.

The work of construction professionals by its nature includes all the elements of the model so job characteristics theory has little to add for the enrichment of the jobs of construction professionals, rather that the enrichment of jobs in other fields has just awoken to ways in which construction professionals have always worked (except, perhaps, for contacts with clients). Historically, this was always the role of senior partners in private practice or senior officers in public service, but has become increasingly more widely distributed amongst employees over the years as clients themselves have adopted less hierarchical organization structures.

The theory has been well researched and most of the evidence is supportive that the characteristics affect behavioural outcomes. Beyond GNS, other variables, such as employees' perception of their workload compared to the workload of others, may also moderate the link between the core job dimensions and personal and work outcomes (Robbins and Judge, 2008).

Huczynski and Buchanan (2007) observe that job enrichment was successfully applied during the 1960s and 1970s. They refer to an application to mainly white-collar groups in the late 1960s, including sales representatives, design engineers, foremen and draughtsmen and also state that job enrichment methods have enjoyed renewed popularity since the 1990s based on teamworking approaches. They continue that applications of job enrichment methods no longer carry novelty value and thus pass unreported; that the concept has been reinvented by Butler and Waldroop (1999) under the label of 'job sculpting'; and that practice in many organizations have gone beyond the enrichment of individual jobs to encompass teamworking, organizational culture change and forms of empowerment. These features have been characteristic of construction organizations for many years due to the organizational demands of originating, designing, tendering and constructing construction projects, which goes to further confirm the informal familiarity of job characteristics theory to members of construction teams.

5.5 Empowerment

Empowerment can be perceived in terms of motivation, which is discussed here, and also in the context of power which is dealt with in Chapter 7. The literature seems reluctant to provide a clear definition of empowerment but Neilson's (1986) is useful:

> Empowerment is giving subordinates the resources, both psychological and technical, to discover the varieties of power theythemselves have and/or accumulate, and therefore which they can use on another's behalf.

Tuuli and Rowlinson (2009a) agree with the lack of specifity and find that empowerment means different things to different people but also point out that empowerment can be conceptualized as a structural concept and as a psychological concept. Structural empowerment (also called empowering acts/practices and empowerment climate) relates to job design and other formal organizational techniques aimed at granting individuals greater control over their work and the decisions associated with it. Psychological empowerment sees empowerment as a collections of cognitions in which individuals have a sense of freedom and discretion, a personal connection to the organization, confidence in their abilities and the ability to make a difference (Speitzer and Quinn, 2001).

However, Huczynski and Buchanan (2007) believe that empowerment has become a broad term applied to any organizational arrangement which passes decision making responsibilities from managers to lower-level employees. They point out that Claydon and Doyle (1996) argue that empowerment is more myth than reality, as organizational changes introduced under this heading are often cosmentic because managers are reluctant to see a reduction in their power.

Empowerment is a progression from employee participation. This in itself grew out of the human relations movement of the 1930s which led to managers recognizing that the input of employees could be valuable in improving organizational effectiveness. Empowerment of workers in decision making within their job area is seen to aid job satisfaction resulting in increased performance. Instead of being told how to do their job, employees are provided with the resources to make their own decisions about how to carry out their work on the basis of their own knowledge and experience, which leads them to improved performance. This process and outcome should increase their self-esteem and their motivation. Depending on the level of interdependency of tasks with those undertaken by others, decisions on administrative matters and flexible working schedules could be within the control of the worker, as could such matters as quality control. If successfully implemented in the right circumstances, the benefit to the company would be better motivated employees with greater self-esteem and sense of accomplishment and hence the use of the full capacity of its employees. Empowerment can also be effective by empowering teams when the tasks being undertaken are interdependent and the teams are of specialists who need to collaborate to achieve their objectives. Empowerment will have no benefits if an organization uses it as window dressing and is insincere in its attempts to introduce it. If this is the case, employees will soon see it as a charade and become cynical, with counter-productive effects.

Liu et al. (2007) examined whether work empowerment of quantity surveyors is an antecedent of organizational commitment. They saw work empowerment as multi-dimensional in nature involving delegation of power by managers and how individuals perceive and internalize such power. It is seen to enhance self-efficacy and, through motivation and commitment, leads to increased performance and effectiveness. It was found that 'when the perception of empowerment increases, organizational commitmenet increases accordingly'. Of the two dimensions of organizational commitment (affective and continuance), work empowerment is related to affective commitment (the employee's emotional attachment, identification with and involvement in the organization) rather than continuance commitment (the employee's costs associated with leaving the organization). In the Hong Kong sample used, male quantity surveyors showed less continuance commitment and the longer they had worked for the organization the less they had. Professional qualifications and nationality were positively corellated with both dimensions of organizational commitment. Chinese chartered quantity surveyors were more commited to their organizations.

Greasley et al. (2005) examined how empowerment is perceived by individuals employed on construction projects. They found that project-based work with its fragmented nature, changing teams and leaders make it difficult for a consistent empowerment strategy to be used. Also the distance both geographically and psychologically between employees and senior managers means it is difficult to promote and realise empowerment. This isolation means that employees tend to have frequent contact only with their immediate superior so their superior's attitude to empowerment is vital to its success. A significant initial finding is 'the notion that empowerment is not a permanent, fixed reality that is shared by all, but rather is something that results in variation in how it is experienced from individual to individual … Thus, only through examination of the individual is it possible to measure the level of empowerment that may exist within a project or organization'. The same team (Greasley et al., 2008) examined the meaning of empowerment from the perspective of non-managerial employees from four construction projects and whether employees wanted to be empowered. They conclude that 'empowerment operates through individual variations in experience and is a perception that individuals hold. All of the employees were able to describe in their own words how they were and were not empowered, "the process of empowerment" and the meanings they associated with this, that is, "the outcomes of empowerment"'. And also: 'The general consensus is that employees typically want some empowerment in the form of control, authority and decision making but the level and form of the desired empowerment varies. Notably they do not associate themselves with the term "power", perhaps because they do not recognize this in themselves or in their role in the team. The limits of empowerment vary according to the individuals' willingness and ability to be empowered and should this line be crossed then empowerment soon becomes exploitation.'

In an examination of empowerment in project teams Tulli and Rowlinson (2009a) found that empowerment climate (structural empowerment) and psychological empowerment play complementary roles in engendering individual and team performance behaviours and are therefore not mutually exclusive. In their subsequent paper, Tulli and Rowlinson (2009b) address the relationship between psychological empowerment and job performance in the construction industry and whether the three performance determinants of motivation, ability and opportunity to perform explain the empowerment–performance relationship. They find that this was the case. They say, 'Indeed, the opportunity to perform actually emerged as the stronger mediator in the psychological empowerment–contextual performance behaviours relationship than ability to perform' but also that the findings 'demonstrate that the relationship between empowerment and performance is more complex than previously thought' and that 'psychological empowerment clearly emerges as a valuable path for organizations to pursue in their search for performance improvement in project settings …'.

Most references to empowerment in the construction literature deal with its relationship to power rather than to motivation which is dealt with in Chapter 7. For example, Tuuli and Rowlinson's (2009a, 2009b) work, which uses data from the construction industry, is essentially concerned with the individual and contextual factors that engender feelings of empowerment and how the roles of empowerment climate and psychological empowerment engender individual and team performance behaviours. But Tuuli and Rowlinson's (2010) paper moves on to consider the performance consequences of psychological empowerment. They state that: 'Practically, by clearly demonstrating that empowered employees exhibit positive performance behaviours, psychological empowerment clearly emerges as a valuable path in the search for performance improvement in project settings' and that 'the findings provide preliminary evidence in support of a comprehensive model of work performance that takes into consideration not only motivation and ability to perform but opportunity to perform. Indeed, opportunity to perform actually emerged as a stronger mediator in the psychological empowerment-contextual performance behaviours relationship than ability to perform'.

5.6 Rewarding employees

Most major books on organizational behaviour contain a section on pay and other extrinsic rewards for work. They cover such things as pay structures, piece-work rates, performance-based pay, bonuses, profit sharing, stock ownership plans, etc., but the focus of these accounts is generally on semi-skilled and skilled workers rather than managers and highly skilled professionals. Nevertheless, certain aspects are of relevance to construction industry professionals. Perhaps the most relevant are performance-based pay plans which rely on performance appraisal ratings with the intention that people thought to be high performers are given more. Rewards are seen to be best provided through bonuses rather than pay rises as, if performance falls off, reducing salaries would create disquiet. However, bonuses are more frequently based on a company's overall performance and simply divided between employees based on salary and/or seniority and not on individual performance, or perhaps based on both. Whilst performance-related pay and bonuses based on an individual's performance is intuitively seen as an equitable and sensible way to approach pay, major problems have been identified with this approach, not least of which is the design of acceptable performance measures and their quantification. In criticizing performance-based schemes Huczynski and Buchanan (2007) cite Kohn (1993) to good effect. He said that such schemes were doomed to fail because:

- Money helps us meet our needs, but research reveals that money is not an overriding concern for most people
- Pay that is dependent on performance ('If they have to bribe me to do this') is manipulative, and heightens the perception of being controlled

- Competition for rewards can disrupt relationships between individuals whose collective performance would be improved by cooperation and is damaged by rivalry
- Dependence on financial incentives to improve productivity diverts attention from attempts to understand and solve the underlying problems facing an organization
- Incentive schemes discourage risk-taking, experiment and creative exploration by sending the signal 'do exactly what you are told'
- Rewards that are contingent on particular levels of performance undermine interest in the job itself, whereas intrinsic motivation is usually the real basis for exceptional work.

He continues:

> The more we experience beingcontrolled, the more we will tend to lose interest in what we are doing. If we go to work thinking about the possibility of getting a bonus, we come to feel that our work is not self directed. Rather it is the reward that drives our behaviour. Anything presented as a prerequisite for something else – that is, as a means toward another end – comes to be seen as desirable.

However, bonuses in general, which were widely seen as a sensible way of linking pay to performance, have become discredited as a result of what the wider public sees as the unacceptable face of bonuses as paid to bankers in the run up to and even during the credit crunch of 2008. What is seen to drive bonuses is an unethical propensity to take risk without responsibility and, whilst this is manifest particularly in investment banking, it can also occur in all sectors of business to varying degrees. Nevertheless, bonus schemes which are equitable should be capable of being designed, for example based on net profits which could not have been known when basic salary was established and/or on the basis of mutual respect between employee and employer and confidence that each party 'will do the right thing'.

All organizations, whether construction related or otherwise, need a well-designed pay system in which employees have confidence. In the construction professions the pay structure is generally based on qualifications, experience and seniority which taken together are expected to define competence. So promotion within an organization or moving to another company is generally for an increase in status which would normally be accompanied by an increase in salary and other benefits, or very possibly for greater intrinsic rewards as discussed previously. Information on salary levels and other extrinsic benefit levels in general in the construction industry is not too difficult to obtain and whilst organizations may not have transparency of salary information, speculation amongst colleagues usually results in some idea of a person's salary becoming 'known', although it is not necessarily accurate. Hence, motivation within organizations may reflect equity theory. The dangers of performance-related pay are stated above and as professional construction firms often award bonuses, they can produce feelings of inequity and demotivation if they are differentiated between employees who do not have confidence in the performance measures adopted. These problems may be avoided if the bonus is declared as proportionate to salary for everyone but, of course, this approach does not recognize individual differences in performance.

5.7 A critique

The myriad of theories of motivation with their overlaps and inconsistencies have lead to many criticisms of their value. Wilson (1999) says that job satisfaction and motivation are difficult subjects to study as there are so many views on offer and so many of them offer contradictory positions. Fincham and Rhodes (2005) comment: '... research and writing covered here does seem to explain and account for the major causes of differences between jobs and individuals in the amount of motivation and job satisfaction they cause and experience' which is followed by '... but given the insights provided by this research, the obvious question to ask is why so much of what characterizes people's experience at work is at odds with the contents of this chapter...' which further illustrates the paradoxes inherent in the subject. They continue by pointing out that the massive amount of academic activity in the area has not been reflected in improvements in people's working lives.

Fineman et al. (2005) are critical but from a different perspective. They believe that people in organizations '... learn motivational scripts and apply them to ourselves or impose them on others'. So they create motives in others in the organization 'in ways that can be helpful, or restrictive and manipulative.' They also believe that we over-simplify complex situations or behaviour into 'a motive' so they become more manageable. This rather Machiavellian view is perhaps close to reality in many organizations. But Judge and Church (2000) found a more apathetic attitude amongst managers whose interview responses included: 'Job satisfaction is virtually never discussed in the senior staff meetings I attend'; 'It's never mentioned'; 'The term job satisfaction is not used here'.

So what are we to do, should we discard the ideas associated with motivation? Whilst it may be difficult to reconcile the contradictions and complexities of motivation, it is too important a subject to disregard. The problems seem to arise due to the complexity of people as a result of their vast range of individual personality and attitude profiles. Their individuality creates many dimensions by which they can differ in their motivation which is then compounded by both the range of motivational influences generated in their workplace and the depth to which they impact. The variables generated are proving to be too many to reduce to widely acceptable theories as they present huge difficulties in research design, methodologies and data collection. So for the present we are left with an array of interesting concepts, some of which are more detailed than others, which we are able to apply, perhaps not in a rigorous research manner, but which may help us to make some sense of our working milieu.

An application of motivation theories to construction (which are sparse), generating insights which may not otherwise have been acquired, is a major piece of work by Bresnen and Marshall (2000a) directed at motivation, commitment and incentives in partnerships and alliances. They say that the lack of any real research effort into the effects of financial incentives in partnering is reflected in 'much of the prescriptive literature about partnering: namely that it tends to conform to the fairly simplistic idea that financial incentives (positive and negative) have a more or less direct effect on the motivation and ability of individuals and groups to achieve particular goals or standards of performance (and to work more closely together in the long term)'. They see this approach as a 'crude' attempt at behaviour modification (OB mod) which ignores the role of cognition and individual differences in motivation. They develop their argument by recognizing that the subjective perceptions and preferences of individuals place different values on rewards and the different perceptions they have about the probability of achieving them, as reflected in contemporary motivation theories. As a result, they highlight the process theories of motivation, in particular expectancy theory. They also introduce goal-setting theory and equity theory as relevant to their theoretical approach to incentives in partnering and they distinguish between extrinsic and intrinsic rewards.

They identify the important issue that clients and contractors are seldom individuals but are instead complex social organizations. Such recognition is important for incentives in partnering and other collaborative initiatives, as the intention is to motivate organizations. They believe that 'The main implication here is that it is important to understand not only that attitudes and beliefs may conflict within an organization (between departments and/or hierarchical levels) but also that collaboration, motivation and commitment may vary across different levels of analysis (for example, being high at the organizational level but low at the level of the individual...)'.

They conclude that the application of contemporary theories of motivation was sucessful in showing that formal financial incentives had serious shortcomings as the only motivator for partnering and that 'it was clear that other sources of motivation, particularly the prospect of further work, were much more important to the companies and individuals concerned' and 'more intrinsic factors (such as achievement and autonomy) are important realized sources of motivation and commitment for individual staff in partnerships and alliances...'. The significance of this paper, apart from its valuable insights about partnering, is to show that construction professionals are motivated by the factors inherent in contemporary motivation theories and, hence, that these theories can usefully be employed to better understand the motivational dynamics that are at play in both teams and individuals in construction. The paper's conclusions, which directly reflect aspects of contemporary motivation theories, include 'the importance of understanding that participants' evaluations of rewards, expectations of performance and perceptions of equity are highly subjective and may differ' and that 'it has demonstrated that motivation and commitment may be the result of intrinsic, as well as extrinsic, rewards'.

Whilst these ideas may be readily understood in the mainstream organization behaviour literature and in other industries, they are not generally explicitly recognized in construction.

The problems with motivation theories is well demonstrated by Moorhead and Griffin (2001) in relation to expectancy theory, probably the most sophisticated of the motivation theories, of which they comment:

> Because expectancy theory is so complex, it is difficult to apply directly in the workplace. A manager would need to figure out what rewards each employee wants and how valuable those rewards are to each person, measure the various expectancies, and finally adjust the relationships to create motivation. Nevertheless, expectancy theory offers several important guidelines for the practicing manager. Some of the more fundamental guidelines include:
> Determine the primary outcomes each employee wants.
> Decide what levels and kinds of performance are needed to meet organizational goals.
> Make sure the desired levels of performance are possible.
> Link desired outcomes and desired performance.
> Analyse the situation for conflicting expectancies.
> Make sure the rewards are large enough.
> Make sure the overall system is equitable for everyone.

And finally, Wilson (1999) gives a thought-provoking commentary on motivations based on Taylor and Walton (1971) and Taylor (1972) by asking how we know the account of motives articulated by people are real, how do we know that motives are not produced to 'create a "case" or "position"? People are likely to give motives which they believe will be acceptable to the receiver; they will say what they believe others want to hear. These views challenge researchers into motivation who use peoples' own accounts of their motivation for the purpose in empirical research'.

6 Communications

6.1 Introduction

Communications are the lifeblood of human activities – social, commercial or any other kind. The word 'communication' immediately brings to mind the activity of passing information, and whilst this is a vital process, the subtleties of meaning, innuendo and emotion which are incorporated into communicating, particularly in face-to-face meetings, are the most important in the critical areas of human communications. All business sectors rely on efficient and effective communications for success, and construction is no exception. In fact, the complexity of the construction process both technically and in terms of interdependency of relationships places great demands on members of the industry to communicate effectively. The first to examine this issue in the modern era were Higgins and Jessop (1965) in *Communications in the Building Industry*, followed by *Interdependence and Uncertainty: A Study of the Building Industry* (Tavistock Institute 1966). Their seminal work was the first to recognize the need for overall coordination of design and construction to be exercised by a single person or group, but the major focus of their work was the formal communications needed to coordinate project documentation. They recognized that the amount of information in the form of drawings, specifications, contracts, etc., generated and transferred on even modest construction projects is enormous and needs a formal system of coordination. Subsequent official reports on the industry have continued to urge improvements in communications, coordination and associated matters (Walker, 2007). Whilst the improvements demanded by the reports were essential and whilst the implementation of some of them has contributed to advances, nevertheless, Emmitt and Gorse (2003) believe that there has been little, if any, noticeable enhancement in communication performance in recent years. Interestingly, however, the reports did not emphasize the more informal interpersonal communications which are central to the design and construction process.

The advances in information technology, in all its guises, since the reports have made huge improvements in the efficiency of communications in

Organizational Behaviour in Construction, First Edition. Anthony Walker.

construction, if not necessarily in its performance, as it has to communications in practically all walks of life. These advances have opened up the idea that people do not need to meet but can communicate electronically from remote locations so that going into the office or on site on a frequent basis is no longer seen to be necessary. But whilst electronic communication technology has permanently changed the way we communicate, face-to-face communications continue to be vital for businesses to prosper. This is reflected in early definitions of communication as 'the social process in which two or more parties exchange information and *share meaning*' (O'Reilly and Pondy, 1979), and 'the primary purpose of which is to achieve *co-ordinated action*' (Baskin and Aronoff, 1980) or more recently 'the sharing of meaning to reach mutual understanding and to gain a response – this involves some form of interaction between the sender and receiver of the message' (Emmitt and Gorse, 2007a). Sharing meaning, achieving coordinated action, receiving feedback and reaching mutual understanding are essential to creative activity, decision making and planning in stochastic situations and can only be maximized as a result of free-flowing face-to-face discussions. This is nowhere more necessary than in construction projects for which the building has to be perceived and designed in concept and detail to satisfy the client's requirements. thus necessitating the interaction of many specialists followed by the incorporation of contractors and subcontractors into the process. A myriad of decisions are necessary, most of which are not routine, requiring members of the project team to share information, be allowed to express feelings and often strongly held views and achieve coordinated action in defining and developing the project. The construction industry is not homogenous; it is made up of a vast number and variety of organizations from large construction companies and the wide array of specialist professional practices and subcontractors, which may range from large national firms to small jobbing firms. Coupled with the complexity this represents is that practically all construction projects are unique and project teams are temporary organizations in which members do not work together on a regular basis; all of this means that communications within construction are amongst the most difficult of any industry.

Experience and intuition lead members of construction project teams to believe that face-to-face communication ranks highly in terms of effectiveness. This view is supported by the work of Gorse et al. (1999), Carlsson et al. (2001) and Shohet and Frydman (2003), cited by Dainty et al. (2006), whose work confirmed that face-to-face communications were preferred and perceived to be the most effective medium of communication within the industry. A survey by Emmitt and Gorse (2003) supported these findings by ranking face-to-face communications first in a list of eight communications media with electronic transmission being towards the bottom of the list – but no doubt the ranking of the latter will have moved up with the passing of time. For these reasons and due to the orientation of this book, this chapter focuses on the behavioural characteristics of communications in interpersonal situations rather than the breadth of formal, technical and systematic aspects of communication of information in construction which is extremely well catered for elsewhere (cf. Emmitt and Gorse, 2003, 2007a; Dainty et al., 2006). So, although reference will be made to written communications, orientation will be towards the nuances of communication in face-to-face and non-verbal forms and other behavioural aspects of communication.

A remarkable paradox is that more technology in communications has not led to a lessening of the need for interpersonal contacts. Following the devastation of 9/11 it was widely expected that people would travel less, both for business and for pleasure, leading to far fewer business meetings and hence face-to-face communications. But that appears not to have been the case; rather, the complexities of modern businesses appear to demand increasingly more face-to-face meetings.

The complexity of face-to-face communication is underpinned by a sense of ambiguity of the words used to describe it. According to *Collin's Dictionary*, one meaning of the word 'verbal' means 'relating or using words, especially as opposed to ideas etc.'; another is 'oral rather than written' and the meaning of oral is 'spoken or verbal'. In all cases, spoken communication can be face-to-face or by telephone or similar device. Therefore, in order to avoid any confusion, this book will use the rather clumsy term 'face-to-face communication' for the spoken communication which is its focus.

It is well recognized that most people in management positions spend much of their time in meetings or in conversation, maintaining information channels, discussing issues informally, monitoring, receiving feedback, working towards decisions, etc., in which behavioural aspects of communication play a large part. However, it is felt that information technology driven communication can allow many to overcome the intimidation often felt in interpersonal communication, that more openness in communication can result and that the communication process is more democratic and subversive, but that the use of powerful language styles in such situations can lead to inappropriate openness and work against the development of groups (Huczynski and Buchanan, 2007). Over recent years the lure of information and communication technology has been put into perspective. Even at the simplest level it is recognized that email can lead to misunderstandings due to the truncated method usually used and without the benefit of face-to-face communications, leading to the realisation that hardware and systems will not be a panacea without the solution of people problems associated with complex construction projects but will, nevertheless, continue to transform communications in every field, including construction.

6.2 The communication process

The communication process has a structure and series of features whatever the medium of communication. The process is described here in terms of face-to-face conversation. The *source* originates the message and *encodes* it, which in this case means putting it into spoken words. It is then *transmitted* in a way that the receiver can *decode* it so that the message is fully understood. However, communication – particularly face-to-face communication – is subject to error; in particular, people put different meanings on words and interpret gestures and nuances differently. *Transmission* can take many forms. For face-to-face communication it is sound waves but may be electric impulses, electronic media or paper in many forms, including mass media such as newspapers, magazines and, of course, TV. Choice of transmission media can be significant in 'getting the message across'. For example, in establishing contact with a prospective client, the opportunity of engaging in face-to-face discussion to promote one's services is extremely valuable. *Decoding* is the interpretation of the message by the receiver. In order to decode the message accurately the receiver must assign the same meanings to the words used as used in the encoding by the source. At the professional level in construction there is a reasonably well understood professional and technical language. Nevertheless, misunderstandings can occur, for example, once communications move into the world of specialist engineers, specialist subcontractors and the law. In face-to-face conversations not only do words have to be ascribed the same meaning but so do non-verbal signals which can be more difficult to interpret. The receiver has to decide whether to make an effort to understand (decode) the message and whether to respond. In perverse cases the receiver may deliberately choose to misunderstand. The key skill of a receiver is to be a good listener. Both sending and receiving require high levels of concentration. Non-verbal aspects in face-to-face communication powerfully reinforce the message through, for example, words said with anger, urgency and humour, and may influence the way in which the receiver responds. Gorse and Emmitt (2003) examined communication during construction project progress meetings and found that exchanges could include emotional outbursts which significantly affected group behaviour. The way in which individuals express themselves is strongly influenced by their personal characteristics so that if knowledge of the source's personal characteristics are known by the receiver, the receiver will decode the message in the light of those known characteristics, but if the receiver is not familiar with the source, decoding becomes more difficult.

Feedback is the receiver's reply to the message. This will show the extent to which the receiver has understood the message and on the basis of that understanding what the receiver's view is. In our face-to-face focus, feedback will likely be spoken but it could be non-verbal, for example a shrug of the shoulders which could indicate a number of meanings and is not very helpful. Silence also represents feedback. For other than face-to-face communication, any medium could be used for feedback.

Noise in the communication process interferes with the process. In its simplest form it can be seen to be interference such as static associated with a radio signal and even noisy machinery on site which makes it difficult to hear face-to-face communication. More subtle, and perhaps more difficult to deal with, are emotions which may distort the meaning intended to be conveyed. Huczynski and Buchanan (2007) focus on this aspect and identify '*perceptual filters*' which can interfere with accurate decoding so that they affect what we say and how we interpret what we hear as a function of our perceptions arising from our values, traits, biases and prejudices. The effect of the prejudices of the parties in the construction process was recognized many years ago by Higgins and Jessop (1966) and, whilst not so strong now, they do surface from time to time. In the UK they are to a large extent formed by the professional institutions which, through their separate educational systems, inculcate strong allegiances to their profession and prejudice against others. Culp and Smith (2001) give an example of how engineers approach communications as a result of their different personality types. Using the extrovert–introvert scale of the Myers-Brigg Type Indicator (see Chapter 2), they showed that extroverts are sociable and expressive and prefer to communicate by talking and dislike getting only written feedback on project performance, whereas introverts dislike interacting with others in person and getting frequent verbal feedback and prefer to communicate by writing. Understanding each others' style improved team cohesion. Filtering may be subconscious, as a reflex to our perceptions and prejudices, or conscious if we purposefully suppress or limit information to be transmitted or purposefully misinterpret that which is received. Huczynski and Buchanan (2007) also point to physical, social and cultural contexts of conversations which may affect understanding of transmitted and received messages as a function of the relationships which exist between the parties in conversation (e.g., senior–junior, colleagues on the same level, client–consultant), their relative social standing and the location of the exchange (e.g., a formal meeting vs. the office party).

Whilst all this may seem to present no particular difficulties, nevertheless errors and misunderstandings in face-to-face communications are rife, as they are in all other mediums of communications; it is not the channels of communication which cause the problems, it is the people. McShane and Von Glinow (2003) illustrate this with a quote from George Bernard Shaw who wrote: 'The greatest problem with communication is the illusion that it has been accomplished' and a quote from a company CEO, John Loose, who said, 'When there is a message, to most people, you have to say it six times. You cannot have a meeting one day and check it off. You have to keep going back.' This became know in his company as the 'rule of six'.

At this point it is worth mentioning the famous phrase 'the medium is the message' (McLuhan, 1964) which means that the chosen medium for transmitting the message says as much or more about the status of the message and the people involved as the contents of the message itself. For example, a formally arranged face-to-face meeting is usually seen to be highly significant, particularly when called by a senior member of the firm, whereas a 'post-it' stuck on an office door does not carry status. It is therefore important to recognize and respond appropriately to the medium in conjunction with the message.

6.3 Communication channels

6.3.1 *Face-to-face communication*

Face-to-face communication is the major method of communication for a number of reasons. The foremost are, firstly, that it is the richest medium in terms of the range and complexity of messages which can be transmitted; secondly, that it allows the most rapid transmission; and, thirdly, that it allows the swiftest feedback which enables the fastest confirmation that the message has been received and understood but, if it has not, allows swift clarification and rectification. Its major disadvantage is that, unless it is recorded, there is no definitive record of what was said.

Whilst written communication can be rich given a good vocabulary and writing skills, most people are better able to express themselves fully through speech, particularly as speech can be accompanied by non-verbal signals. However, if a message has to be passed through a succession of people it is likely to be distorted – remember the party game (known as Chinese Whispers) in which a message is passed around ring of friends in which what the last person believes the message to be bears little resemblance to the original? But face-to-face communication is vital when decisions have to be taken in complex situations where the answer is not obvious, and particularly in cases where creativity is necessary. Rapid exchanges of views which lead to solutions or creation can only be achieved by the face-to-face interchange of ideas as individuals in a pair or a group bounce ideas off each other. The opportunity to persuade others to one's own way of thinking or to take on board the ideas, concerns and positions of others is best achieved by face-to-face exchanges.

The need for face-to-face exchanges in the construction industry as projects are defined, designed and built is clear to see and was confirmed by Pietroforte (1997). He found that professionals in the USA interacted through an informal structure in which they exchanged small amounts of information to clarify issues rather than using the formal structure laid down by contracts, reflecting the informal organization structures recognized by mainstream management research (Walker, 2007). The way in which project teams make progress is through many hours of meetings as ideas, design and technical and contractual matters are discussed, options considered and decisions confirmed, so the ability to articulate one's opinion is paramount. However, the problem which surfaces in these situations is that the more articulate and forceful personalities often hold sway even though their contribution may not be the best. To avoid such an outcome, project managers must be particularly articulate and have the most persuasive personality if they are to balance the competing opinions in the best interests of projects as a whole. Frequent face-to-face exchanges also arise from the need for clarification due to the often erroneous assumption that drawings, specifications, and other contract documents are complete and not contradictory but the reality is that this is not usually the case (Gorse, 2002).

Different ways of speaking are used in different circumstances in which variations are usually achieved by changes in tone, pitch, speed of delivery, volume and choice of words. Contrast, for example the types of delivery used for formal speeches and brainstorming sessions, both of which are face-to-face communications but which vary enormously in presentation. The range of situations in which construction professionals have to speak is very wide. For presentations to prospective clients the style has to be persuasive and with sufficient gravitas to impress and show that your organization has the capability to undertake the project. Design origination meetings can be very excitable with little restraint on the manner of expression. Negotiations between clients' and contractors' representatives can vary between being aggressive, if such a style is adopted by both parties, to conciliatory if common ground is being sought. Project team meetings with the fellow professionals from other practices and the client tend to be very restrained and conducted with a measured professional delivery. A host of other face-to-face circumstances occur on construction projects, each of which may generate a different style of presentation. Additionally, a whole set of different circumstances will occur within a person's firm, not related directly to projects but concerned with running the firm, which may require different types of face-to-face contribution; for example, staff problems may need a level of sensitivity in conversations not normally necessary in more professional matters related to projects. Other examples of 'in-house' matters requiring distinctive face-to-face treatment are motivation of staff, disciplinary matters and social chit-chat. However, the ability to vary one's approach across all the circumstances is an extremely rare talent. An illustration of the importance of face-to-face meetings is voicemail; this is a face-to-face communication but there is no immediate response and confirmation that the message has been received and understood (or not), which can lead to great frustration and lack of progress.

6.3.2 Non-verbal communication

For clarity, the above discussion of face-to-face communication has taken place in isolation from its most important counterpart – non-verbal communication; that is, communications not using words, either spoken or written. Whilst the manner of speech, tone, etc., may be seen as non-verbal communication, physical signals, such as gestures, expressions and postures, are the most potent non-verbal forces and can convey more meaning than words. The range of non-verbal communication is much greater than we realize as most of us give non-verbal signals subconsciously by reflex. Huczynski and Buchanan (2007) identify the dimensions of non-verbal behaviour which may be summarized as (the examples of meaning will not be common to everyone):

- Eye behaviour, e.g., regular eye contact (but not continuous) shows interest, narrowing the eyes shows doubt.
- A facial expression, e.g., a grimace shows discomfort, a smile shows contentment.

- Posture, e.g., fidgeting can show boredom, a shrug shows indifference, sitting back with arms folded shows an unwillingness to continue the discussion.
- Limb movements (kinesics), e.g., palm upwards to show sincerity, a handshake depending how it is done and by whom can portray a wide range of meanings.
- Tone and pitch (paralanguage) as discussed earlier.
- Distance (proxemics), e.g., leaning forward slightly to show interest, invading personal space can be seen as threatening.

They also quote Guirdham (1995) as listing 136 non-verbal behaviours and Pease (1997) giving 17 examples of women's repertoire of non-verbal courtship behaviour. Perhaps the most symbolic non-verbal communication is silence when one would expect someone to speak which is often interpreted as a sign of disagreement in the West, although not necessarily so elsewhere. One of the major differences between cultures is the use and meaning of non-verbal communication, the meaning of handshakes, silences in conversation and the amount of personal space and many more can all have different meanings. Whilst it is not within the capacity of this book to cover this minefield, anyone managing in a cross-culture situation must take the matter seriously.

A good illustration of the importance of non-verbal communication is the blandness of verbatim minutes of a meeting. The meeting may have been animated and controversial with forceful exchanges but much of this may have been expressed non-verbally which cannot be (or is not) recorded in formal minutes; so much of the impact is lost. The frequency of use of non-verbal signs in meetings, including those related to construction, will vary to an extent as a function of the personality of the individuals. Extrovert types are always more likely to use non-verbal signs but the frequency of the use of non-verbal signals will also vary depending on the type of meeting and the people present. Formal presentations to clients and formal speeches may well include a high incidence of non-verbal activity as stress is laid on certain aspects (akin to an actor's performance?), formal meetings of professionals for progress and other reasons and meeting for the purpose of negotiation will probably see little use of non-verbal signs unless some crisis point has been reached but design meetings and other creative exercises are likely to see much animation.

The best known technique for classifying face-to-face communication is Interaction Process Analysis (IPA) devised by Bales (1950) which allows classification of interactions into 12 sub-categories and so identifies socio-emotional or task-based interaction in order to show individuals' profiles, as discussed further in Chapter 9. Other techniques which can be used to track the pattern of communication are 'communication network analysis' producing the 'communigram' and 'communication pattern analysis' leading to a 'communication pattern chart'. In construction Gameson (1992) successfully used the IPA technique for classifying interactions during the first meeting of clients and professionals. Gorse and Emmitt (2009) also found IPA valuable in mapping interaction in construction progress meetings which provided rare insight into the informality of face-to-face interaction within a formal project setting. Considerable difference was found between the 'normal' range of socio-emotional acts found by Bales (1950) and those found in construction which showed very low use of socio-emotional acts. Socio-emotional interaction is seen to be important because it is used to build, develop and maintain relationships and recover from conflict, hence it is often termed 'relationship communication'. It is considered that a lack of socio-emotional interaction leads to the build-up of latent tension that can threaten relationships. However, whilst the level of socio-emotional interaction was low in construction compared with previous studies, they were generally of social groups not work groups which can be expected to have lower socio-emotional interactions. Nevertheless, it was concluded that professionals in construction do not have appropriately developed communication skills to discuss issues in a more socio-emotional interaction than task-based interaction.

6.3.3 Impression management

Impression management is the manipulation of the signals given out to others so that the face-to-face and non-verbal messages are not subconscious actions, as they usually are, but are controlled to manipulate the impression and perception others have of us. It is akin to acting; it is a performance and raises the question of whether it is deceit, but it is claimed that most people do it to some extent most of the time by the way they dress and conduct themselves. Taken to extremes it is difficult to sustain but can be tutored and seen as a component of the 'spin' culture of business and politics.

6.4 Listening

It seems obvious that in a conversation listening is as important as talking, but in practice the vast majority of people do not behave as though it is. Most people are not good listeners and those that think they are often hear without listening; that is they take in the words spoken face-to-face but make no real attempt to understand them or their underlying meaning.

In the communication process described earlier, the receiver is shown as decoding the message; that is, receiving and attempting to understand it. Decoding involves two actions, sensing and evaluating, which then lead to responding or feedback. The receiver senses the message and pays attention to it, then listens to the words face-to-face and should also sense the tone of voice, emphases and non-verbal signals in order to receive the complete message. Receivers should attempt to see things from the sender's point of view so that they can decode the message in the way the sender intended.

Receivers should concentrate on what is being transmitted and not make assumptions about what the sender is transmitting without listening to the whole message. The worst failing in this respect is for the listener to respond before the sender has finished speaking; such interruptions break the flow of the speaker and inhibit the delivery of his message and the ability of the listener to comprehend the full message. Good listeners do not respond until the speaker has finished and even then they pause to gather their thoughts before replying. They also maintain interest even when it is difficult to do so and concentrate on the speaker both visually and aurally to ensure all the non-verbal signals are received.

In evaluating the message a good listener should empathise with the speaker which requires the listener to 'get under the skin' of the speaker, that is to be sensitive and sympathetic to the points being made by the speaker and not to instinctively reject the speaker's words because of some bias or lack of interest. Most people have something useful or interesting to say. Barrett and Stanley (1999) make this point in relation to the construction project briefing process which they see as a strongly interpersonal communications undertaking. It is necessary for listeners to organize the information being transmitted and listeners are strongly aided in this by being able to process information three times faster than the average rate of speech; the downside of this is that they are easily distracted. The outcome for good listeners is that by the time the speaker has finished the listener has organized what has been said and is ready to respond. Responding requires receivers to give their point of view which may involve them first clarifying that they have a proper understanding of the message and so the cycle continues until full understanding is reached, a decision made or an impasse reached.

6.5 Questioning

Discussions are concerned with exchanges of information and understanding which are achieved by asking questions of those involved. The way in which we ask questions can play a significant part in how the response is given. Huczynski and Buchanan (2007) give an account of different types of questions and their purpose, summarized here briefly as:

Closed question: to obtain a precise answer or factual information; to control the conversation.
Open question: to encourage discussion.
Probing question: to be specific after an answer to an open question; to show interest.
Reflective question: to show concern and interest regarding feeling and emotions.
Multiple questions: to give a choice of questions (which may confuse).
Leading questions: to encourage the answer you expect to hear.
Hypothetical question: to encourage creativity.

The distinction between closed and open questions is an important one. Closed questions should get a factual reply; simple and efficient when that is all that is required. Closed questions can also be used to establish the pattern of a conversation as a basis for open questions later and to establish who is in control. Probes are another type of open question and show interest. The reflective question maintains rapport, encourages disclosure of feelings and emotions, shows interest and concern and encourages further disclosure of information. Interestingly, they point out that multiple and leading questions are often heard on radio and television, the former frequently directed at politicians and the latter at suspects in police dramas. Hypothetical or 'what if?' questions are particularly valuable in brainstorming sessions and at the various stage of design of construction projects.

Emmitt and Gorse (2003) point out that high-status professionals are reluctant to ask questions in situations where such action will show that they do not understand something as this may reflect on their credibility. They draw on Gameson's (1992) work which showed that 'interaction during client briefing found that the construction specialists would rely on their own limited knowledge rather than suggesting that the contribution of other specialists would be useful' in order to protect their status. Whilst saying that 'professionals tend to avoid asking questions because help-seeking behaviour implies incompetence and dependence', they also point out that Gorse's (2002) work 'found that during site-based progress meetings the contractors' representatives considered to be most effective asked more questions than those considered to be less effective' which in the former seems to point to an issue of confidence within the professionals and in the latter to the status of contractors' representatives who are expected to have to ask questions.

6.6 Barriers and issues

Information overload refers to a situation in which information received exceeds a person's capacity to process it. Whilst the focus of this chapter is on face-to-face communication, it is felt necessary to refer to information overload because, even though overload is caused by the sum of all types of information, it impacts significantly on face-to-face communication as it leads to stress and other causes of dysfunction in face-to-face communication. People have an information processing capacity which varies between individuals. The major cause of the massive increase in information overload is, of course, electronically-transmitted information, particularly email. The vast majority of managers cannot deal effectively with all the information they receive and have to be selective in what they process. It has been reported that more that 40 per cent of managers in the USA say that 'receiving so much information weakens their decision making ability, delays important decisions, and makes it difficult to concentrate on their main tasks' (McShane and Von Glinow, 2003). Suggested solutions include speed reading, scanning documents and better time management, but whatever techniques are used people need the ability to integrate the information they receive to arrive at decisions and too much information inhibits such ability. Dainty et al. (2006) point to the plethora of different types of communication which impact concurrently on members of the construction industry, particularly through ever-expanding information and communication technologies (ICTs) which they may be ill-equipped to manage and for which they need training to help them identify information that is important to them in their work, otherwise integration of ICTs will be inhibited.

In dealing with *noise* earlier, which is a major barrier to communication, reference was made to *perceptual filters* but *perception* and *filtering* are also dealt with separately (McShane and Von Glinow, 2003). That people have different perceptions of the same thing was dealt with in Chapter 2. This phenomenon has a significant impact on communications as receivers of messages hear selectively based on their needs, motivations, prejudices, expectations and other personal characteristics; that is, they hear what they want to hear. We interpret what we see and hear and call it reality. Filtering refers to the purposeful manipulation and misinterpretation of information. The more levels in an organization, the more opportunity to manipulate information as each level tells the next level what they think they want to hear or what they want them to hear. The idea of perpetual filters combines these two phenomena which is frequently the case as filtering is carried out on the basis of perceptions. A further type of noise was identified by Dainty et al. (2006) who see it as occurring frequently at project meetings when only a small number of members speak whilst many do not. What was happening was that those who did speak also spoke on behalf of those who did not. The latter would pass information to the spokesperson which creates noise as messages travel from one person to the other before being delivered and may become distorted and lose meaning in the process.

Jargon and *ambiguity* are modes of expression which can lead to confusion and lack of understanding. Jargon refers to reserved terms which have highly specialized meanings usually understood mainly by specialists in a specific field of endeavour. An extreme example in construction is the use of Latin terms by construction lawyers. Jargon is often referred to in a derogatory manner when it uses pseudo-specialist terms purposefully to confuse, to attempt to try to show the user to be superior to the receiver or to hide a lack of knowledge. In such cases it will seem to the receiver to be no more than gobbledegook and presents a serious barrier to understanding. However, in many specialized fields reserved terms are essential, for example in medicine and most branches of science and engineering. In such cases, specialists will confer using specialized terms, but wherever possible when conversing with non-specialists every endeavour should be made to explain matters in non-specialist language; nevertheless, in many instances it will be impossible to explain these terms in this way with any degree of accuracy. Construction provides a good example as professionals will converse using their own specialist terms but should do so only in closed meetings; in meetings with clients they should endeavour to use terms in language understandable by the client but if this is not possible full explanations should be given. The key is not to use jargon just for the sake of it. If confronted by jargon one should not be afraid to ask the meaning.

Ambiguity should generally be avoided as it leads to misunderstanding and abortive work at best and to seriously inappropriate decisions at worst. Ambiguity is not the same as differences in perception; it is lack of clarity due to flawed expression. Senders should ensure that their message cannot be misconstrued, so careful choice of words is essential. The purpose of the receiver paraphrasing a message and feeding it back to the sender is to confirm understanding and avoid ambiguity. Ensuring that no ambiguity occurs often requires a blunt statement such as a straight yes or no. To avoid bluntness in the cause of politeness, some people soften a message to the extent that the receiver is not entirely sure what the message is. However, ambiguity can be used on purpose to confuse or as a cover-up. Semantics is closely associated with ambiguity and arises due to people attributing different meanings to the same words or language forms, particularly in a language as rich as English in which words can have subtly different meanings. A phrase heard from time to time is 'it is only semantics' meaning that there is no real difference between what two people are saying even though they are using different words or language forms; often there is a real difference in meaning but one party does not realize it.

Situational aspects of conversation can create barriers to understanding but which, if understood by both parties, can be overcome. *Power difference* is a classic situation in which both parties may converse in a manner which reflects their view of themselves. Employees may distort the upward communication by telling their superior what they think they ought to know and superiors talk down to subordinates often because they do not fully understand the working situation of their subordinates. *Gender differences* are significant to the way men and women behave in conversation and have become more so as increasing numbers of women enter the workforce and attain senior positions in business and the professions. Sociolinguist Deborah Tannen (1991, 1995) provided important findings on differences in conversational styles between men and women that have led to barriers between them in conversation. Tannen believes that women generally have a linguistic style that focuses on relationships whilst men generally have a linguistic style which focuses on status as a result of the rituals learned in childhood. These positions can be seen to be at the opposite ends of a spectrum so that they do not apply to all men and to all women; nevertheless; there are linguistic characteristics which better fit men and those which better fit women. Women use a language of relationships, intimacy and modesty. Conversations build closeness to give understanding and support, tend to be indirect in approaching an issue, are disinclined to brag, take others' feelings into account, avoid putting others down, attempt to save face for others and tend to apologise frequently, even when it is not their fault. Men generally use a language of power and independence to reinforce status, and use conversations to that end, they tend to be direct and attempt to overpower others in conversation to achieve 'one-upmanship' by putting others down, they do not readily say sorry and look for opportunities to criticize. The general outcome is that women are likely to appear less self-assured than men. They tend to play down their certainty and express doubt more openly. The benefit of such an approach is that a more questioning approach can lead to more measured consideration of the matter under discussion rather than the more forceful approach of men which can lead to higher risks if all issues have not been explored due to men's more macho linguistic characteristics.

With an increasing number of women joining the construction professions, such differences in linguistic style, particularly if extreme, could have a significant impact on project team and other meetings. Men's style could mean that women may have less opportunity to contribute to discussions as men attempt to dominate and interrupt more readily than women. When female project team members represent specific professional skills, the danger is that the contribution of the skills they represent may not be given sufficient weight in meeting outcomes, leading to imbalanced design and progress of the project. On the other hand, men tend to focus on exchanging information in an impersonal style which contributes to efficient meetings but the conflicting styles can lead to irritation. Even though the extremes described here may not commonly occur, differences may be sufficient for project managers, chairs of other meeting and individuals to need to recognize and understand them so that meetings can be managed to ensure that they are positive forces and not detrimental to the development of the project.

Political correctness can impact on the effectiveness of communication, both face-to-face and written. It is essential to be sensitive to the feeling of others and not use words which offend, stereotype and intimidate. The use of such language angers or releases other emotions in those it is directed against and so, as well as offending, can reduce their contribution to discussions. The other side of the coin is that political correctness also complicates conversation by removing words from the vocabulary and replacing them with words which are not so precisely defined and commonly understood. Language is evolving continuously and the use of politically correct language is part of this process and yet another challenge for managers. Dainty et al. (2006) refer to politically incorrect language being common in certain sections of the construction industry and suggest that it helps to sustain informal interaction and temporary networks. Whilst it is seen as a strong part of the 'culture' of construction, it is responsible for much of the poor external image of the construction industry which is likely to have a detrimental effect on the recruitment of women.

Education of the major contributors to construction projects in the UK has created barriers between them which are reflected in the way they communicate with each other. This arises from each profession being educated through specialist undergraduate courses from the time they enter university from which they form strong allegiances with their profession rather than with the construction industry and construction projects. The result of such strong sentient feelings is that each profession may 'fight for its corner' rather than seek to understand the position of the other contributors and see the project as a whole. This position is reinforced by the professional institutions which have vested interests in maintaining the status quo. This barrier is unlikely to exist should all undergraduates be able to share a common undergraduate course from which they could develop shared values and a common language and from which they could progress to specialize as architects, engineers, surveyors and project and construction managers at postgraduate level. If having to obtain two degrees is seen to be expensive an option alternative structures are available such as part-time study which could be, for example, by conventional means or on-line. Instead we have project teams comprising members who have difficulty speaking in the same 'professional language' and we wonder why the construction industry has more than its share of conflict.

Cultural differences arising from societal norms in different countries create significant communication barriers. Fully understanding these differences is beyond the scope of this book so reference to them is limited to considering high and low context cultures to give a brief indication of the issues and differences involved. In high level cultures people take into account the context of the person they are dealing with as much as, if not more than, the detail of what is being said or written. They rely more on non-verbal and situational clues (So and Walker, 2006). In low level cultures people rely on words to give meaning and take far less account of non-verbal signals and status. High level cultures are seen to be located in the East – China, Japan, Korea – low level cultures in the West – Europe and North America. High level cultures traditionally relied on trust rather than formal written contracts and trust is built up slowly through relationships developed through meetings and social gatherings involving what often appear to be casual and insignificant conversations, so face-to-face communications are significant. Issues are approached indirectly. The need for high ranking personnel to be present at relatively insignificant meetings is part of this ritual, reflecting the importance attached to relationships developed through face-to-face communications as trust is established. Low level cultures rely on written enforceable contracts and directness in communications and negotiations. But as the high level cultures of the East have become increasingly exposed to Western business methods through globalisation more reliance on written contracts, particularly in construction, has taken place when a Western company is involved. This brief illustration hides a vast complexity of different behavioural attributes in the many countries of the world not mentioned here. The basic message is to assume people from other nations behave differently from you until shown otherwise. This will help to reduce error and embarrassment.

Ineffective communication is frequently at the root of conflict in the construction industry. Whilst the ineffectiveness of any of the types of communication used in construction may contribute to conflict, face-to-face communication is most frequently the means of clarification where information and meaning are not clear. Face-to-face communication may clarify or further mislead if it is not precise. Lavers (1992) states that clarification constitutes good practice and is part of the legal obligations of construction professionals. Decision making defines the structure of construction project organizations (Walker, 2007) and whilst project objectives, professional issues and technology define the decision choices, arriving at these choices will involve clarification and debate, much of which will be carried out through face-to-face communication. It is therefore essential that such conversations are effective. Similarly, arriving at decisions which are other than procedural are usually carried out in face-to-face meetings in which the clarity, balance and objectiveness of the spoken contributions will determine the effectiveness of the communications and hence the decisions.

Communication breakdown must be avoided no matter how trivial it may seem. If some members of the project team cease speaking to each other, escalating conflict and ineffective management of the project is likely to be the result. Project and specialist managers should be vigilant in identifying such breakdowns and act swiftly to diffuse problems.

6.7 Aspects of organizational communication

Project organizations can take many organizational forms (Walker, 2007) through which the vast amount of information that defines a project and instructs how it is to be constructed is coordinated. But, whatever the organizational form, in the process of generating this information many hours of interpersonal communication will be spent discussing and defining its content. The style of these discussions will be determined largely by the characteristics of the project manager. An autocratic project manager will act formally and limit the extent of face-to-face communication so the full abilities of the professional team would not be allowed to contribute to the project. At the other end of the spectrum a more democratic project manager would have an open approach to discussion. Project team members would be able to make a fuller contribution and all the characteristics of interpersonal communications discussed so far would come into play. Such an approach would favour members skilled in face-to-face communications and require project managers to be able to correctly interpret spoken messages. The issue for the project manager in this position would be to control discussions so that progress is not hindered whilst at the same time full value is gained from members' contributions. Hence, knowledge and understanding of how people contribute in face-to-face communications is an important skill for project managers.

In terms of company organization (professional firms and contracting companies), the classical pyramid organization chart portraying an organization's hierarchy is still generally taken as the way to manage a company even though many new models have emerged (Leavitt, 2005); it is also seen as showing the lines of communication within an organization. Such communications tend to be formal and documented and have become more so with the increase in size of organizations. However, in managing the output of a company, e.g., designs, contract documentation, etc., communication channels are horizontal between the various units of the company and, particularly in construction, between the various external contributing companies. Such communications comprise many informal face-to-face contacts, as well as the formal technical output. Informal communications frequently lead to the creation of informal organization structures (Walker, 2007). Interpersonal communications are continually taking place within the company and between it and other companies as well as in project organizations, emphasising the importance of understanding styles and nuances of face-to-face communication.

Communication from the top of companies down the hierarchy on matters of organization significance is invariably formal except in the case of very small companies. Company policy, developments and news are transmitted in writing, either electronically or by hardcopy or more usually a combination of the two. The nature of communications from above has a major effect on determining an organization's culture. The edicts which are passed down either explicitly or implicitly indicate how employees should behave in order to succeed within the company and include such items as mission statements and statements of corporate values. Whilst this formality may be necessary for the efficient running of a business, making the business effective requires employees to contribute not only to production but also to the company as a whole by providing feedback to the hierarchy. Achieving this requires the hierarchy to cultivate an open communication climate in which employees are encouraged to make suggestions on the running of the company and are able to respond to formal company communications and have their responses considered seriously; that is communication up the hierarchy which is often best by way of face-to-face meetings in informal settings. Organizations with a highly hierarchical structure or limited opportunities for interaction across levels frequently have communication problems because no feedback is given (Shriberg et al., 2005).

Open communications rely to a large extent on face-to-face communication, described by Moorhead and Griffin (2001) quoting Peters and Waterman (1982) as 'a vast network of informal open communications' that fosters trust which minimizes the effects of status differences. Peters and Waterman further describe communications in effective companies as chaotic and intense, supported by the rewards structure and the physical arrangement of facilities. This means that the performance appraisal and reward system, offices, meeting rooms and work areas are designed to encourage frequent, unscheduled, and unstructured communications throughout the organization. The same ideas could also be applied to projects as well as companies. Many may disagree with the argument but it certainly gives food for thought. The creative aspects of design in the construction process may well benefit from such a setting but the more formal structured aspects of project development may prefer a less frenetic setting.

This scenario raises the issue of the use of workplace design to encourage interpersonal communication in both project and corporate settings, particularly open-plan offices (on which there are mixed views). Obviously, office walls prevent employees from talking to each other. However, whilst it is also obvious that if walls are removed people will communicate face-to-face more often, it is felt that open-plan offices increase employee stress due to a loss of personal space and hence privacy. Again these ideas can be applied to projects. Dainty et al. (2006) use the design and development of the Millennium Dome project built in London in the late 1990s as an example of a project team housed together in an open-plan office. A further aspect of openness is for members of the hierarchy to be seen around the company by employees and engage them in conversation. This can contribute considerably to openness and can also add to the charisma of the members of the hierarchy but becomes increasingly difficult to do successfully as organizations become larger and because of the tendency of employees to focus on detailed personal matters in such conversations rather than company-wide issues. The wide range in the size of professional practices in the construction field illustrates the issues referred to here and, as those who have worked in firms of different sizes will recognize, the complexity of managing communications in firms is likely to increase the larger firms become. In a survey across a range of industries, Gallie et al. (1998) found that: processes for informing employees of organizational issues were widespread but involved distributing information rather than holding meetings; the higher the skill level of employees the more ready access they had to information, which was particularly the case for professional and managerial staff; however, contradicting what was said before, the larger the organization the better the communications. The reason for larger organizations performing better was said to be because they had dedicated units implementing communications policies. And finally a note of warning: the danger arising from mismanaged company communications which attempt to 'spin' messages, lack understanding of the employee's point of view, are irrelevant, avoid important issues and do not allow feedback is likely to lead to cynical and uncooperative employees.

The *grapevine* is probably the oldest company communication mechanism and also exists on projects. It operates to spread unofficial information about the company and about projects, and exists because it enables employees to hear news (or more probably rumours) about the company and project before the news is officially issued by the company or project manager. It is said that rumours are spread about matters that are important to employees, not about trivial matters, because they cause anxiety (Rosnow and Fine, 1976). A small number of individuals (rumour mongers) tell people in their social sphere in the company the latest rumour who then spread it further. Most rumours are founded in truth but not necessarily the exact truth as stories become distorted in the telling. McShane and Von Glinow (2003) believe that the grapevine has both advantages and disadvantages. One benefit is said to be that the grapevine enables sense to be made of the workplace when information is not available through formal channels but this is not advantageous if the information is distorted to the extent that it is misleading. A further benefit is claimed to be that the grapevine relieves anxiety but, if information is distorted, anxiety will increase, particularly in uncertain times when the rumour is not confirmed or denied. Grapevines can be mechanisms for reducing company and project morale if they are transmitting false rumours on issues which adversely affect employees, the project or the company, such as impending redundancies. Managers have to keep their ears to the ground to know what the grapevine is saying and respond rapidly to issue formal factual statements of the position to minimize negative consequences. Grapevines also transmit company and project stories and reputations and so feed into corporate culture but they are out of the hands of the senior managers and may not contribute to the culture they are trying to establish.

Traditionally, grapevines were transmitted by word-of-mouth and are impossible to stop, so need to be nurtured to ensure they do minimum damage. But the word-of-mouth grapevine is being complemented or even overtaken by email and instant messaging as the main medium for gossiping which has developed on a global scale and particularly affects companies with world-wide representation. However, non-verbal nuances are not available to qualify electronically transmitted written messages.

6.8 End piece

The focus of this chapter has been face-to-face communication but cannot be allowed to pass without reference to electronic information processing and telecommunication. We all recognize the power of email, instant messaging, intranet and extranet links, video-conferencing, virtuality and phone messaging and future developments which have and will further reshape company and project communications, but treatment of these topics and how people behave when using them can fill many books and cannot conceivably be covered adequately here. But face-to-face conversations will continue to be the greatest force for important decision making, which is why the world's leading politicians, businessmen, professionals, academics and others will continue to fly thousands of miles to meet.

And finally a reference needs to be made to a lighter side of face-to-face communication (which can also have a dark side) – joking. Dainty et al. (2006) identify 'banter' as establishing camaraderie in interpersonal relationships in construction project teams and firms. Irony, sarcasm and cynicism are typical ways in which groups relate internally but can alienate other groups at whom it is directed. Fineman et al. (2005) have a chapter on 'Serious Joking' – joking in an organizational context; a powerful communication tool. Their message is put across here by quoting their conclusions:

> Humour and jokes, far from being inconsequential, are important features of organizational life. They break the organizational routine and enable people to cope with boring or alienating jobs. They generate trust and affection for those sharing a laugh and a joke and permit the venting of unacceptable views and emotions (like aggression or contempt), by offering a moral amnesty which permits the breaking of taboos.
>
> The targets of organizational jokes are varied, but most workgroups have an individual or another group which serves as the butt of disparaging humour. When directed against superiors or outsiders, jokes strengthen the solidarity of a group and enable the group to score symbolic victories against

their psychological adversaries. When directed against members of a group, jokes may be part of joking relations, highlighting the intimacy and trust between group members, or they may serve to reinforce group norms and force compliance. They can also be a means of teasing or bullying some members, but as with many of the common forms of language joking, the jokers may end up excluding themselves rather than those they intended to hurt. Humour may also be exploited as a management tool.

Finally, the target of jokes may be the organization itself, whose foul ups and absurdities are celebrated because they undermine the façade of rationality and seriousness. Such jokes, for a brief moment or two, explode organizational order and restore the human factor, in all its fallibility and unpredictability, at the heart of organizations.

7 Authority, Power and Politics

7.1 Introduction

Power in organizations has proved difficult to pin down and define. There is a tendency amongst managers to want to deny its existence even though its reality is readily apparent (Pfeffer, 1992). Power can take a number of forms but the main division is between formal power – better known as and referred to in this chapter as *authority* which is enshrined within the rules and structure of organizations – and informal power – referred to in this chapter as *power*, which operates outside the authority structure and arises from a person imposing his or her will on others within an organization. Whilst the authority structure is, of course, extremely important, it is the power structures (which are not made explicit) which can be insidious if not used in support of an organization's objectives. The exercise of power generates organizational politics.

Organizational politics is seen as distasteful by those who believe that organizations should exercise rational decision making. Whilst organizational politics can be a force for good, its greater propensity is to work against an organization's best interests. Fineman et al. (2005) characterie organizational politics at work as 'people talking about others behind their backs; pub chat that makes (or breaks) opportunities; rumours and alliances among powerful people; overheard conversations; attempts to impress powerful people. Indeed it is hard to think of organizing without politics'. All of which points to a need to understand power and politics if organizations are to be understood.

The concept of power has always been difficult to understand in theoretical terms but, although a commonsense grasp of the term seems easy enough (Fincham and Rhodes, 2005), it is clear that power is not always simply a matter of the bosses being able to do what they want (Fineman et al., 2005). The sociology of power was first explored by Weber (1947 trans.) who stressed the ambiguity of the idea. Lukes (1975) provides a widely accepted model known as the 'three faces of power' which offers a framework within which other approaches can be placed. The first face is an account of actual power behaviour; the second is about the ability of powerful groups to prevent choices from

Organizational Behaviour in Construction, First Edition. Anthony Walker.

being considered or excluding certain interests from the centres of authority and decision making; and the third face is not about excluding interests but in preventing interests from ever forming in the first place. Fincham and Rhodes (2005) give an example of the latter as 'the way in which the social horizons of many working class people are narrowed and constrained by education and upbringing'. Foucault (1977) is another significant theorist who considers that new forms of power are concerned with activities such as surveillance, controlling technologies and structures, and Wilson (1999) draws attention to the micro-techniques of power and disciplinary practices used in modern organizations. She gives as an example what she calls 'the disciplinary gaze' as exercised by organizations in, for example, employee appraisal systems. In this category, Fincham and Rhodes (2005) refer to 'the specific languages of occupations and professions – the sales talk, the technical jargon, the bedside manner, and so on. Professional discourses are an independent source of power, such that experts, through their talk, can define situations and create new areas of knowledge'. All of this has resonance for construction professionals. Theories of power are drawn from society at large, not from the specific situations in business organizations, but their application has resulted in greater understanding of the forces at play in business.

7.2 Authority

Authority is intrinsic in achieving objectives through organizations. Authority, which is vested in a person through the position held in the organization and, hence, their right to make decisions upon which others are required to act, is the most common understanding of authority in organizations. Such authority is therefore seen as essential in order to get things done. This simple view of authority describes what is often referred to as *formal* or *legal* authority and rests on three basic assumptions (Cleland and King, 1972):

- The organization chart is a realistic descriptive model of an organization.
- Legal (or line) authority is delegated down through the chain of command. Therefore, if one has legal authority, one can demand the obedience of others.
- Given sufficient authority, an individual can accomplish organizational objectives regardless of the complexity of the forces that are involved (Ries, 1964).

But the concept of authority is insufficient in today's complex organizations and informal patterns overlie the formal structure. But before moving on to examine such phenomena, the origins of authority are worth exploring.

The original source of authority in society can be traced back to private property rights which were held by the crown and the church. They owned the land from which food was produced and hence had the ability to enforce their authority. This was the model from which authority in industrial processes arose during the industrial revolution. Independently of the rise of this aspect of traditional management 'theory', Max Weber, the influential German sociologist, produced his typology (Weber, 1968 trans.):

Traditional authority – resting on an established belief in the sanctity of immemorial traditions and the legitimacy of those exercising authority under them.

Rational–legal authority – resting on a belief in the 'legality' of patterns of normative rules and the right of those elevated to authority under such rules to issue commands.

Charismatic authority – resting on devotion to the specific and exceptional sanctity, heroism or exemplary character of an individual person, and of the normative patterns or order revealed or ordained by him or her.

Traditional authority has its roots of feudalism; rational–legal authority underpins Weber's notion of bureaucracy. He believed that the charismatic forms arose in periods of instability and crisis when people turned to people they believed (rightly or wrongly) could resolve crises. Thus to a large extent a formal view of authority was enshrined within early management thinking.

The work of Barnard (1938) took the understanding of authority significantly forward. His view that goals are imposed from above whilst their achievement depends on willing cooperation from those lower down the organization, led to his view of authority in which he states that it is a 'fiction that authority comes from above' and that 'the decision as to whether an order has authority or not lies with the person to whom it is addressed and does not reside in persons of authority or those who issue these orders'. These ideas gave root to the concept of *informal* authority which is now explicitly recognized as transcending authority and is enshrined in the concept of power.

These ideas were developed by Dornbusch and Scott (1975) who identified three types of authority: endorsed, authorised and collegial, which essentially represent controls on authority. They actually referred to these as endorsed power, authorized power and collegial power but the use of 'power' in this context is confusing as explained later when the differences between authority and power are explored. Originating from Barnard's work they identify authority by endorsement as that arising from a subordinate group acting as a coalition which limits and regulates the exercise of authority over them by a superior. They then go on to recognize that an important characteristic of formal organizations is that each superior will (in the vast majority of cases) also have a superior in the hierarchy. Subordinates will have the opportunity to appeal to their superior's superior with the hope that their immediate superior's authority will be curbed. They term the authority of the superior's superior as authority by authorization. A further source of enforcement of authority norms is seen to be the colleagues or equals of the superior in question. They believe that this source may be expected to be of particular importance in professional organizations and is referred to as authority by collegiate.

Authority has evolved, but Leavitt (2005) believes it 'has been softened, even cloaked in cordiality. But it is still there, and it's still the central reality of organizational life'. Whilst authority is modified by power structures much of the management literature tends to overstate the absolute influence of informal aspects. In business people are always likely to respond to a command from their superior. However, they may not carry out the order effectively or may seek to dodge it. This latter possibility is high when a subordinate's knowledge of his or her task is greater than that of the superior; conditions which frequently exist for construction projects. Pfeffer (1992), in an account which demonstrates the strength of authority, states that 'obedience to authority is conditioned early in life and offers, under most circumstances, many advantages to both society and the individual'.

Whilst authority is given by the ability to impose sanctions, the greatest of which in a business sense is the withholding of a salary rise or ultimately dismissal, there are many constraints on authority. Formal limitations are contained in laws, contracts, etc., and informal limitations are provided by morality and the capacity of the person so ordered to carry out a task. In fact, absolute authority does not exist as it depends upon the sanctions being sufficiently high to make the person so commanded obey the order. In a business sense an employee may accept the sanction of dismissal rather than obey the order. In a more extreme example, authority imposed by a state over its people may not be accepted and so revolutions begin and the ultimate sanction of death is not considered great enough to command obedience.

7.3 Power

Power is the influence individuals have over the people with whom they interact to the extent that what they require is done. It operates with subordinates, superiors, peers and friends and has to be earned. If misused, it can be counterproductive. Cleland and King (1972) identify the following talents as contributing to what they term an individual's informal authority but which is now usually referred to as power:

- superior knowledge
- an ability to persuade people to his or her way of thinking
- a suitable personality and the ability to establish rapport with others
- a favorable reputation with peers and associates
- a record of accomplishments which lends credence to his or her experience and reputation
- an ability to build confidence in peers and associates
- patience to listen to the problems of subordinates and peers and a willingness to help out when asked or when the need to help is sensed
- an ability to resolve conflict between peers, subordinates, and associates.

The great strength of a person in a position of authority who also has these characteristics is not difficult to imagine nor are the connections between effective leadership, authority and these characteristics.

It is important to distinguish between authority and power in organizations. Power is a much broader concept than authority (Weihrich and Koontz, 1993) and many pages have been devoted to its definition. Many stem from Weber's (1947 trans.) definition that 'power is the ability of a person to carry out his own will despite resistance'. Hence 'the ability of individuals or groups to induce or influence the beliefs or actions of other persons or groups' (Weihrich and Koontz, 1993), 'the capacity of individual actors to exert their will' (Finkelstein, 1992) and the simple definition 'the ability to get things done' (Kanter, 1983) used by Lovell (1993) which will have a significant appeal in relation to those in the construction industry.

Scott (1992) relies on the approach of Emerson (1962):

> It would appear that the power to control or influence the other resides in control over the things he values, which may range all the way from oil resources to ego-support, depending upon the relation in question. In short, power resides implicitly in the other's dependence.

He believes that Emerson's approach means that power is not to be viewed as a characteristic of an individual but as the property of a social relation. This means that in defining the power of an individual it is necessary to specify over whom he or she has power. That is, an individual cannot have power generally. Nevertheless, the possession of power is generalized by Pfeffer (1992) in defining power as 'the potential ability to influence behaviour, to change the course of events, to overcome resistance, and to get people to do things that they would not otherwise do'.

Pfeffer defines organizational politics as the exercise or use of power. That is, politics and influence are the processes, actions and behaviours through which potential power is utilized and realized. Scott comes to a not dissimilar conclusion in combining the definition of power and its use as:

> We will define interpersonal power as the potential for influence that is based on one person's ability and willingness to sanction another person by manipulating rewards and punishments important to the other person.

Whether such a narrow definition of power is relevant in a construction context is open to question.

7.4 Relationship between authority and power

Most dictionary definitions of authority contain a reference to power. For example:

> Legal power or right: power derived from office or character or prestige (*Chambers Twentieth Century Dictionary*)

We know that authority stems from a structural position in an organization and can be enhanced by the characteristics of the person in that position. The person in such a position has authority over his or her subordinates which can be conceived, according to the dictionary definition, as legitimate authority or legitimate power. In this conception the terms 'authority' and 'power' can be seen to be synonymous but in this chapter power is taken as illegitimate.

The situation in which power gains its wider meaning is when it is used outside the authority structure such that the person exercising power imposes his or her will on others in the organization. To repeat; such power is seen as illegitimate. For this reason there exists a general feeling of disquiet about its presence in business organizations (it appears much more acceptable in overt political systems). This leads Pfeffer (1992) to believe that we are ambivalent about power and to quote Kanter (1983) who believes 'Power is America's last dirty word. It is easier to talk about money – and much easier to talk about sex – than it is to talk about power'.

However, power can be a 'positive' or 'negative' force. In its positive form those with power use it to further objectives which are directly compatible with the organization's official objectives. In its negative form power is used to achieve objectives which do not subscribe to organizational objectives. Even in what may appear to be its positive form it can have negative aspects as control of such forces is by definition difficult as they are not explicit and whilst on the face of it may appear positive they may in fact be exerting a negative effect.

Positive power has been described by French and Bell (1990) as 'a balanced pursuit of self-interest and interest in the welfare of others; viewing situations in win–win (non zero-sum) terms as much as possible; engaging in open problem solving and then moving to action and influencing.' They see negative power, on the other hand, as 'an extreme pursuit of self-interest; a tendency to view most situations in win–lose (zero-sum) terms; and predominant use of tactics such as secrecy, surprise, holding hidden agendas, withholding information or deceiving'.

7.5 Sources of power

There have been many perspectives on the sources of power in organizations, some of which are complementary, some of which overlap. Together they help to explain the nature of power in organizations.

Emerson (1962) sees power as relational, situational and potentially reciprocal. That is, it occurs between people in specific circumstances and may work in both directions. Power can have many foundations but writers often refer to power being based on the resources that can be employed in the attainment of desired goals. In this context, resources are defined extremely broadly, for example money, skills, knowledge, strength, sex appeal. What types of resource will function as sanctions will vary from one individual to another and from situation to situation (Scott, 1992). Emerson considers that power relations can be reciprocal in that an individual may have power over another in one matter but power may work in the opposite direction between the same people in another matter. Pfeffer (1992) recognizes the link between power and interdependency in organizations. Interdependency means that the dependency of one person or unit on others exists and is the basis of power relations. Such power relations are likely to be significant in highly interdependent areas of business such as those found on construction projects in which specialists frequently cannot complete their work without the performance of other specialist team members, and whilst it may not seem to be sensible to exercise power in such circumstances, there may be people working on other agendas in which they use power with a view to a pay-off in other circumstances in future (Walker, 2007).

Within this context there is a reasonably consistent view of the sources of power (Weihrich and Koontz, 1993; Newcombe, 1994) in the categories of reward, coercive, expert and referent power. Reward power refers to the power to offer enhancement (in many forms), for example, pay, position, conditions of service. Coercive power is closely related to reward power and is the power to punish by way of deprivation of benefits, for example pay, status. Whilst these categories are frequently seen to be legitimate (i.e., arising through authority) and organizational they do not have to be so. Many indirect opportunities to reward or punish exist outside the authority structure and can be seen as the exercise of power (illegitimate).

Expert power arises from skill, knowledge and, increasingly, information. Finkelstein (1992) believes that the ability of top managers to deal with environmental contingencies and contribute to organizational success is an important source of power. Managers with relevant experience may have significant influence on a particular strategic choice and are often sought out for their advice. However, power tends to accrue when a manager's expertise is in an area critical to an organization.

Referent power is the influence that people exercise because people believe in them and their ideas. Finkelstein refers to this as prestige power. He believes that managerial prestige promotes power by facilitating the absorption of uncertainty from the institutional environment both informatically and symbolically. He also believes that prestige also provides power through suggesting that a manager has gilt-edged qualifications and powerful friends.

Whilst these sources of power are accepted, there are attempts to subdivide and extend them. For example, Kakabadse et al. (2004) use reward power (which includes coercive power), role power (which is authority), personal power (referent power), knowledge power, network power, information power (which can all be seen to be part of expert power) and corporate memory power (which may also be seen as expert power).

These sources of power are seen as personal although they can operate within the authority structure, and can have a great effect as confirming its legitimacy. They possess the ability to operate freely outside the authority structure and be potent sources of power which can undermine an organization's objectives if used negatively.

Finkelstein's paper also deals with structural power which is conceived as the equivalent of the authority structure. He also identifies ownership power which he defines as 'power accruing to managers in their capacity as agents acting on behalf of shareholders'. He believes that the strength of a manager's position in the agent–principal relationship determines ownership power. The strength of managers' ownership power depends on their ownership position as well as on their links to the founder of a firm so that managers who are also shareholders are more powerful than those who are not.

Whilst accepted definitions of power see it as a personal relationship, Huczynski and Buchanan (2007) draw attention to power embedded in organization structures through the factors which give power to specific departments of an organization, not individuals. Nevertheless, departmental power gives individuals power through their access to decision making, information and budgetary control. A department becomes powerful if other departments depend upon it to carry out their functions effectively, particularly if the department controls financial resources, hence the power of finance departments and hence finance directors. Further departmental strengths include criticality of its work to the major output of the company, a department's work not being capable of being substitutable by others inside or outside the company and a department's ability to reduce uncertainty. For example, a situation could arise in an architectural practice with an extremely high reputation for design in which the design department gains dominant power over all other departments as the practice believes that without continuity in exemplary designs it could founder. They also point out that when such a perception of a department's power becomes so embedded in the organization structure, power becomes virtually invisible, taken for granted and difficult, if not impossible, to challenge. If such power is unequally distributed it can be to the detriment of the long-term prospects of the organization and may lead to its downfall. McShane and Von Glinow (2003) use similar ideas in connection with an individual's power (as opposed to a department's) when discussing the contingencies of power; that is, the conditions necessary for the ability of an individual's power sources to generate power. The four conditions are non-substitutability, centrality, visibility and discretion which, if not in place, will prevent power translating into influence; that is:

- a person's power source must not be able to be substituted by someone or something else;
- a person's centrality and hence power increases with the number of people affected by his or her actions and with the speed with which others are affected;
- a person with potential power must not be 'invisible' but must ensure others are aware of their potential; and
- a person must have the discretion to make decisions without permission.

Information is an increasingly important source of power and whilst it is a fallacy that it is a new source (it has always been important), it is increasingly relevant in a knowledge-based economy with rapid electronically-based transmission of information. In organizational terms McShane and Von Glinow (2003) believe that information power derives from authority and expert sources through the ability to interpret and control the flow of information provided to others and the ability to reduce uncertainty by providing information to this end. The position of information gatekeepers who rely on information power is increasingly challenged by the need for knowledge sharing required by knowledge management and team-based organizations, pressures which are keenly felt in the team-orientated structure of construction firms and projects.

7.6 Personality and power

Personality or charisma is often referred to as a source of power but is better perceived as a reinforcing agent to the sources described. The strength gained through expert or referent power will be increased by an appropriate personality. Huczynski and Buchanan (2007) point out that, paradoxically, organizations' recruitment, appraisal, training and promotion policies are usually focused on recognizing ambition and the innovative talents possessed by people who have personalities appropriate to power seekers and political behaviour.

McClelland (1961) identified that people culturally acquired three types of need: power, achievement, and affiliation. The strength of each need varies between people. Those with a strong need for power need to influence and lead others and are likely to engage in political behaviour. McClelland and Burnham (1995) subsequently identified that the power needs of individuals can be either personalized power or socialized power. People with a need for personalized power need power for its own sake and use it for their own purposes, for example to advance their careers. Those with a need for socialized power use their power to help others, hence such a trait is valuable in politicians, other public servants and corporate leaders. They have a high sense of social responsibility, altruism, moral standards and ethics in order to develop trust and respect in others which will reinforce all aspects of their power, particularly referent power.

An extreme form of personality associated with power plays, and one which has entered the public psyche more than any other, is Machiavellianism, named after Niccolo Machiavelli, a 16th-century Florentine philosopher who wrote guidelines for rulers to secure and hold power, and published in a book called *The Prince.* He promoted the primary methods of achieving and keeping power as manipulation, opportunism and deceit. Machiavellian personalities prefer to be feared, manipulate using persuasive methods, not always seen as legal and certainly unethical, and believe that achieving the desired ends justifies any means used to do so. It is often referred to in situations when someone is seeking revenge and has assumed a significance in the public mind that is rarely realized in practice.

7.7 Politics in organizations

Organizational politics is the exercise or use of power (Pfeffer, 1992). Politics has to do with power, not structure (Mintzberg, 1989). Whilst Mintzberg agrees that political activity is to be found in every organization he believes that politics act to the detriment of the effective functioning of organizations by 'disordering and disintegrating what currently exists'. He unequivocally states that 'I am no fan of politics in organizations' but he recognizes that no account of the forces at play in organizations can be complete without a consideration of politics. In agreeing with this view, Moorhead and Griffin (2001) cite a survey by Lenway and Rehbeim (1991) which showed that about one third felt that political behaviour influences salary decisions, 28 per cent felt it affects hiring decisions, and a large majority found it more active at high organizational levels. Over half believe that politics is unfair, unhealthy and irrational yet believe that it is necessary for executives to act politically to get ahead as it is a reality of organizational life. They use the results to show that political behaviour is pervasive in organizations.

Mintzberg's (1979, 1989) configurations illuminate the use of power in organizations and use an analogy of politics as an organizational illness which needs to be understood. He sees it working both for and against organizational effectiveness as it can undermine healthy processes but can also strengthen an organization by acting as a symptom of a more serious disease, enabling early action by provoking adaptive mechanisms. Mintzberg identifies 13 political games which by his definition involve use of power but many of which use authority as a part of the play. Some co-exist with strong authority and could not exist without it, other usually highly divisive games are antagonistic to authority and further games arise when authority is weak and substitute for it. Two of the games will be more easily recognized by construction people:

1 *Expertise game*: non-sanctioned use of expertise to build a power base, either by flaunting it or by feigning it; true experts play by exploiting technical skills and knowledge, emphasising the uniqueness, criticality, and irreplaceability of the expertise, also by seeking to keep skills from being programmed, by keeping knowledge to selves; non-experts play by attempting to have their work viewed as expert, ideally to have it declared professional so they alone can control it.
2 *Rival camps game*: played to defeat a rival; typically occurs when alliance or empire-building games result in two major power blocs, giving rise to a two-person, zero-sum game in place of an n-person game; can be most divisive game of all; conflict can be between units (e.g., between marketing and production in manufacturing firm), between rival personalities,

or between two competing missions (as in prisons split between custody and rehabilitation orientations).

Mintzberg approached the study of organizations on the basis of the configuration of five basic types to which two others were added at a later date. One of the latter is the political organization which he describes as:

> What characterizes the organization dominated by politics is a lack of any of the forms of order found in conventional organizations. In other words, the organization is best described in terms of power, not structure, and that power is exercised in ways not legitimate in conventional organizations. Thus, there is no preferred method of co-ordination, no single dominant part of the organization, no clear type of decentralisation. Everything depends on the fluidity of informal power, marshalled to win individual issues.

He identified political organizations as having four forms:

Confrontation: characterized by conflict that is intense, confined, and brief (unstable)
Shaky alliance: characterized by conflict that is moderate, confined, and possibly enduring (relatively stable).
Politicised: characterized by conflict that is moderate, pervasive, and possibly enduring (relatively stable, so long as it is sustained by privileged position).
Complete political arena: characterized by conflict that is intense, pervasive, and brief (unstable).

The latter (complete political arena) appears to be an extreme type of organization rarely found in its absolute form whilst the other three are less uncommon but still rare. The shaky alliance form is the one which is most easily recognized in construction terms. It exists when 'two or more major systems of influence or centres of power must co-exist in roughly equal balance'. Mintzberg uses the symphony orchestra to illustrate this form but those in construction will find it familiar, e.g., architect/engineer, designer/constructor.

However, the political organization is not the only type subject to political influences. Mintzberg recognized that politics may exist in each of his other basic types of organization to varying degrees but less extensively than needed for them to be classified as political organizations. Of these, the professional and innovative types are particularly relevant to the management of construction projects. He identifies a professional organization as relying on the standardization of skills, which is achieved primarily through formal training. It hires professionals for the operating core and then gives them considerable control over their own work. The innovative organization relies on 'adhocracy'. Mintzberg believes that 'sophisticated innovation requires a very different configuration, one that is able to fuse experts drawn from different disciplines into smoothly functioning ad hoc project teams'. The uniqueness of construction project management organizations is that they often combine Mintzberg's professional and innovative configurations.

What is significant is that Mintzberg believes that there is considerable room for political games in each of these configurations as both have relatively weak systems of authority, though strong systems of expertise. When combined as in construction project management their ability to convert to political organizations must be even greater.

One can do no better than quote Mintzberg:

> The professional configuration may have a relatively stable operating core, where activities are highly standardized, but its administrative structure, where all kinds of professionals and managers interact to make choices, is hardly stable and, in fact, very supportive of power games. The innovative configuration is far less stable, generally having a highly fluid structure throughout that literally promotes games.

Whilst taking a firmly negative view of politics in organizations Mintzberg does recognize that politics can serve a functional role:

- As a system of influence it can ensure that the strongest members of an organization are brought into positions of leadership
- It can ensure that all sides of an issue are fully debated
- It can stimulate change that is blocked by legitimate systems of influence
- It can ease the path for the execution of decisions.

Whilst reluctant to endorse any political organization as functional he does acknowledge that a shaky alliance that reflects natural balanced and irreconcilable forces in the organization could be functional. He uses the example of the differences between research and manufacturing people in a firm which needs the two in balance and considers the alliance functional in such a situation. This is a situation which is directly analogous to design and construction. He goes on to say, 'This is because the organization could not function if it did not accommodate each of these forces. It has no choice but to take the form of a shaky alliance. Some conflict is the inevitable consequence of getting its work done.'

Whilst Mintzberg's analysis and illustrations remain amongst the most cogent there are many other accounts of politics in organizations that are valuable but which mostly cover the same ground even though it has been held that 'writers, researchers and managers are not able to agree on the true meaning of the term politics' (Kakabadse et al., 2004). They describe the situation within organizations as a competition 'for resources, for attention, for influence; there are differences of opinion concerning priorities and objectives; clashes of values and beliefs occur frequently. All these factors lead to the formation of pressure groups, vested interests, cabals, personal rivalries, personality clashes, hidden deals and bonds of alliances'. Within this milieu, politics is a process to influence other individuals or groups towards one's own views and impact in different ways within construction-related firms and within construction project teams.

Certain conditions are generally accepted as fertile ground for politics to take root and prosper in organizations. Most revolve around resources. A scarcity of resources often requires political manoeuvring in order to maintain or improve the position of a group or a person in the organization. Uncertain, fudged, unstructured and complex decisions about resources also provide opportunity for political activity as parts of the organization represented by a person or group of people manipulating the position for their own ends. Organizational change also provides similar opportunities, usually for the same reasons, as similar conditions apply. Much also depends on the hierarchy of the organization as their tolerance or even support of political activity will give acquiescence and encourage such activity, particularly if they take part themselves and 'political animals' are seen to be promoted to more senior positions.

7.8 Types of political activity

McShane and Von Glinow (2003) identify various types of political tactics which OB scholars have grouped into six main categories, a number of which have been alluded to previously.

Briefing against others (or attacking and blaming others) is probably the most common and insidious form of political activity. It can vary from explicit condemnation of others to the more common subtle innuendoes and laying of false stories and spreading of rumors. The phrase 'briefing against others' is frequently found in connection with politicians but is also prevalent and equally effective with businessmen.

Controlling information. Power is gained by the ability to control and manipulate information. Not only is power gained through 'spinning' information to suit the transmitter but the creation and timing of information dispersion through setting agendas, controlling who attends meetings and arranging meetings to suit their personal interests give power to the person in control of these events. Not releasing information is a frequently-used tactic used to disadvantage others.

Forming coalitions. A person or group may seek to broaden their scope of influence by recruiting other people or groups to its cause and so increase its power base. Pooling power in this way towards a common objective gives a sense of broad appeal to which others will then give their support and a 'bandwagon' then begins to roll.

Networks and obligations. Cultivating social relationships both within and outside the company for whom an individual works is a political activity, particularly if at some point it involves creating obligations for which network members owe favours to each other in return for favours which have previously been granted. Networking can supply information which may increase expert power; it also increases referent power if your network is seen to include people of high standing. By excluding those who are not in the network, power can be seen to increase for those in it; an example can be seen in the 'glass ceiling' frequently encountered by female employees. The obligation to act in response to a previous favour can be a strong force to maintain one's place in the network, so once a favour has been done the giver can be seen to have power over the receiver. This sense of networking and obligations is nowhere more developed than in the Chinese system of 'guanxi' (So and Walker, 2006) which is a powerful force in Chinese society, both commercially and socially. Closely knit groups of professional firms from different disciplines in the construction industry are not uncommon. They often seek to work together on projects as they are familiar with each other's style of working. However, with increasing fee competition this mode of operation has reduced, particularly on public sector projects, although it can still be beneficial as fee levels are not usually the only criteria in competitions and is frequently beneficial on private sector projects.

7.9 The ethics of political behaviour

An instinctive reaction is that all political behaviour in organizations is unethical but organizational behaviour texts recognize both ethical and unethical political behaviour (Robbins and Judge, 2008; Moorhead and Griffin, 2001). Unethical behaviour is that which disregards the ethical principles identified in Chapter 4. If the ethical principle of utilitarianism (the greatest good for the greatest number of people) is disregarded in carrying out political activity, the perpetrator follows a narrow path of self-interest. On the other hand, if political activity is, for example, designed to preserve the largest number of jobs in a recession it can be said to follow this ethical principle. The ethical principle of individual rights, such as freedom of speech, can easily be infringed by political behaviour. If someone in the organization is prevented from legitimately putting their point of view as the result of politicking by another, this can be construed as unethical. On the other hand, if someone ensured that another could 'have their say' this would be seen as ethical. In organizational terms the distributive justice ethic requires all to have equal access to higher positions and if this is the case inequality is acceptable. Hence an example of unethical political behaviour would include 'spinning' stories and rumours against colleagues with the intention of making them appear unacceptable for consideration. Ethical political behaviour would be putting forward for a position someone who had been overlooked but was eligible and qualified.

7.10 Power and leadership

The concepts of power and leadership are, of course, closely related. Chapter 11 on leadership examines the characteristics needed in leaders and the circumstances in which different characteristics and leadership styles are appropriate.

Leadership is directly associated with both authority and power. The link between leadership and power is that the authority of leaders is reinforced by their power and personal characteristics.

This raises the issue of the relationship of power to the functions of leadership. If a leader has strong power sources it is possible to conceive the idea of a leader operating outside the authority structure. Whilst there may be some good served by such a situation in terms of compensating for weak or misdirected legitimate leadership in others, the potential for serious dysfunctional conflict is extremely high. Therefore, leadership patterns and performance in organizations should always be considered within the context of both the authority and power structures.

7.11 Authority and power in the construction process

Construction companies, and perhaps to a lesser extent engineering and quantity surveying companies, may be more formally structured with more rigid hierarchies than are design organizations such as architects and specialist interior designers. The members of the former group of firms may therefore have a higher respect for and respond more to authority than design organizations. It could be thought that members of the latter may be more inclined to conspire to override authority in pursuit of idealistic ideas which could be contrary to the objectives of their firms and their clients but maybe this is stereotyping; that is not to say that other types of firm will not have members who have specific objectives that they will pusue through the use of power and organizational politics. An example could be a specialist engineer who attempts to manipulate his appointment as a director of his company by exaggerating the significance of his specialism over another. Another could for a team leader to attempt to divert higher calibre staff resources to his or her project to the disadvantage of other team leaders' projects in order to raise his or her standing within the firm.

However, the use of power and politics in professional firms may often work for the good of both firms and projects. Expert and referent sources of power have to be earned through performance and it is in the best interests of the firm that people with these attributes influence and advise others in the firm. On the other hand, reward and coercive power sources are capable of misuse if used outside the authority structure and can lead to staff dissatisfaction, low morale and departure of good staff. Whilst power and politics can be important to the wellbeing of staff and the effectiveness of firms, for clients it is the manner in which they impact on projects which is significant.

The extent to which authority is explicitly stated in project structures varies considerably both between and within projects. The ultimate in explicitness is contained in standard forms of contract between clients and contractors for construction work which lay down the rights and obligations of all parties under the particular contract. However, for the same project, and in sharp contrast, the project team may be appointed by a short inprecise letter with no explicit statement of either their authority or responsibility. On the other hand, the project manager and the consultants may be appointed using formal agreements which are interrelated to show the authority and responsibility of all contributors to the client and to each other. The authority structure within the client organization itself is of great importance to the project team as it will have a significant effect on the establishment of the brief for the project and will determine in the first instance who in the client organization the project team should be listening to.

An issue on construction projects which differs from many other types of organization is the lack of sanctions available to a project client. If a project is wholly in-house to the client (e.g., a development company) then the sanctions available to enforce authority will be similar to those of other organizations. However, when all or some of the elements of the project team are external to the client organization the likelihood of sanctions being used is remote as it would be a dramatic and potentially disastrous course of action for a client to fire any of the project's professional and contracting firms, particularly when significant resources have been invested. Whilst this has occurred, it would be an absolute last resort as a client is increasingly locked into the project as it progresses. In such a difficult situation the client and project team's member firms, contractors and consultants will therefore have to negotiate from their relative positions of strength in order to achieve completion of the project. Intermediate sanctions are essentially only those available under the construction contract and other formal agreements but will not necessarily get the project back on track.

Little has been written about power and political structures in the construction process itself yet their use and their potential for influencing the output from the contributing firms and hence projects is well recognized by those directly involved. In his paper on sociological paradigms applied to client briefing, Green (1994) uses a political metaphor to explain the nature of multifaceted clients. Newcombe (1994) applies a power paradigm to the procurement of construction work contrasting the nature of power in traditional and construction management systems of procurement; Poirot (1991) examined a matrix system and found it to be a power-sharing, power-balancing organization and Bresnen's (1991) paper purported to show how goal and power differentials can affect project outcomes but, whilst describing how complex construction project organizations are, did not explicitly analyse the power structure.

As all construction-related firms are focused on projects, power issues tend to be exposed in the project management process, often, but not always, to the detriment of the project when the performance of firms is affected. Thus the nature of the project organization has a significant effect on the opportunity for the use of power. At one end of the spectrum the project may be totally in-house to the client including design and construction capabilities; examples are a large developer or a petrochemical company which constructs its own plant.

In such circumstances the potential for the use of power would be high as political structures are capable of being established through the long-term relationships which will be present. At the other extreme is the client company which builds very infrequently and hence employs external consultants and perhaps a project manager. In such situations the power structure is likely to be less well developed and hence weaker due to fragmentation of the organizational structures involved. Between these two extremes is a whole range of different configurations of clients, consultants and contractors. Each project organization structure will generate different power structures and the ideas relating to power discussed previously will need to be applied to analyse the appropriateness and effectiveness of the use of power in each case.

7.11.1 Power within client organizations

The exercise of power within client organizations can have major effects on a project's objectives and how they are defined. Power struggles within client organizations can distort objectives leading to the construction of a project which the client considers unsatisfactory on completion (Cherns and Bryant, 1984). The reason for dissatisfaction will stem from how the client initially defined the project (often as a result of a power-driven political process within the client organization) but in many cases this will not stop the client believing that it was due to some fault on the part of the project team. Such an outcome can affect the credibility of members of the project team such that the strength of their reputation is reduced and so may create problems for them on future projects.

An analysis of the construction of a new university showed that the exercise of power within a complex client structure and a sensitive wider political context meant that the project was commenced without a full understanding of the type of university to be constructed (Walker, 1994; Walker and Newcombe, 2000). Whilst the project was completed extremely quickly, the final cost was three times the original estimate; nevertheless, ultimately the project was considered to be good value for money as the university which was finally built was of a much higher quality and scope than that originally proposed and was seen to be worth the eventual cost. That the original conception of the university changed dramatically during its development was as a result of a complex client structure in which members conducted highly political games. Nevertheless, the members of the project team were perceived to have created 'a massive cost overrun' and suffered from this reputation resulting in some of them being rejected for future commissions. The 'overrun' was all that was remembered. The final cost would not have been seen as an overrun if the people involved in the power plays within the political context had recognized the changes taking place in the university's design and if they had not committed to an extremely short period for design and construction.

Whilst the above illustration shows that the outcome was to an extent negative for the project team, the exercise of power was positive in delivering a high class university three years earlier than expected to the benefit of the community as a whole. The potential for the use of power within public client organizations is probably even greater than for private sector projects. The Hong Kong Convention and Exhibition Centre took ten years from conception to the start of design due to competing power plays by government departments. Only when a chairperson held in very high regard in the community and reinforced with charisma was appointed to the Trade Development Council (TDC) was sufficient authority vested in the TDC to enable it to break the deadlock (Walker and Kalinowski, 1994).

A distinctive power source on construction projects is that of patronage which can be seen as a form of reward (or coercive) power. This accrues to whoever has the power to influence and/or make the appointment of professional and construction firms to a project team. In a simple view this would be the client but clients are rarely unitary but take complex forms with many people being involved in the appointment process of project teams in both the private and public sectors, all of whom accrue reward (or coercive) power over the firms being considered to a greater or lesser degree. Anyone who has been involved in this process may also develop what could be termed secondary reward (or coercive) power as others may ask for their opinion about the firms which they have been involved in appointing and which are being considered for appointment to other project teams. Whilst generally appointments are awarded on the basis of competitive bids, particularly in the public sector, nevertheless in the private sector commissions are frequently awarded on the basis of recommendations. Even for competitive bids, recommendations are often needed in order to be admitted to the list of bidders. Traditionally architects had the gift of selecting the consultants with whom they would work. In contemporary times, this gift, when available, has in many cases passed to the project manager. Such a situation gives the project manager or the architect reward or (coercive) power over the consultants as they will be unwilling to go against the desires of those who may hold the gift of future work. Even on projects where consultants have been appointed in competition the position is little different as the process may introduce consultants to project managers or architects with whom they have not previously worked. The potential of future commissions from new sources will act to give power to the project manager or architect.

7.11.2 Power and project team members

The influence of the client on the expert and referent power of the project team members is paramount such that any opinion which the client makes explicit about any member of the project team, particularly the project manager, will have a profound effect on that individual's influence within the project team. The potential for using power is likely to be much greater on wholly in-house projects and may well operate in such situations in ways which are comparable with the way in which it operates in organizations generally but it is in circumstances where projects are not carried out wholly in-house where special features of power specifically related to the organization structure of construction projects arise.

The sources of power described earlier (reward, coercive, expert, referent) are all to be found in construction. In the management literature, reward and coercive power are seen to be legitimate (authority) as they offer benefits or sanctions which operate within the formal structure. However, it has been argued that this may not always be the case for construction projects. The impact of many construction projects on their environment, socially, economically and technically, is such that a project often has many and varied stakeholders with a wide range of interests who exert pressure on project teams, their members and clients and in doing so may exercise reward and coercive power over them.

Expert power is a major force on construction projects. The specialisation of professional skills contributes to the effectiveness of expertise as a power base. Projects are so technologically and managerially demanding that a reputation as an expert counts for a great deal. Contradicting the expert advice of a specialist contributor is difficult even though the advice may be believed to be inappropriate. A corollary is that such experts are likely to gather support from colleagues against less expert members of the team who may disagree with the expert.

Referent power is the influence that people exercise because people believe in them. Although closely related to expert power it is more associated with a record of substantial achievement. This source of power is particularly reinforced by charisma. It is a particularly potent force on projects at a number of levels. At the initiation of projects people with referent power are able to force through the acceptance of projects 'against the odds', for example, the selection of one airport site as opposed to another. Project managers with referent power are naturally in a strong position as are other members of the project team, for example, an architect with an international reputation for the design of famous buildings. Project managers and others who worked on the Mass Transit Railway in Hong Kong, which was an extremely successful project, gained great referent power and went on to obtain other important project management and other posts as a result.

At another level, a play which relies on referent power is when one of the professional firms sends along a director to a project meeting, which would normally be attended by a project level staff member, for the purpose of obtaining a particular outcome. The intimidation of other members of the meeting by the presence of a respected director can be seen as the use of referent power. Architects as a profession gain referent power from their historical social position. This certainly still occurs in the UK and no doubt in other countries, though probably less than in the past. In the early part of the last century architects emerged as the elite social class associated with construction stemming from the patronage of their clients as described by Bowley (1966). Architects continue to derive referent power from this historical context.

Finkelstein (1992) refers, in passing, to prestige (referent) power also being provided through a manager having powerful friends. This facet is significant in the construction process due to the high interdependence of the elements of the process. The use of 'powerful friends', rather better put as 'contacts with influence', plays a large part in moving the project forward at all stages. Such influence can include, for example, arranging finance at the early stages, contacts with government, both nationally and locally, to ease the approvals process and good contacts with suppliers and subcontractors during construction. People who can get things done through their contacts acquire strong referent (prestige) power.

7.12 Politics, projects and firms

Earlier it has been suggested that construction project organizations can be categorized (using Mintzberg's taxonomy) as shaky alliances of professional and innovative organizations. The shaky alliance is one of Mintzberg's four types of political organization which is said to be characterized by conflict that is moderate, confined, and possibly enduring yet relatively stable. This configuration can be seen to apply to projects for which the project team is not wholly in-house; that is an alliance of independent professional firms and contractors. Project teams which are wholly in-house could perhaps also be categorized as a combination of professional and innovative organizations but with elements of machine organizations which if they became political organizations could fit any of Mintzberg's other three forms – confrontational, politicised or complete political arena. If they became confrontational or a complete political arena, the development of an effective project could be difficult to achieve as such organizations are seen to be unstable. However, this would be less likely if they became a politicised organization which is said to be relatively stable, as long as it is sustained by privileged position.

Not only do these typologies have relevance for understanding the effect of organizational politics on projects, they are also valuable for understanding organizational politics of client organizations which could be any of these forms at the time the project is being initiated and objectives are being defined. The possibility of obtaining a suitable brief must be extremely limited in the case of client organizations which are political organizations of the confrontational and complete political arena forms but if of the shaky alliance and politicised organizations forms then obtaining a rational brief should be less difficult if the nature of these forms is understood by project teams.

The other important elements which have the potential to become political organizations are the individual professional firms which make up a project team. Mintzberg would classify them as professional organizations, which he believes is one of the organizational types in which there is considerable room for political games. If so, they have the propensity to become political organizations of a confrontational, politicised or complete political arena type. If a professional firm transformed itself from a professional organization to a political organization during the course of a project, the effect on the project could be severe. The likelihood would be that rather than focusing on the project, the attention of members of the firm allocated to the project would be deflected to the internal problems of the firm. In an activity as closely interdependent as a construction project, this could be to the detriment of the firm's contribution to the project.

Whilst the illustrations here use Mintzberg's descriptions, many other accounts of politics in organization are also good bases for analysis (cf. Kakabadse et al., 2004; Schermerhorn et al., 2004; Robbins and Judge, 2008). An interesting point is that not all writers believe that organizational politics are a negative force. Some believe politics enhance the achievement of organizational goals and survival and are good for career advancement (Schermerhorn et al., 2004). The important aspect for the construction industry is to recognize that these forces exist within client organizations and will probably have an impact on project definition and client involvement over which members of the project team (whether in-house or not) will have no, or very little, influence. Similarly such forces may be present in the professional firms from which team members are drawn, and may influence their attitude to the project and their performance. These forces could lead to the politicising of the project team.

7.13 Empowerment again

In dealing with authority and power comprehensively it is appropriate to return to empowerment (see Chapter 5). At first glance it appears to be an elaboration of delegation but close examination shows that it is far more than that. Stacey et al. (2000) appear to see empowerment as integral to complexity theory by stating that self-organization is another word for empowerment but then cloud the issue by stating that empowerment is 'really a more idealistic word for delegation'. The literature seems reluctant to provide a clear definition but Neilsen's (1986) is useful:

> Empowerment is giving subordinates the resources, both psychological and technical, to discover the varieties of power they themselves have and/or accumulate, and therefore which they can use on another's behalf.

It can be seen that empowerment revolves around the need to provide psychological support and stems from an understanding of powerlessness. According to Kanter (1977):

> People held accountable for the results produced by others, whose formal role gives them the right to command but who lack informal political influence, access to resources, outside status, sponsorship, or mobility prospects, are rendered powerless in the organization …

One view is that of power sharing/empowerment (Rudolph and Peluchette, 1993) which is seen as a 'motivational process of increasing the degree of control which the followers perceive themselves to possess'. The model they develop incorporates the sense of empowerment and motivation which, ultimately, is seen to affect performance. People who feel powerless do not respond well in an organizational setting, do not provide the level of effectiveness required and often have a negative effect on the achievement of organizational goals. To overcome the negative effects of powerlessness a positive policy of empowerment is necessary.

Conger and Kanungo (1988) identify leadership practices which are empowering and which are consistent with those of Kanter. They include:

> ... the expression of confidence in subordinates including giving positive emotional support during experiences associated with stress and anxiety, the fostering of opportunities for team members to participate in decision making, the provision of autonomy free from bureaucratic constraint, the observation of others' effectiveness, i.e., providing models of success with which people identify, the setting of inspirational and/or meaningful goals, and, above all, the establishment of a trusting and cooperative culture.

The benefits of empowerment are claimed to be that it motivates people to face greater challenges than they would if they felt powerless. People are likely to accept higher performance goals and hence leaders are able to put such challenges before subordinates with a reasonable expectation that they will respond. People will be motivated to persist in the face of difficulties due to their increased level of confidence in their own abilities to influence events. In a study of 111 work teams, Kirkman and Rosen (1999) found that: 'More empowered work teams were ... more productive and proactive than less empowered teams and had higher levels of customer service, job satisfaction and organizational and team commitment.' Nevertheless, Conger and Kanungo believe that empowerment could have negative effects. They believe that empowerment might lead to overconfidence and hence misjudgments on the part of subordinates. Because of false confidence in positive outcomes, organizations may persist in efforts which are inappropriate to their aims.

Liu et al. (2003) identify the situation in which some leaders may feel threatened by empowered subordinates because there is some loss of control, and also the corollary of some employees not wanting to be empowered as they are non-assertive and cannot accept responsibility. The question of responsibility does not seem to be clearly addressed in the literature on empowerment. A key issue would appear to be to ensure that those who are empowered also accept responsibility for their actions otherwise the false confidence referred to by Conger and Kanungo could lead organizations into unwise commitments.

7.13.1 Empowerment and projects

Empowerment is an important issue in many organizations which have powerless and disadvantaged groups. However, on the face of it, in the process of managing construction projects, empowerment of a project team could be expected to occur naturally due to the characteristics of the task which requires highly skilled professionals to be self-motivated and accept high performance goals. This they have always been required to do due to their professional training and traditions so the concept is not new although it has not been clearly identified and given the name empowerment. It is generally assumed that project managers instinctively empower their project teams as little effort should be needed because professional contributors should generally empower themselves. Nevertheless, these assumptions may not be correct for every occasion so the concept may need to be understood and applied to further enhance the project team's contribution.

Newcombe (1994) claims that the construction management procurement system produces a different power configuration from the traditional procurement system. He believes that the former system allows a better opportunity for the exercise of expert power and charisma coupled with the empowerment of all the parties involved, especially the specialist trades contractors. Newcombe (1996) extended these views by contrasting what he saw as the hierarchical traditional system of project organization with the 'potentially democratic' construction management form. He believes the latter is based on empowerment or power equalization and that such power structures are to the benefit of clients, as they will not produce the fragmentation and friction created by the traditional system. He argues, 'If the empowerment approach is adopted then skill in building networks of contacts such as designers, trades contractors, clients, suppliers and stakeholders with an interest in the project will be necessary.'

These ideas reflect the current preoccupation of the construction industry with various forms of partnering but empirical work is required to substantiate their value. Illustrating the unevenness of empowerment in traditional construction organization structures, Hammuda and Dulaimi (1997) state that construction managers were always 'traditionally empowered' on their sites by making decisions in relation to their autonomous project within their wider companies' boundaries but that this is becoming less so as projects become more complex. In comparing construction with service and manufacturing industries they compared companies not projects, so in finding that empowerment in the construction industry was very much behind other industries, they were referring to actual construction work undertaken by contracting companies not to the much wider work involved in the project management process in which empowerment at the professional level is generally seen to be the way professionals operate. At the level of site management, the application of empowerment may be very limited although it is at this level that further empowerment of workers, subcontractors and specialist contractors may have the potential to be of the greatest benefit but may require more stability in the workforce.

An interesting perspective is taken by Liu et al. (2003); in dealing specifically with the power of the project leader, they point to the 'power gap' which may arise and say:

'... as a project becomes increasingly complex, both project leader and project members may feel pervasively more powerless as a result of a widening gap between the amount of power granted by the position and that actually required to get the job done.'

They continue,

'... not all leaders are capable of manipulating the power gap successfully. Certain leaders may feel threatened by an empowered subordinate or an empowered group of subordinates because there is some loss of control and, hence, will not contemplate power sharing. On the other hand not all subordinates want to be empowered.

The assumptions about empowerment in construction are reflected in and challenged by Tuuli and Rowlinson (2010) who point to the emerging view that empowerment can be seen as either a structural (empowerment climate) or a psychological concept. The former is deeply rooted in job design and is akin to that described earlier. The latter refers to individuals feeling that they are empowered. Both concepts were identified in the construction setting of the research and were seen to need the creation of an empowering climate for the full benefit of empowerment to be achieved. Also examined is empowerment in construction as a multi-level concept at both the individual-level and the team-level simultaneously. A significant finding is that 'Managers seeking to engender ... cooperative acts which are particularly needed in the high interdependence contexts of projects, can therefore not selectively empower individuals, but must ensure that all team members are empowered if full benefits are to accrue from empowerment' (Tuuli and Rowlinson, 2009a). A further study found that psychological empowerment was a valuable contribution to the improvement of job performance in projects (Tuuli and Rowlinson, 2009b). This is only a brief review of a few ideas emanating from this set of papers which have shown that empowerment is not the relatively simple idea that is has been seen to be in construction organizations but is a complex concept worthy of serious study.

However, Williams (1997) tempers enthusiasm for empowerment by pointing to the conflict between it and risk management. He believes that 'unfettered empowerment in a modern project setting ... can lead to high risks He continues:

'... some Risk Managers, in projects which have embraced the management philosophies of both empowerment and risk management, have found themselves in an almost schizophrenic role, required to manage risks others are empowered to effect or enhance. What are needed, are compromises, empowering project workers and teams whilst recognizing the holistic nature of the projects.

Whilst not referring to construction specifically, nevertheless his message is clear.

8 Culture

8.1 Introduction

The most common meaning of the word 'culture' probably relates to the characteristics which are seen to typify the inhabitants of different countries and tribes – its anthropological meaning – but is also used to describe the characteristics of a refined person or in connection with cultivating plants or breeding animals. Dictionaries provide many more meanings which have led to it being described as one of the two or three most complicated words in the English language (Williams, 1983). Culture is also used freely in casual conversation with a vague, unspecified, all-embracing meaning, for example, a culture of greed, a culture of dumbing-down, popular culture and so on. Its anthropological meaning is the one which is of value to management studies. Tylor (1871) is believed to be the first to use the term in an anthropological sense in describing culture as 'that complex whole which includes knowledge, beliefs, art, morals, law, custom and any other capabilities and habits acquired by man as a member of society'. In those days the focus was on understanding the way by which people in different groupings in society distinguished themselves from other groups. At the higher levels of society, such as nationalities, there is a rich source of factors on which to draw to identify and describe cultures in meaningful ways; however, it is when these ideas are used to identify and describe the cultures of industries and business organizations that the identification and description of cultures is problematic.

Nevertheless, organizational culture came to eminence in the early 1980s when four popular books appeared promoting the idea of a strong organizational culture as a way of guiding the behaviour of workers to the advantage of their companies, together with the possibility of managers being able to change a company's culture to its benefit. These books were Ouchi's *Theory X* (1981), Pascale and Athos's *The Art of Japanese Management* (1982), Peters and Waterman's *In Search of Excellence* (1982) and Deal and Kennedy's *Corporate Cultures* (1982). It has been suggested that the idea of organizational culture originated in the work of the human relations movement in the early part of

Organizational Behaviour in Construction, First Edition. Anthony Walker.

the 20th century which was interested in non-logical rationalisations for actions and the replacement of the so-called scientific or classical management school by a more humane way of understanding people's values, beliefs and feelings (Huczynski and Buchanan, 2007).

Organizational culture is seen as the web of common meanings in specific organizations (the dominant culture) but in which different managerial units may acquire different cultures (subcultures) in order to perform effectively. The dominant culture arises unconsciously from the common elements that influence the way members of an organization think, which delineates what is seen to be the correct way for the organization to do things. A comparable process may take place in the different managerial units to create subcultures. In both cases, members of an organization take them for granted and follow them unconsciously. These hidden meanings are created over time and sustained by 'surface manifestations of culture' (Schein, 1985) which are the observable behaviour patterns and objects reflecting a culture.

Whilst many texts on OB accept organizational culture as a given and have applied it to analysing many diverse organizations, there are many scholars who are critical of the application of the ideas of culture to organizations and/or do not find it useful for analysis. There have been limited attempts to apply the ideas of culture to whole industries – one such is of the wine and museum industries in California (Phillips, 1994) – but the construction industry seems to have become transfixed with the idea that its culture is the root of all its problems, of which more later.

Whilst research on the value of organizational culture to organizational performance (which is where its benefit is claimed to lie) is unconvincing, business tends to have maintained a greater interest in organizational culture than some other topics of OB, encouraged in part by the enthusiasm of management consultants for the topic.

8.2 Definitions

The elusiveness of the idea of organizational culture is shown not only by the lack of a widely-accepted definition but also by the wide variety of definitions which have been proposed. They vary from the broad to the specific and from the succinct to the rambling. Amongst the more well known are:

- 'The way we do things around here' (Deal and Kennedy, 1982).
- 'The pattern of basic assumptions that a given group has invented, discovered or developed in learning to cope with its problems of external adaptation and internal integration' (Schein, 1985).
- 'The collective programming of the mind which distinguishes the members of one organization from another' (Hofstede, 1993).

Palmer and Hardy (2000) give a masterly comment under the heading of 'Endless definitions':

> A range of management definitions of culture can be found; some of these definitions mention beliefs and values; others emphasize 'knowledge'; some highlight shared meanings; others point to an organization's 'ethos'; some mention myths, symbols and rituals. Some writers maintain that agreement exists around the fact that culture is holistic, historically determined, anthropological, socially constructed, soft and difficult to change. Others, however, question even this degree of consensus arguing that culture has become a buzzword meaning 'many different and sometime contradictory things' (Wilkins, 1983). Even definitions that are widely used, such as that of Schein, have been the source of dispute. ... Sathe (1983) argues that it is pointless to argue about which definition of culture is correct because it does not have a 'true and sacred meaning that is to be discovered.'

This lack of congruence of definition points to imprecision in understanding what organizational culture really means and hence inhibits meaningful empirical research. This led to Barthorpe (2002) believing that '... the word culture has been particularly recalcitrant in this regard [definition], to the extent that some argue for the abandonment of the word altogether'.

8.3 Critiques of organizational culture

The imprecision evident in the wide range of definitions of organizational culture is reflected in the wide-ranging criticisms of the concept which has lead to scepticism, particularly on the part of academics, as to whether it is useful for the analysis and understanding of organizations. Huczynski and Buchanan (2007) believe that organizational culture remains a controversial concept and point to Needle (2000) who wrote: 'The treatment of culture at the level of the firm varied considerably, ranging across the banal, the simplistic, the misleading, the highly complex, the impenetrably academic and the highly critical.' Other critical voices include:

- Grint (1995): 'Culture is rather like a black hole: the closer you get to it the less light is thrown upon the topic and the less chance you have of surviving the experience'.
- Brown (1998): 'Organizational culture may be thought of as a reformulation of existing models and theories to satisfy changing views of how organizations work, and which is likely to enjoy its pre-eminence for a finite period of time'.
- Palmer and Hardy (2000): 'One debate here is whether culture has had its fifteen minutes of fame' and 'One view attributes the death of culture to its limited practical relevance – so many attempts to change it and improve organizational performance have failed'.

Huczynski and Buchanan (2007) draw attention to the contrasting perspectives between consultants and writers of self-improvement books and many management textbooks on the one hand (the managerial or functional perspective) which is said to be both normative and prescriptive and, on the other hand, mainly academic social scientists. They use Needle (2000) to make the point that 'It [the managerial approach] makes assumptions about employees that it does not explore and constitutes a set of beliefs and values that are deliberately created as part of a management strategy, and which are used to guide behaviour and processes within the organization'. This they describe as the culture '*has*' view which holds that every organization possesses a culture, just as it has a strategy, structure, technology and employees. Critics believe that the managerial approach does not seek to explain what an organization's culture is, nor does it assess its significance; rather it prescribes what it should be to achieve greater efficiency and how it should be changed to further this. They set against this view of the mainly academic social scientists who believe that the culture *is* the organization and the organization *is* the culture and also that organizational culture is a term that is 'overused, over-inclusive, but under-defined'. They continue 'It thus rejects the notion that culture possesses any objective, independent existence which imposes itself on employees' and 'Culture is produced and reproduced continuously through the routine interactions between organizational members'.

Under the illuminating heading of 'In Search of Culture: Methods Galore!' Palmer and Hardy (2000) provide a critique of research into organizational culture in which they say: 'A menagerie of different organizational features has been targeted in attempts to identify manifestations of culture.' They continue by saying that the issue revolves around:

> ... whether culture is seen as a variable, something which the organization *has*, much like a structure or a strategy; or whether it is seen as a perspective where the culture *is* the organization and the organization *is* the culture. The former view assumes culture can be identified, measured, analysed and managed. It has an objective existence, independent of the people who wish to study or change it. Users of quantitative methods are most likely to share this attitude. The latter view argues that culture cannot be reduced to independent or dependent variables. Moreover, it maintains that what is seen as culture is a product of how the person is 'doing' the seeing. In other words, the frameworks and perspectives used to identify culture shape what is perceived to be the culture. This treatment of culture denies any 'objective' existence and instead focuses attention on the interwoven nature of language, meaning, symbols and rituals. Users of qualitative methods are most likely to share this view.

Notwithstanding such conflicting views of organizational culture, its relevance and importance need to be considered as it features prominently in the OB literature. If one accepts that it exists, the question is what, if anything, does it contribute to the effectiveness of organizations, or does it work against effectiveness?

Fincham and Rhodes (2005) recognize that 'culture is clearly a complex concept and one we need to explore in some detail. It refers to the totality of knowledge in an organization or society. But if almost anything is part of culture, the concept runs the risk of becoming vague, a dumping ground of unexplained factors in organizations' but continue 'Culture is basically substantive; it reflects the beliefs and techniques that characterize *specific* organizations'.

On the other hand, Robbins and Judge (2008) are in a less critical mode, accepting that organizational culture does exist but treating it in a non-judgemental manner and, in summary, state that it performs a number of functions by:

- Creating distinctions between one organization and others
- Conveying a sense of identity for organization members
- Generating commitment larger than individual self-interest
- Stabilising the social system which holds an organization together by providing standards for what employees should say and do
- Serving as a sense-making and control mechanism that guides and shapes the attitudes and behaviour of employees.

They continue by recognising that organizational culture can become a liability when, because of its entrenched nature, it acts as a barrier to organizational change and progress. In this connection, an organization's culture can lead to resistance in hiring staff who have characteristics different from the general workforce but who are needed to aid change. Also, lack of compatibility of organizational cultures is seen to a major obstacle to successful mergers.

8.4 Organizational climate

Organizational climate should be distinguished from organizational culture to avoid confusion although the distinction is far from clear. Organizational climate, which pre-dates the interest in organizational culture, is said to refer to the prevailing atmosphere which exists in an organization. It focuses on the beliefs individuals hold regarding the organization and is a product of individual psychological processes. This contrasts with organizational culture which results from shared assumptions, values and beliefs that arise from an organization's traditions and underlie and determine what is seen as the correct way for the organization to do things.

Although the terms are sometimes used interchangeably and are similar, they do differ is some fundamental respects. The roots of climate are psychological whereas those of culture are anthropological and sociological. Culture is seen as the way people learn what is acceptable and unacceptable in an organization; climate does not deal with values and norms but with current atmosphere; culture is more difficult to alter in the short term than climate as it is deep-seated. Drawing on Ashforth (1985), Palmer and Hardy (2000) comment, 'Despite such confusion, and the acknowledgement by some that "the concepts do slide greyly into one another", a valiant fight is still fought by those who wish to retain the distinction between the two.' Rousseau (1990) is one who supports retaining the distinction. However, they also cite Denison (1996) stating that '... the convergence between studies of both culture and climate means that they are not particularly different. These "two research traditions should be viewed as differences in *interpretation* rather than differences in the *phenomenon*" being studied'.

8.5 Organizational culture, types and performance

Its proponents believe that a strong organizational culture improves an organization's performance and leads to success and lower staff turnover. This represents a managerial perspective which concentrates on the contribution cultural elements make to organizational unification and control (Scott, 1998). However, research on the impact of culture on organizational performance is mixed, depending on how the research is done and what variables are measured (Moorhead and Griffin, 2001).

A strong culture is one in which the organization's core values and beliefs are intensely supported by its members and in which this support is institutionalized and hence long lasting. A weak culture is at the opposite end of a cultural spectrum. McShane and Von Glinow (2003) believe that 'companies have weak cultures when the dominant values are short-lived and held mainly by a few people at the top of the organizzation.' Not all strong cultures are seen to be good as a company's organizational culture needs to be appropriate to its environment or employees will have difficulty anticipating and responding to the needs of customers and other stakeholders (McShane and Von Glinow, 2003). Moorhead and Griffin (2001) give examples of cultural traits which can hurt performance: 'customers are too ignorant to be of much help' and 'innovation is not important'. A further potentially serious drawback to a strong culture is a propensity to resist change, particularly in the fast moving business environment of the 21st century. Also Scott (1998) warns that a strong culture can develop into an authoritarian system leading to abuse. Possibly one of the most spectacular example of the dangers of strong cultures is the financial crisis of 2008/2009, in which the strong culture of financial institutions resulted in them paying no attention when some of their employees warned of the severe dangers associated with their lending and investment practices. OB texts do not seem to refer to any benefits of a weak culture, leading to the impression that it is a 'bad thing' but, in situations in which free and independent thought and innovation is needed, the lack of a straightjacket can be advantageous.

Huczynski and Buchanan (2007) observe that whilst Ogbonna and Harris (1998) say that some researchers still support the idea that strong cultures improve performance, they do so only with caveats such as the linkage is dependent on a firm also having certain traits like adaptability. Rather, competitive advantage is seen to arise from firms having superior competencies to their competitors which are not able to be copied so that 'if a firm's culture is to represent a source of competitive advantage, it must be rare, adaptable and non-imitable (Barney, 1991)'. But also that it is in the interests of both management consultants and managers to maintain that 'culture makes a difference'.

Newcombe (1997) gives the advantages of a strong culture in construction projects as: enhanced effectiveness; contributors learn to live together; are less parochial; and that it reconciles conflicts. He sees the limitations of a strong culture as: difficulty in introducing change; conservatism inhibiting innovation; and isolation of a project from its environment. He implicitly adopts a managerial perspective by stating that 'culture is built slowly by committed leaders with a strong vision or mission for the project'.

Palmer and Hardy (2000) assemble powerful critics of the idea that strong cultures enhance performance. They point out that:

- 'the term strong has been used in a myriad of different ways' (Yanow, 1993) [which contributes significantly to the difficulties of acquiring a body of convincing research]
- 'researchers challenge claims to have found support for strong cultures (Hofstede et al., 1990) because of the impossibility of operationalizing cultural strength'
- 'the strong culture argument assumes a causal direction in which good performance results from a strong culture [whereas] an alternative argument is that a strong culture only emerges in companies which are experiencing good performance' (Kotter and Heskett, 1992)
- 'strong cultures may initially be high performers, but barriers to the emergence of new and innovative ideas may replace co-operativeness' (Hilmer and Donaldson, 1996).

Regarding research in this area, they continue by saying, 'A fundamental problem lies in the methodological difficulties concerning how both culture and performance are operationalized, making it unlikely that definitive evidence, either supporting or refuting the link between culture and performance, will ever be found' but Ogbonna and Harris (1998) state, 'However, whilst a conceptual understanding of the intricacies of the managing culture debate has provided many worthwhile theoretical contributions, many theorists have noted the need for further empirical work in this area.'

8.5.1 Types of organizational culture

Classifying organizational cultures into defined types is an attractive idea. It aims to makes analysis simpler, it makes establishing relationships with other organizational variables a possibility and comparisons between the experiences of those working in each type also become possible. However, none of the attempts to achieve classification have been sufficiently convincing to have been widely adopted. Although it is possible to describe different types and give each a name, the narrowness of the description and the choice of a single word to describe them is insufficient to convey the complexity and uniqueness of an individual organization's culture. An organization's culture will be a complex mix of types and although it may be possible to discern some elements of the types which have been identified or, at best, describe the most obvious traits of the culture, definition of a fully specified and undisputed type is unlikely to emerge. As Fincham and Rhodes (2005) say:

> A number of accounts have been at pains to stress organizational culture not as a monolithic concept, but as encompassing the idea of multiple cultures. Organizations are made up of different occupations and professions, different social classes and sexes, and may be spread over different geographical areas. All these can form the basis of distinctive subcultures and counter-cultures that compete to define the reality of the organization.

In addition to the very broad classification of strong and weak types, probably the typology most frequently referred to is that of Handy (1993) which comprises briefly:

- *Power culture.* Depends on a central power source with power and influence spreading out from the central source.
- *Role culture.* The role or job description is often more important than the individual that fills it.
- *Task culture.* The whole emphasis is on getting the job done.
- *Person culture.* The individual is the central point. If there is an organization or structure its emphasis is on serving the needs of individuals.

Fincham and Rhodes (2005) comment that Handy's cultural types seem to add very little to the account that is provided by structure. The range of different types of culture if not infinite is enormous, so defining and classifying them could be insuperable. For example, Palmer and Hardy (2000) identify 23 types drawn from many writers on organizational culture and euphemistically state that 'this approach has some shortcomings'.

8.5.2 Adaptive cultures

The managerial view believes that strong cultures perform better when an organization's culture and an organization's environment are compatible (often referred to as cultural fit). This view reflects the contingency theory of organizational design and its underpinning of systems theory (Walker, 2007). The wrong type of strong culture would be counter-productive. The idea of 'tailor-made cultures' is attractive but operationalizing such an idea seems impracticable.

A major question is: if a cultural and environmental fit is established how does the culture react to changes in external factors? It is suggested that organizations adopt an adaptive culture in which employees focus on the changing needs of customers and other stakeholders and support initiatives to keep pace with these changes (McShane and Von Glinow, 2003). They continue by stating that employees in adaptive cultures pay as much attention to organizational processes and their improvement as they do to organizational goals (which again shows convergence with the systems theory of organization); that employees believe that 'it's our job' rather than 'it's not my job'; and that they are quick and proactive.

A further significant aspect of adaptability and cultural fit occurs when mergers and joint ventures between organizations take place. McShane and Von Glinow (2003) believe that in these situations corporate leaders are usually focused on financial and marketing issues to such an extent that they fail to consider the compatibility of the organizational cultures involved, which are likely to differ. They point out that two-thirds of US firms which merged between 1997 and 2000 underperformed their industry peers in the following years and they give interesting real world examples. Palmer and Hardy (2000) use Cartwright and Cooper (1993) to show the relative difficulty of merging power, role, task and person culture types. An example is mergers where the dominant party has a role culture and the other organization has either a task or a person culture, in which circumstances managers in the junior organization resist the bureaucracy of the role culture. Fellows (2006) points to similar difficulties in construction joint ventures which are magnified if the project is international with partners of different nationalities.

8.6 Corporate image

Fincham and Rhodes (2005) contrast corporate culture and corporate image and point out that they are not the same thing. Often, that which managers popularly and loosely refer to as organizational culture tends to be no more than the corporate image of the organization that top management would like to project. They draw attention to the research methods of populist management writers who carry out surveys amongst only senior managers to elicit the 'organization's culture' and which exclude other levels of employees, to produce an elitist version of culture which is no more than a wish of top management. As they go on to say: '... the contrast to the original anthropological idea of culture, as the gradual emergence of social beliefs and practices in a community of people, could not be greater.'

8.7 Applications to construction organizations

Whilst in recent years there has been an increasing literature on culture related to construction organizations, there has been little empirical research reflecting problems in construction of producing convincing research studies. Zhang and Liu (2006) aimed 'to develop a culture – effectiveness model of [Chinese] contractors' motivated behaviour towards performance'. As a contribution towards the achievement of this aim, their paper specifically sought to identify the organizational culture profiles of Chinese construction enterprises. Their valiant effort served to show the difficulties of achieving meaningful outcomes in this field, difficulties faced in equal measure by mainstream management studies. They recognize that by choosing to use just four types of culture (clan, adhocracy, market and hierarchy) they used only a small sample of the many types referred to previously and that, due to the problem of definition of types, those they used are not mutually exclusive. They start from the basis that 'Given that organizational culture plays a significant role in work performance and effectiveness, the apparently low effectiveness of the construction industry may be related to the culture of the contractor's organization' but the assumption underlying this proposition is subject to considerable debate as empirical evidence in support of there being a relationship between organizational culture and performance is not forthcoming in mainstream management. In fact, their paper recognizes that the proposition needs further investigation and that 'at present, culture research findings are not entirely consistent with one another [because of the complex nature of the constructs of organizational culture and effectiveness]'. Based on the assumption that organizational culture and effectiveness are, in fact, related they make the interesting statement that '… the postulation is that the relationship between organizational culture and effectiveness is reciprocal and that organizational culture is both an asset and a liability, depending on its positive, or negative, impact on organizational effectiveness'.

A further study set in the Hong Kong construction industry (Phua and Rowlinson, 2004) sought to operationalize culture for the purpose of construction management research; an ambitious aim, as mainstream management researchers have said that operationalizing culture and performance is a fundamental problem and may never be solved. The study was based on culture in partnering and used a social identity framework. It pointed out that writing and research on partnering culture has been prescriptive without being related to:

- context
- the dimensions of culture that are important to project success
- how they are brought about, and
- in what way they are related to performance.

They state that their results suggest that the relationships between culture and individual behaviours are far more complex than alluded to by simple normative generalizations and underscores an urgent need for future research to adopt a more comprehensive framework for defining and measuring culture in construction management research; a comment which could apply to research on culture in any management setting.

A research project which demonstrates the intransigent nature of organizational culture as a subject for empirical study and, in particular, its application to construction is by Riley and Care-Brown (2001) who attempted to compare the culture found in a construction company with that found in two manufacturing companies. The issues begin with recognition that 'to manage and influence culture, it is necessary to first define and then develop a conceptual model of what culture is ...' but then continue by stating that an accepted definition does not exist. Undertaking research from this base raises questions of validity which face all empirical research on culture. The model of culture used is that devised by the National Economic Development Office (1990) (NEDO) embodying eight characteristics but they point out that, as with definitions, there is no universally accepted list of characteristics. Also the NEDO questionnaire incorporates ten 'key areas' which are said to define culture, some of which could be disputed. Significant differences were found between the cultures of the construction and manufacturing companies. The culture within construction was found to be a 'project culture' and manufacturing was found to be a 'company culture' but these categorizations are extremely broad and are difficult to specify. Members of the construction industry are unlikely to be surprised by this outcome. However, the paper goes on to say that a typical construction company appears to have two cultural identities, a head office corporate culture and a distinctive site-based or project culture, whilst manufacturing industry appears to have only a corporate- or company-based culture. The subcontracting structure of construction projects and the cultural complexity this engenders through subcontractors each having its own subculture is highlighted. Although it is not acknowledged, this situation mirrors the subculture ideas of mainstream management but in a more resolute form as each subcontractor is an independent firm; hence the concept of a dominant culture becomes so much more problematic. It was found that shared values are quite different between construction and manufacturing industries as in construction the soft issues and the beneficial effect of managing the soft issues are less understood. An interesting finding shows how the construction industry is often misunderstood by those remote from it:

> The results show the construction industry being less innovative than the manufacturing industry. This may not be a true reflection of reality because the nature of a construction site is one of continually producing new solutions to unique site production situations. This sort of innovative thinking may be taken for granted by site staff and not thought of being anything special and hence not scored as such.

However, it should be added that such innovation is more likely to take place in connection with some trades and subcontractors more than others as certain technologies lend themselves to on site innovation.

8.8 Observable aspects of culture

It is the nature of culture that it cannot be seen. Though managers and employees may believe that they are able to describe the values and beliefs on which their firm's culture is based, it is when they take their firm's culture for granted without the need for articulation that it is seen to be at its most telling. Nevertheless, manifestation of an organization's culture is said to be observable through the visible things which a culture produces. Such symbols and actions project a firm's vision of its culture which it wishes to show to the outside world. Many have been identified and can be loosely categorized as:

- *Physical objects* which include the style of buildings occupied by a firm, particularly its head office, office layouts, e.g. open plan, furniture and clothing.
- *Rites and rituals* include: corporate 'days out' for bonding activities; routine activities such as company chants and songs which employees are expected to join in with; catchphrases, mottoes and slogans; and symbols which represent the company.
- *Induction events and courses* which can be seen as the indoctrination of new employees with the ways of the company and how they are expected to behave in order to fit in. Further courses can be seen as maintaining commitment to the prevailing culture.
- *Language and gestures* are the ways in which spoken and written words and accompanying non-verbal gestures are used to convey specific company attitudes or technical meaning. This includes the way in which employees address other employees at the same and different levels of the organization, customers and other stakeholders.
- *Stories* which come in many forms in organizations, such as simple narratives of events which are seen as symbolic within the organization, legends, sagas and myths about historical or recently invented events (often embodying heroes who personify the firm's beliefs and values). They may or may not be based on the firm and are intended to inspire employees to subscribe to the firm's culture. Jokes with an underlying message could also come under this category.

The majority of the visible aspects of organizational culture identified here are not likely to be observable in professional construction firms and other professional organizations. They are more familiar in large-scale commercial organizations, particularly in the retail sector and others in the public eye (McDonalds would be a prime example). But nevertheless, certain aspects could be found in construction, the most likely of which would be the design of offices. For example architectural practices would wish to display their design credentials: stories about firms' and individuals' past achievements and heroes of which construction abounds; and the technical language which professions frequently have to use. Professional firms would be expected to seek to establish and transmit the type of professional culture in which their clients would have confidence through, for instance, rational, thoughtful, measured and quietly confident advice.

8.9 Creating, sustaining and changing organizational culture

Taking a managerial view of culture, a firm's founders establish the initial culture of their firms as start-up firms are invariably small and the founders have a huge influence. As firms grow, however, new forces come into play and new managers and employees may effect shifts in a firm's culture. In the first instance, founders will usually recruit employees who think as they do; they will then further influence them to continue to think in this way and become role models for their employees. As a firm grows, the values and beliefs this process develops become embedded as its culture. Once a firm has been in existence for a number of years, new entrants at managerial level need to learn its culture if they are to use it to the benefit of the company. They must learn its values and the behaviours and actions those values support, focusing on developing a deep understanding of how organizational values operate in the firm, which is not an easy task (Moorhead and Griffin, 2001).

The employee selection process is fundamental to maintaining an organization's culture. Whilst recruiting people with appropriate skills and experience is vital to success, compatibility with the firm's culture usually looms large and can work to the firm's detriment if allowed to compensate for a shortfall in ability. Part of this process is also acquainting applicants with information about the organization so they can decide whether they will be comfortable working in the organization, although this may be difficult to achieve as the organization may project an image of itself which is not a true reflection of the reality of its culture. A continuation of this process can be seen to be the requirement for new recruits to attend induction courses which reinforce the expectations of the organizations in terms of the values and beliefs employees are expected to subscribe to and how they should behave to reflect these beliefs and values.

The next step in the process is organizational socialization through which employees develop their understanding of the organization's culture through socializing with longer-standing employees. By working and mixing with others they come to know over time what is acceptable and what is not, how to communicate their feeling to others and how to interact with them. New employees learn through observing the behaviour of experienced people and by understanding the requirements of top management. Huczynski and Buchanan (2007) point out that, whilst formal training, corporate statements, etc., about organizational culture are not unimportant, the socialization process is primarily dependent on people's close observation of the actions of others. They also make the important observation that sometimes the description of a firm's culture in pronouncements by top management and in training courses is in conflict with the values of that organization as expressed in the actions of its employees. They give as an example that a firm may say that employees are their greatest asset but then treat their employees badly. In such situations new employees will soon learn that the rhetoric of formal announcements and training has little to do with the real culture and come to accept the real culture rather than the 'official' one, leading to cynicism in the workforce.

The OB literature tends to imply that an organization seeks to have employees acting in entirely predicable ways in order to reflect its culture and, whilst it may be in the interests of many organizations to have such cohesion, dangers such as 'groupthink' can emerge which inhibit creative thinking as a result of brainwashing. A culture which results in standardized behaviour is increasingly less relevant to today's working environment and has never been relevant to many companies in the construction industry. A culture of creativity is essential to many professional firms as it is to high-tech industries where innovation not standardization, is their lifeblood. So, rather than recruit in their likeness, such firms may be looking for a 'grain of sand' to inspire change. Firms which have a strong inflexible culture will need to change as the environment in which they exist changes. However, as they grow and more top management is recruited they will need to bring in new ideas even though they may have been selected in the image of the firm's founder and the original senior management. Such changes may occur in an unconscious manner but a time may come when a conscious decision is made to attempt to change an organization's culture, invariably in the expectation of improving the firm's performance, which may not be realized.

There are mixed messages about whether managers can change an organization's culture. Fincham and Rhodes (2005) see it as the central problem arising from the managerial approach to organizational culture. Moorhead and Griffin (2004) say that 'Organizational culture resists change for all the reasons it is a powerful influence on behaviour'. Because the values and beliefs in an organization's culture are deeply embedded within firms, the stronger the culture the more resistant it is to change and, even if achieved, is likely to take years rather than months. Robbins and Judge (2008) suggest that conditions which are receptive to change are: a dramatic crisis; a change of leadership; a young or small organization; and a weak culture. Huczynski and Buchanan (2007) see the issue of culture change as a complex matter. They point to Thompson and Findlay (1999) who state that there is confusion about culture change because it is not clear what is being changed. This is demonstrated by two different views of culture: one is seen to be a set of shared values to guide change and the other a set of practices or behaviours. The former has resulted in some large organizations attempting generalized change programmes requiring employees to question their current beliefs, behaviours and values. The latter approach has resulted in organizations introducing specific initiatives, for example in customer care, rather than generalized culture change. Empirical evidence from the previous ten years reviewed by Thompson and Finlay (1999) showed that 'management were not yet able to "govern the souls" of their employees' and that staff responses to attempts to change organizational culture led to 'distancing behaviour, cynicism, deep acting, and resigned behavioural compliance rather than internalization of values or attitudes'.

It seems that the results of attempts to change cultures are limited at best, largely unpredictable and not within the control of management. Thompson and Finlay (1999) suggest the reasons for this are that 'the majority of companies experience enough difficulties managing their operations successfully let alone managing their employees' "hearts and minds", that there is a high degree of employee awareness of management's motives in instituting these programmes, much of it translated into dissatisfaction and distrust and that since the 1990s the supporting conditions within the workplace environment which encourage long-term commitment have been absent or are reducing'. They make a telling point in relation to professionals generally, which has particular resonance for construction professionals, when they identify that professionals have seen their standing in society reducing and their level of expertise being challenged leading to a reduction in respect and greater scrutiny both by their organizations (particularly where they are managed by non-professionals) and by the public. This has led to less autonomy, greater pressure and a reduction in loyalty and commitment to their companies compounded by contract employment, temporary positions, downsizing, outsourcing, etc., which do not encourage company loyalty and hence are not conducive to developing or changing organizational cultures.

8.10 Cultures, subcultures and construction

Subcultures underlie the dominant culture of most organizations. The larger and more complex an organization, the more likely that subcultures will form. They tend to form to reflect specialisations, common experience and problems and are often defined by departments and geographical locations. Whilst the comfortable view of organizational culture sees the core values of the organization being retained despite the existence of, or even enhanced by, subcultures, this is not necessarily the case. Subcultures can become countercultures and directly oppose an organization's core values.

Countercultures can create conflict and dissenion and detract from the development of a dominant culture but can also have sound benefits. They can act as an irritant which counters complacency; help to maintain ethical behaviour; generate creativity; and encourage constructive debate; and so aid the organization in moving forwards. The adaptation of an organization's dominant culture to suit changing environmental conditions is enabled by countercultures. However, the validity of organizational culture as an explanatory construct can be weakened if the strength of subcultures are such that they do not allow a dominant culture to emerge because there are no shared values to determine what is a uniform interpretation of appropriate behaviour (Robbins and Judge, 2008). Palmer and Hardy (2000) drawing on Sackmann (1992) emphasize that what constitutes a subculture is much more complex than it is often portrayed in the literature and, drawing on Rose (1988), that the term subculture can have different meanings. They can be unique cultures of meanings and values 'not connected to a core umbrella culture' or 'an array of distinct subcultures that exist in relationship to a core umbrella culture'.

These ideas have implications for the proposition that construction project organizational cultures exist (Kumaraswamy et al., 2002) alongside the organizational cultures of professional and contracting firms. Kumaraswamy et al. (2002) state that project cultures comprise four overlapping subcultures: organizational (as referred to above), operational, professional and individualistic. The latter appears to refer to national cultures. Professional subcultures are seen to be delineated by the sentient differences between professions (Walker, 2007) which can be conceived as a number of subcultures, one for each profession. The idea of sentience in construction was developed by the Tavistock Group (Miller and Rice, 1967). A sentient group is one to which individuals are prepared to commit themselves and on which they rely for emotional support. It is a particularly strong force in the construction industry as a result of members' allegiance to their profession. Operational subcultures are said to 'deal with issues such as safety, quality and organizational learning, how people react to claims and disputes, their approach to risk management and planning and control systems'. From this challenging array, Kumaraswamy et al. (2002) see the project culture, which should be the dominant culture, being determined by the relative importance of subcultures to the project but they do not recognize the likelihood of the subcultures being so strong that a dominant project culture is not allowed to emerge. If that was to become the case, the outcome would be that the subcultures would be of the type 'not connected to a core umbrella culture' referred to earlier.

Culture is said to be a powerful force in the construction industry and is particularly complex on projects as a result of subcultures. The overriding image of the industry is macho, uncompromising, uncaring, opportunistic and adversarial to which Green (1998) adds a culture of 'control and command'. Although some construction firms and their workers may still exhibit such cultural traits, they are certainly not common to all firms (contracting and professional) contributing to construction projects but therein lies the complexity of the culture of project organizations. Put simply, architects and other designers may be perceived to have a predominantly aesthetic culture (Ankrah and Langford, 2005); engineers a culture of inflexibility; quantity surveyors one of pedantry and conservatism, and contractors of practicality and adaptability, but this suggests stereotyping. The culture of client organizations complicates this scenario, as their culture is usually unknown to the project team members at the outset of a project. These are no more than illustrations but the point is that each group of specialist contributors to projects will bring their own subculture to the table even when all contributors are in-house to the client organization. The differences are likely to be greater if they are from separate organizations. There also appears to be a belief that all subcultures have an adversarial component in their cultural profile. Such attitudes raise the issue of the relationship between culture and trust in the construction industry. Generally, it could be said that mistrust is a major traditional position between clients and contractors and between professional firms and contractors. However, efforts to generate trust have gained momentum with the advent of partnering and other relational initiatives which seek to change the culture of the industry through arrangements for specific projects.

Cultural differences illustrate the difficulty of the task of the project manager in integrating project teams and raise the question of whether it is possible to develop a project culture that is constructive in producing effective project teams. The acceptance of a managerial view of culture implies that culture can be managed. That is, a project manager, given appropriate leadership attributes is able to develop an appropriate culture within a project organization. The type and duration of a project are likely to have a significant effect on the opportunity for a project manager to develop a project culture. Organizational cultures are seen to be formed over long durations originating from an organization's founder's values and beliefs and are acted upon subconsciously by organizational members. Most construction projects do not have the benefit of such long durations. A high management profile may give the project manager the chance to imbue a strong sense of purpose in the project team and the longer the overall project duration, particularly the period from establishing the project brief to starting work on site, the more time the project manager will have to work on cultural aspects if so inclined. A major attribute would be charisma and the consequential respect of the project team but, even given the best talents, they would need to be sufficient for the project manager to overcome the forces working against the establishment of a project culture. Partnering is seen as presenting more fertile ground for project cultures to be created as an appropriate culture of collaboration is seen as fundamental to achieving the benefit of partnering. Bresnen and Marshall (2000c) recognize that partnering is essentially not an organizational structure but a manifestation of organizational culture. Nevertheless, the parties to the partnering agreement will bring with them their unique subcultures which may be difficult to mesh into a project culture, as discussed earlier relating to mergers and joint ventures. But the partnering concept gives a better chance of success, particularly on a long duration rolling programme of projects, that which, if achieved, would make an immensely valuable contribution to project success.

The perspective outlined above and related to construction is essentially a managerial view of culture which is the orientation taken by most people in the industry. The alternative social science perspective is not one to which people in construction readily relate. However, the social science approach to culture appears to explain culture in a manner which is more readily related to professional organizations, particularly those in construction. The independence of professionals does not lend itself to them subscribing to values and beliefs incorporated in an organization's view of its culture. Their behaviour will generally reflect their professionalism. Huczynski and Buchanan (2007) contrast the consensus view of the managerial approach with that of social science which sees organizational culture as differentiated or pluralistic. This view reflects the system's approach to organizational structure and is a more convincing profile of construction project culture. They see differentiation as arising from divisions such as management vs. labour, staff vs. line and marketing vs. production leading to a lack of cultural consensus and, whilst the categories of differentiation may be conceived differently in construction, e.g., on the basis of skills, technology, location (Walker, 2007), the common aim is understanding differentiation and the way in which organizational reality is constructed and deconstructed. They summarize:

> Thus, the differentiation perspective sees cultural pluralism as a fundamental aspect of all organizations; seeks to understand the complexity and the interactions between frequently conflicting subcultures; and therefore stands in direct contrast to the managerial integration or unitary perspective.

Culture is seen as a product of group experience which will be found wherever there is a definable group and as there are many different groups within organizations, particularly within construction, there will be many interacting subcultures differentiated laterally (architectural, engineering, contracting, subcontracting) and vertically (junior professionals, team leaders, partners and all the levels within contracting and subcontracting firms). Of particular interest to construction professionals is the recognition that organizational subcultures have different sets of values which are associated with the concept of sentience. They also draw attention to earlier work by Gouldner (1957) who distinguished two social identities, cosmopolitans and locals, for which construction professionals are akin to cosmopolitans:

> Cosmopolitans had low loyalty to their employing organization, had a high commitment to their specialized role skills and were likely to use extra-organization reference groups. Locals, in contrast, were high on company loyalty, had low commitment to specialized role skills and were likely to use an in-company reference group.

Social science's approach to understanding culture in construction organizations, particularly in relation to project organizations would appear to be much more likely to be productive than the prescriptive managerial approach.

8.11 Culture and ethics

Ethics and culture are connected in two ways. One is through the ethical behaviour of members of an organization contributing to culture and the other is whether it is ethical for managers to attempt to impose an organization's culture on employees.

In the first case, the expectation of the standard of ethical behaviour of members of an organization is transmitted from the top of an organization to the bottom and embedded in an organization's culture by example and by the way in which unethical behaviour is handled by the hierarchy. What constitutes ethical behaviour is defined by the hierarchy of an organization by their statements, behaviours and actions and by individuals' personal ethical standards and are a major component of the shared assumptions, values and beliefs of an organization's culture. Posner and Schmidt (1992) point out that 'managers reported that the action of their supervisors/managers were the most important factor that influenced ethical and unethical behaviour.'

Standards of ethics do not have an absolute measure but can be placed on a scale from totally ethical to totally unethical. The ethicality of an organization will be only one aspect of a company's culture but nevertheless by far the most important in ensuring its credibility. Examples of unethical cultures are all too prevalent.

An aspect of the construction industry which makes a major contribution to the culture of professional contributors' personal ethical standards are the rules of conduct laid down by members' professional institutions. Whilst many members of the professions may see their ethical responsibilities as much wider than the rules of conduct, those rules do form the bottom line. Professional ethics are one element which distinguishes construction organizations from many other commercial organizations and hence is a distinguishing feature of their subcultures. Fellows (2006) gives a good account of professional ethics and believes that ethics 'constitute a vital behavioural link between culture, climate and behaviour of members of an organization'.

The second case concerns what Palmer and Hardy (2000) refer to as the 'dark side' of cultural change and asks to what extent is it unethical for managers, in attempting to change an organization's culture, to coerce employees to change the way they think so that managers increase their own power? They see this area as being subject to continuing debate, illustrated by Sathe (1983): 'It is in the nature of the manager's job to influence organizational behaviour in a responsible and professional manner, so it is his or her job to conscientiously shape organizational beliefs and values in the appropriate direction'. And they say that: 'In taking this stand, he distinguishes between shaping organizational values and beliefs from shaping personal and political ones, the implication being that attempts to shape the latter are unethical'.

8.12 Industry cultures

Whilst the main unit of the study of culture in business has been the organization, some attention has been paid to an industry as a unit of study, particularly in the 1990s, but has not been intensive, perhaps because of the amorphous nature of industry culture. Even claims that industry cultures do exist (Philips, 1994) have led to others questioning the value of applying the construct of culture to industries (Chatman and Jehn, 1994). Indeed, an early example relating to industry culture (Gordon, 1991) was not concerned directly with industry culture but rather with the influences industry exerted on organizational culture. He believed that 'within industries, certain cultural characteristics will be widespread amongst organizations, and these most likely will be different from the characteristics found in other industries.' He also considered that 'a corporate culture, as a product of the company's adaptation to its environment [industry], will resist change, but change in its environment may necessitate a cultural change in order for the company to survive and prosper'. Whilst not referring to industry cultures explicitly, Abrahamson and Fombrun (1994) examined the determinants and consequences of what they term 'macrocultures' in the auto and steel industries in the USA. Macrocultures, seen as beliefs shared between managers across related organizations, are analogous with industry cultures.

In empirical work associated with these ideas, Philips (1994) stated that the existence of industry cultures is explored through her study of two industries, wine and museums, which presented evidence of industry cultures. Chatman and Jehn's (1994) empirical study of 15 firms across four service sector industries related to industry culture in assessing the relationship between industry characteristics and organizational culture found, predictably, that there was less cultural variation within industries than across them. They see their work as a first step towards generating more systematic assessments of an industry's effects on organizational culture and also that research that can address the causal chain from industry characteristics to organizational values and norms is essential. Their practical advice to firms was that they 'should consider the benefits of imitating the cultures of successful players in their industries'. Philips (1994) also has suggestions for managers: do not assume managers in other industries have the same mindset as you; knowledge of other industries' mindsets should help communications and assimilation; and that managers need to have available a variety of cognitive lenses through which to view organizational life. Abrahamson and Fombrun (1994) suggest that an understanding of macrocultures may facilitate negotiations due to common assumptions: increase trust; reduce transaction costs; and that managers may have a greater capacity to engage in collective strategies necessary to counter threatening events and trends. On the other hand it is also said to lead to inertia and prevent inventiveness and innovation.

It seems that interest in industry culture is not firmly established in mainstream management or, at least, is not seen to have much analytical value, but it has been the subject of much interest in the construction industry. A series of official reports (Murray and Langford, 2003) stretching back decades has pointed to the industry's culture being the cause of the industry's poor performance, poor reputation and other deficiencies. However, whilst many reports on the construction industry may not use the word culture in its formal, if ambiguous, sense, it has been used in connection with reports in a more colloquial manner and over recent years attention has been paid to the issues raised through actions taken by industry bodies implementing practical initiatives. These actions do not always go under the heading of culture, but the issues they deal with and their approach are designed to impact on the culture of the industry (Strategic Forum for Construction, 2009). Although construction academics have taken to writing about culture, particularly industry culture, as with mainstream management, there has been little empirical research.

Fox (2003) is a notable exception. His work was consolidated in his paper (Fox, 2007) in which he 'accepted' the existence of a construction industry culture and whilst he argues strongly in favour of its existence, the problem of definition continues to perplex in the same way as it does for organizational culture.

Underlying and compounding the difficulty of defining construction industry culture is that of defining the construction industry itself. The vast scope of the industry makes defining the boundaries of subcultures problematic. Subcultures, many of which are extensive, have differing beliefs and shared values. A simple example would be differences between the subcultures of an industry defined as comprising construction companies and professionals working internationally and one comprising locally-based small construction and maintenance companies. Within these two extremes, and on a wider view including civil engineering, etc., are a vast number of subcultures. Ive and Gruneberg (2000) have argued that the construction industry comprises several separate industries rather than a single industry. The complexity of such an array of subcultures is compounded by the cultures of the various professions which contribute to the industry (Ankrah and Langford, 2005). This multi-level conceptualisation of a hierarchy of meta-cultures and subcultures is referred to by Fox (2007) and mainstream management writers but whether it is possible to define each subculture in relation to the construction industry and then a dominant culture to encompass them all seems questionable. Maybe each subculture is better conceptualized as a culture. As Fox (2007) says, 'If we cannot define the industry, we cannot define its culture.'

This is not to say that many of the characteristics of culture are not valuable in enhancing understanding of the construction industry. There have been many calls in official reports and pronouncements for changes to the industry which have alluded to culture in a colloquial manner; nevertheless, they have resulted in substantial improvements in the industry. A long-standing major concern has been the adversarial nature of the industry (Phua and Rowlinson, 2003) leading to what Rooke et al. (2003) refer to as 'the claims culture' which is a 'familiar aspect of construction industry culture' and that 'proponents of change must overcome sheer cultural inertia'. One major initiative designed to address this problem has been partnering. Essentially, partnering is designed to reduce transaction costs by, for example, greater collaboration for greater efficiency and elimination of adversarial attitudes and hence the cost of disputes. For such things to happen a change in culture within the partnering organizations is needed to develop the trust necessary to make partnering work effectively. There may be recalcitrant problems in achieving this, even when partnering relationships are intended to extend over a long period and many projects. A shared culture is much more difficult to achieve on a single project, even when it is of long duration. Nevertheless, there is much anecdotal evidence of successful partnering for which a shared culture based on trust has been achieved (Liu and Fellows, 2001). Whilst partnering may be claimed as a phenomenon of construction industry culture change, it is not industry wide, although the propensity for partnering attitudes to spread beyond partnering firms to the wider industry exists. Partnering is again typical of the single issue nature of culture and culture change in the construction industry. Others include the application of total quality management, health and safety, sexual equality, sustainability and, perhaps the most frequently occurring, an adversarial/claims culture. So it seems that a definition of a construction industry culture has not been holistically and homogeneously defined and perhaps cannot be and so remains to be seen as a collection of cultural characteristics which may be relevant only to specific groups and situations.

8.13 National cultures

Recognition of the differences between national cultures has existed ever since people began to travel outside their homeland and has accelerated to its present high profile and rate of momentum due to the rapid increase in globalization. An extensive review of national culture and cross-cultural issues is outside the scope of this book but some comment does need to be made on its relationship to organizational culture. Stereotypes of national characteristics are well known – for example, the English are reserved, the French are romantic, Germans are without humour etc. – all of which are gross oversimplifications relying on one characteristic to categorize a nation. Writers on both national culture and organizational culture attempt to identify traits and classify countries and organizations into types, their objective often being to establish how national cultures impact on organizational cultures in the global activities of companies. Frequently quoted examples are McDonalds and Disney which have strong and distinctive organizational cultures and which have found it necessary to adapt their culture and marketing to suit the relevant national culture with which they are working.

Hofstede (1991, 2001) and Hofstede and Bond (1988) carried out a significant cross-cultural study of employees in the same multinational companies based in 40 countries and identified five dimensions of prominent traits of national cultures, each of which is a continuum on which a country's culture can be placed:

- *Power-distance* – degree of acceptance of unequal distribution of power.
- *Uncertainty avoidance* – extent to which people feel threatened by ambiguous situations.
- *Individualism/collectivism* – extent to which people prefer to operate individually rather than in groups.

- *Masculinity/femininity* – extent to which 'male' or 'female' values predominate.
- Long-term/short-term orientation – extent to which values are for future benefits or early gain.

Whilst Hofstede's work has been highly influential, it has not been without its critics. Barthorpe (2002) refers to limitations due to restriction to one privately owned multinational company which was technically focused and Western orientated and criticisms regarding methodological and sampling issues summarized by Cray and Mallory (1998). Wilson (1999) cites a number of critics, commenting that, 'It is possible to dismiss his project as an attempt to measure the unmeasurable' and 'It is not clear whether something as abstract as culture can be measured with survey instruments at all'. But Hofstede's work is a major contribution and criticisms only go to demonstrate the problematic nature of trying to make sense of the most complex of constructs and the sheer scale of the research projects needed.

The question of whether national culture overshadows organizational culture is considered by Huczynski and Buchanan (2007) who brought together the findings of the following authors:

- Both company managers and employees bring their cultural background and ethnicity into the workplace (Lubatkin et al., 1998)
- National culture explained more of the differences than did role, age, gender or race (Hofstede, 2001)
- More pronounced cultural differences occurred among employees from around the world working within the same multinational company than among those working for companies in their native lands (Laurent, 1983)
- Pressure to conform to the culture of a foreign-owned company brought out employees' resistance, causing them to cling on more strongly to their own national identities to such an extent that a company's organizational culture cannot erase it (Adler, 2002)
- Different cultures had different ways of resolving conflicts, which was problematic in multinational teams (Tinsley, 1998).

Palmer and Hardy (2000) draw attention to Adler and Jelinek (1986) who recommend 'careful attention to societal culture' so that managers can create an organizational culture that is in harmony with societal culture but they say that 'it is not clear, however, how managers can create such harmony …'. And make the telling point that 'Nor is it clear how such alignment benefits organizational performance'.

These issues show that, increasingly with greater globalization, construction firms and related professional firms have had to adapt to the national culture of the country in which they are working. They cannot work in isolation but have to hire local labour, purchase local materials and deal with local or multinational clients, all of which is compounded if working in a joint venture project.

Also, if their workforce is multinational similar issues will arise within their organization and have a profound effect on their organizational culture. Rowlinson (2001), in a study of barriers to change in a construction-based professional government department in Hong Kong, found that a problem at first appeared to be a mismatch between the organizational culture as perceived by the workers and the organization structure that was being implemented. On investigation using Hofstede's cultural dimensions it was found that power-distance and individualism-collectivism were key issues and that barriers were due to deep seated traditional, national, cultural issues. A further application of Hofestede's dimensions is Brochner et al.'s (2002) study of the Swedish construction industry in which low power-distance and low uncertainty avoidance can be traced in the development of specific quality and collaboration practices in Swedish construction. These issues also show clearly the complexity of culture which, whilst useful in terms of understanding situations which may arise in companies, neither provides a theoretical base on which to carry out rigorous research nor a route map for practising managers.

9 Groups and Teams

9.1 Introduction

It is widely recognized in the mainstream management literature that a large part of many people's activities at work, and hence their social behaviour, occur in groups. It is also recognized that this has become increasingly so in the second part of the 20th century and into the 21st due predominantly to the scale and complexity of the issues facing modern-day business and in the expectation that groups and teams are more productive than individuals, particularly when the task to be undertaken requires multiple skills, judgement and experience (Glassop, 2002). So terms such as management team, production team, quality circle and crew are common in a vast range of different business types. But as Fincham and Rhodes (2005) point out, 'There is ... an important gap between the "hype" about teams – exaggerated expectations of what they are able to deliver – and what is often the reality.'

The way in which individuals relate to group membership is explained by social identity theory which is about how elements of our identity arise from our group membership (Tajful and Turner, 1986). Individuals adapt as they move from feeling and thinking as individuals (personal identity) to feeling and thinking as group members (social identity). Our groups and social categories are part of the way in which we see ourselves (self-concept), which affects how we act within a group. People's social identity is how they are defined and evaluated by themselves and by others and prescribes appropriate behaviour for them. To evaluate themselves people not only compare themselves with others with whom they interact but also compare their groups with other groups. This enables them to see the world as 'them and us' and so enhances their understanding of the world around them and maintains and even enhances self-esteem, particularly for members of prestigious groups. Putting people into groups and identifying with some of these groups seems to be a basic human characteristic, showing that human beings are social animals.

The origins of the recognition of the value of group working arose from the Hawthorn Studies (Mayo, 1945) conducted between the 1920s and 1940s which

Organizational Behaviour in Construction, First Edition. Anthony Walker.

coined the phrase 'the Hawthorn Effect'. Huczynski and Buchanan(2007) state '... Mayo went on to propose a social philosophy which placed groups at the centre of understanding human behaviour in organizations. He stressed the importance of informal groups, and encouraged managers to "grow" them'. Mayo's conclusions led to the human relations approach to organization and from this base and from socio-technical theory (Emery and Trist, 1960; Wilson, 1999), team working has grown to become practically a way of life in many organizations. They sum up the contemporary situation:

> The original theoretical developments in the area of group structure and process occurred between the 1930s and the 1950s. Many were conducted in non-organizational contexts, and frequently involved children and university students. Their findings were then applied to companies. The more recent developments have been practical rather than theoretical. They have been accompanied by linguistic change within the research and management literature where the pre-dominant term is now 'team' rather than 'group.' While western companies may have been reluctant to structure their organizations around groups, they have been prepared to train their managerial and technical staff to work more effectively in teams. Thus, team building and team development activities have established themselves as a major element in both management training and organizational development (OD) activities.

9.2 Defining groups and teams

This brings us to the definition of and distinction between the terms 'group' and 'team'. The previous quotation refers to a shift in usage from group to team, but the literature is not consistent in the use of the terms. The major confusion is whether the terms are distinctive concepts or interchangeable. At a fundamental level, Kakabadse et al. (2004) identify a group as any collection of people who actually think of themselves as a group in order to distinguish themselves from a random crowd of individuals. They also state that group members have a common purpose and usually some collective identity, defined membership criteria and have different roles to play. They clarify their use of the term by saying that groups at work are called work groups, committees or teams but do not define them separately from the overall term 'group'. On the other hand, Robbins and Judge (2008) are specific and see work groups as information-sharing groups that do not need to engage in joint effort and work teams as generating synergy through coordinated effort. These references and much of mainstream management literature imply that groups and teams are to consist of employees at a technical, semi-skilled and unskilled worker level rather than at a professional level which is the case in construction project management.

McShane and Von Glinow (2003) state that 'All teams are groups because they consist of people with a unifying relationship. But not all groups are teams; some groups are just people assembled together'. Annett and Stanton (2000) agree with this view. The difference would seem to be that a team has a more specific objective than a group although the former then goes on to say that group and team are used interchangeably in their book. Huczynski and Buchanan (2007), whilst believing that it is important to maintain the distinction, also use the latter citation but recognize that the terms are used interchangeably in general usage. They also point out that 'Authors who are management consultants frequently use the term "team" metaphorically: that is they apply this label to a collection of employees to which it is imaginatively, but not literally, appropriate'. They point to further confusion being created by the idea that groups can transform themselves into teams as they mature, so they arrive at their 'performing stage' once they have organized themselves to fulfil their purposes and ask would it be appropriate for all groups to transform themselves? Those which were information exchange mechanisms and mutual support groups could remain as groups.

The fuzziness of the concepts of group and team are given greater focus by Moorhead and Griffin (2001) who state that, in organizations, groups and teams are not the same thing. They define a group as two or more persons who interact with one another such that each person influences and is influenced by each other person. They continue that group members may be satisfying their own needs in the group and need not have a common goal but, in contrast, team members are committed to a common goal, therefore a team is a group with a common goal. However, classification into groups and teams on this basis, if it is needed at all, is dependent on what one sees as a common goal which can range from being widely drawn to very specific. A definition of a team frequently referred to is 'A team is a small number of people with complementary skills who are committed to a common purpose, set of performance goals, and approach for which they hold themselves mutually accountable' (Katzenbach and Smith, 1993). The problem of finding suitable definitions is illustrated here as construction professionals would find it unacceptable to define a team as comprising a 'small' number of people since the size of a project team is defined by the size and complexity of the construction project and hence the range of specialist skills needed and not by any predetermined number.

Essentially defining and distinguishing groups and teams is a matter of semantics. Groups and teams are defined and described by their composition, purpose and focus and there are as many definitions as there are groups and teams. They can range from informal unstructured groups to highly structured and focused project teams. As Fincham and Rhodes (2005) state '... we will use the terms team and group interchangeably. On close inspection of the terms there seems to be no important difference between the two'; and so will this book.

9.3 Critique

Generally, the claims for the value of teams to organizations include:

- Improved performance: lower production costs, improved quality, better customer service

- Better employee relations: greater participation and empowerment hence better quality of work and less stress, lower turnover and absenteeism
- Organization benefits: enhance problem solving, greater innovation, greater flexibility.

Whilst team working can appear to be a panacea for the solution of all organizational problems, it remains controversial. Huczynski and Buchanan (2007) cite Hayes (1997) who emphasizes the use by managers and consultants of 'team' as a sporting metaphor for cooperation (whilst at the same time being different) to give the impression of everyone committed to the cause and pulling in the same direction, but caution:

> 'The idea of a "team"' at work must be one of the most widely used metaphors in organizational life. A group of workers or managers is generally described as a 'team' in much the same way that a company or department is so often described as 'one big happy family.' But often, a new employee receiving these assertions quickly discovers that what was described as a 'team' is actually anything but. The mental image of cohesion, coordination and common goals which was conjured up by the metaphor of the team, was entirely different from the everyday reality of working life.

They also highlight the detrimental effect groups can have on their members, citing Hampton (1999):

> Yet groups are endowed with a darker side, one which is highlighted in mobs and crowds. They are seen as taking over the individual's mind, depressing intelligence, eliminating moral responsibility and forcing conformity. They can cause members a great deal of suffering and despair and can perpetuate acts of great cruelty. If groups are capable of great deeds, they are also capable of great follies.

Whilst perhaps not so dramatic as the above statement, management can have the power to manipulate groups to keep the behaviour of their members in line with management objectives and so suppress free thinkers which can work against the productivity of a group by, for example, inhibiting innovation. Also, forming a team to address an issue or problem is not always the right answer as it may also be used to delay facing a problem or making a decision. It may be seen as the easy way out but this 'solution' may return to haunt the managers who set up a team if it produces an outcome which is not acceptable to the higher echelons of the organization. Other more straightforward reasons for not forming a team are:

- when the task can be undertaken quickly and decisively by one person with no detriment to the quality of the outcome
- to preserve resources as teams are more time and energy consuming and more expensive than alternatives due to increased communications costs, potential conflict and extra administration

- when individuals with appropriate skills and characteristics for membership and leadership and training facilities are not available.

Wilson (1999) reviews research into teams at the worker level rather than the managerial or professional level which found that not all employees had the same reaction to team working. She pointed out that Ezzamel and Wilmot (1998) on examining the introduction of team working in a company found that younger members with no family commitments and less pressure to maximize income through bonuses 'free-loaded' on those who were working harder. But that 'while team work appeared to deliver universal benefits like cost effectiveness and enhanced profitability for the company, it also concealed a variety of unsavoury features of work reorganization, including coercion masquerading as empowerment and the camouflaging of managerial expediency in the rhetoric of "clanism" and humanization (Knights and Wilmott, 1987).' She cites Knights and McCabe's (2000) examination of the results of team working in an automobile manufacturing company as finding no universal reaction to the experience of team working. Employees had difficulty with the discourse of team working; its intrusion into their lives; colleagues held enthralled by team working; and the psychological warfare waged by management through the ideology of team working, leading to cynicism and resistance to change.

Following further review, Wilson (1999) concludes that 'Despite sustained criticism of team working, however, the concept appears to have survived'. She draws on Buchanan (2000) for the reasons why:

- it is appealing to Western society's sense of being a team player, which is difficult to deny
- it appeals to the sense of solving problems and making decisions together
- it enables us to share our skills and creativity
- it is a flexible concept in terms of, for example, constitution, purpose and size.

One could add that in organizations it is difficult to reject a suggestion that a team be formed for fear of appearing to be autocratic and so rejecting the contribution of others. Notwithstanding the criticisms of team working, it must be said that properly established, maintained and run teams have been seen to achieve outcomes which are far greater than could have been achieved by individuals working alone and, in many instances, teams are the only possible way to undertake certain tasks. Construction professionals would certainly agree that it is inconceivable that a construction project could be designed and built using any other organizational form, yet interestingly the mainstream management literature on teams does not seem to use examples drawn from construction.

9.4 Teams in construction organizations

People employed in the construction industry will find it extraordinary that many writers on OB still see the use of teams and groups as an aspect of modern management practice. Members of the construction industry have always worked in teams due to the practically total focus of the industry on projects necessitating the use of interdisciplinary groups of specialists in project teams. It is inconceivable to members of the industry that projects could be designed, developed and constructed in any other way. However, it is worth examining the ideas arising from the OB literature to see whether there are lessons for construction teams.

Whilst construction is dominated by interdisciplinary project teams, many other types of team exist within the industry's specialist firms. Foremost are project teams within the specialist professional firms which develop the firm's contribution to a project, for example, architectural firms' design proposals and the design-related activities of comparable teams in each of the specialist contributors' firms. Also all firms will have teams to manage other aspects of the business. The OB literature will have something to contribute in these situations. Apart from in-house project teams, groups within firms usually reflect the structure of firms so professional firms may have management teams that comprise a group of partners or directors formed to consider, for example, expansion of the firm, diversification and selection of new partners, in much the same way as most businesses. Another type of group at this level may be a finance and programming group which is concerned with forecasting and particularly with cash flow. The larger professional firms are often structured around business sectors such as health, education, waste disposal, airports. Groups may be formed under the leadership of a director or partner to monitor activity and opportunities in a specific sector. A particular characteristic of sector teams in construction is to invite representatives of client bodies and specialists from outside the firm to become involved in such groups as mutual benefits can be derived.

This account refers mainly to professional firms rather than the majority of contractors but the more expansive contractors, particularly those involved in design and build, public-private partnerships and build-operate-transfer projects are likely to follow a similar pattern. Contractors focusing essentially on conventional construction contracting, but also broader based contracting firms, whilst again being predominantly project orientated, may have in-house groups to examine specific issues such as health and safety, material supply, plant purchases, subcontractor management, etc. A distinctive feature of contractors over professional firms is that they employ site labour: skilled, semi-skilled and unskilled. Much of the mainstream management literature on groups is concerned with work groups of similar grades of employee so is more relevant to contractors' organizations although major differences lie in the fact that contractors' work groups are project based and hence temporary and that much work on site is undertaken by subcontractors.

As stated, the major distinction between firms involved in construction projects and the vast majority of other firms is that they are overwhelmingly and fundamentally project based. In the very early stages of a project, each professional firm involved in a project will form an in-house project team to work up their basic concept for the project. For instance, an architectural practice will have a team of design architects creating the initial ideas for the project and the structural engineer will do the same for the structural design. They will liaise with the client and coordinate their ideas but a full project team may not have been established at that time, although it is most advisable that it should be formed at the very start of the project process to achieve the highest possible standard of project management. Teams operating in-house prior to the full project management team being formed will operate much as described in the mainstream management literature. As soon as possible a full project team under a project manager should be formed which will include all the specialists needed to develop the project. What makes a construction project team different from those described in the mainstream management literature is that its members are drawn from a wide range of specialists, most frequently from independent firms, plus client representatives. Therefore the differentiation between the members is extremely high as they come with allegiances to their profession and to their firm and, although they may have worked with other members before on other projects this is not necessarily the case, but in all cases they have to be welded into a cohesive team. This is a situation which rarely occurs for the majority of companies outside the construction industry but is the overwhelming characteristic of the management of construction projects. This level of differentiation may be reduced if projects are undertaken by multidisciplinary practices and for design-and-build projects but nevertheless the complexity of projects ensures the management of project teams is a demanding task.

Just how much construction project teams have to learn from mainstream management is open to question so its relevance will be commented on as it arises in the text. It may be that firms in other industries have something to learn from construction project management, particularly the management of interdisciplinary project teams.

9.5 Types of group and team

As said previously, when discussing the definition of and factors which attempt to distinguish groups from teams, they are defined and described by their composition, purpose and focus and there are as many definitions as there are groups and teams and for the same reasons there are as many different types as there are groups and teams. Nevertheless writers of mainstream OB texts seem to need to classify groups and teams in an attempt to bring order to a diverse field.

At a very fundamental level there seems to be agreement that formal and informal groups can be distinguished. Formal groups are invariably functional in organizational settings; that is they are formed to carry out organizational tasks and other activities. They will be formed by management to correspond to the needs of the organization, so departments and sub-units can be seen as formal groups but others, which could be interdepartmental or specially create sub-groups within departments, are also constituted as formal groups. Informal groups on the other hand are not established by the organization but occur naturally by the affiliation of like-minded people (albeit the same people as in the formal groups) who for a variety of reasons have an interest in meeting. For example, they can be formed to exchange information as special interest groups such as women's' groups, or simply for friendship and other social needs. Whereas formal groups are embedded in the organizational structure of companies, some informal groups may form to create an informal structure which is sometimes found in organizations in order to overcome ineffective formal structures. Informal groups will develop through the spontaneous interaction of people who are probably already members of formal groups and they may share many of the characteristics of social and leisure groups. Membership of formal groups in business organizations will normally be determined by managers but why do people join informal groups and other groups where membership is not mandatory? Fincham and Rhodes (2005) draw on Sheldon and Bettencourt (2002) stating that they '... found that groups fulfilled both affiliative needs (the need to be included) and the need for relatedness (closeness), and interestingly other, more self-orientated needs (the needs for autonomy and personal distinctiveness). This apparent paradox-groups seeming to fulfil both self and socially oriented needs – are explained by suggesting that in reality these sets of needs are not often in conflict and in adults tend in fact to be complementary. Other self-oriented needs which groups fulfil in the workplace would include self-assertion, recognition, and achievement.

Apart from formal and informal groups there does not appear to be any consensus on identifying and labelling groups and teams. One interesting approach is that of Sundstrom et al. (1990) who identified four types of team: advice, action, project and production. The activities of advice, project and production teams are self-evident; action teams take a form similar to project teams but over a much shorter time frame/work cycle. Types are seen to be differentiated from each other by:

- Technical specialization: a continuum from members having high level technical skills (high differentiation) to general experience and problem solving skills (low differentiation)
- Coordination needs: a continuum from working closely with other groups within the organization and with each other (high coordination) to operating independently (low coordination)
- Work cycles: ranging on a scale from short repetitive work cycles to a long single process
- Outputs: what the team produces.

A category which has received much attention is Self Directed (or Self Managed) Work Teams (SDWTs), also known as Autonomous Work Groups (AWGs), which are groups of employees who perform highly related jobs and take responsibility for their work without frequent recourse to supervisors or senior managers. They were identified by Trist and Bamforth's (1951) study of coal-mining methods which contributed to the Tavistock Institute's work on socio-technical theory with which researchers in construction project management will be familiar (Higgins and Jessop, 1965). In mainstream management literature they comprise skilled and semi-skilled workers who plan and schedule their work, allocate tasks to members, manage the pace of work and make operating decisions, hence the use of autonomous to describe them in AWGs. This represents a shift from assembly line production and is said to lead to higher performance but some results from self managed teams have been disappointing (Robbins and Judge, 2008). McShane and Von Glinow (2003) point out that Volvo's car plants replaced the traditional car assembly lines with fixed workstations of about 20 workers but have amongst the lowest productivity in the automobile industry (Berggren, 1993). These work teams have some resonance with project teams used in the construction industry, particularly in relation to their autonomy which allows them to be self regulating but requiring their output to be approved (in the case of construction by the client). However, the automobile and other manufacturers which use them have a choice between assembly line and SDWT/AWG production but the construction industry does not.

Examples of the nomenclature of other types of team found in the OB literature include the following, the titles of which in many cases are self explanatory:

- Problem-solving team
- Quality circle; investigates quality problems in processes and products that arise in the workplace, they make recommendations
- Management team
- Command group; functional reporting relationships (a group of employees who report to a specific manager), relatively permanent, often shown on organization charts
- Task group/force; perform a specific task
- Affinity group; employees from the same level in an organization who meet to share information, address problems and developments
- Product development teams
- Cross-functional teams: members from diverse areas within an organization, allows specialist contributions, coordinate complex projects
- Interacting groups: interdependent members, need to cooperate and coordinate
- Co-acting groups: members work on a common task but mainly independently
- Counteracting groups: members work together to negotiate and reconcile conflicting demands and objectives.

There are no doubt many other categories but there does not appear to a comprehensive, consistent, widely acceptable and defined list of types. It can be seen that the types listed are not mutually exclusive as one can have, for example, co-acting production development teams, problem-solving management teams and a command group task force. Using these mainstream management terms, a construction project development team would be classified as an interacting cross-functional team, although earlier reference has been made to project teams in mainstream management. Commenting on cross-functional teams, Robbins and Judge (2008) say, 'cross-functional teams are no picnic to manage. Their early stages of development are often very time-consuming as members learn to work with diversity and complexity. It takes time to build trust and teamwork, especially among people from different backgrounds with different experiences and perspectives.' Construction teams do not normally have the luxury of time due to the urgency of delivering projects on time. Whilst construction project managers readily recognize these issues, they take these difficulties in their stride as solving them is essential to successfully delivering construction projects, which may come as something of a surprise to managers in other industries.

Special mention should be made of virtual teams which are anticipated to be used increasingly as business continues to globalize. Rather than meeting face to face, virtual teams use computer technology such as e-mailing, video conferencing and area networks for collaboration. Whilst virtual teams can perform all the things that real teams can, their drawback is in the form of communication. Members of virtual teams lack the ability to express themselves using paraverbal and nonverbal cues so reducing the quality of interaction. This is particularly detrimental to creative work and where complex issues require iterative and subtle debate which benefit from social interaction. Nevertheless, virtual teams enable team work to take place across thousands of miles and many time-zones, which would otherwise be impractical or at least extremely difficulty and expensive. A combination of team working in real time supplemented by virtual team working is a valuable combination on complex construction projects.

In attempting to distinguish between types of group, Huczynski and Buchanan (2007), citing Steiner (1972) examine how teams operate. They identify group tasks on the basis of the type of interdependency required by the task. For an *additive task* all group members do basically the same job and the group's performance is the sum of their combined contributions. Such tasks have low interdependency. An example given is a group pushing a stalled car. In a *conjunctive task* one member's performance depends on another's. There is high interdependency and a group's least capable member determines performance. It is said that groups perform less well on conjunctive tasks than individually. An example given is a relay race. For a *disjunctive task* one member's performance again depends on another's and there is high interdependency but this time the group's most capable member determines its performance. In this case others feed into the most capable performer rather than the outcome being a truly combined effort. Coordination is seen as important but only in the sense of stopping others impeding the top performers. An example is quiz teams.

These categories of tasks are not particularly helpful to construction teams as the complexity of developing and constructing a project generates extremely high interdependency which generally is integrated by mutual equally weighted contributions rather than being determined by the most or least capable contributor. However, it is possible that on high profile projects, a project team's tasks could be of the disjunctive type if, for example, design is of paramount importance and therefore the architect is seen to be the most capable – an example could be Sydney Opera House. Also, a construction team's task could theoretically be a conjunctive type if a member is seen to be below the standard required, but if there is a serious shortfall in performance it is likely that they would be replaced. But in reality the task of construction project teams is of a different order from those described here due to the strength of the interdependencies their tasks generate.

The differentiation of skills in the construction industry can be clearly seen and it is also quite clear that all the contributors are interdependent in carrying out their work of producing the completed project to the client's satisfaction. The network of interdependencies is practically total. It is not that each contributor is dependent on one other but that all contributors are in some way dependent upon all the others. The development of a construction project consists of a series of tasks which are combined to produce the finished project. The contributors are interdependent because on the one hand the various tasks that have to be undertaken to achieve the finished project require an input from a range of different skills, and on the other hand because the tasks themselves are interdependent as frequently a task cannot be commenced until another has been completed or unless another task is undertaken in parallel.

Different types of interdependency exist and have been classified as pooled, sequential and reciprocal (Thompson, 1967).

Pooled interdependency is basic to any organization. Each part renders a discrete contribution to the whole. The parts do not have to be operationally dependent upon or even interact with other parts, but the failure of any one part can threaten the whole and therefore the other parts, for example the decentralized divisions of a large, diversified company. Specific groups and teams would not normally be formed to deal with these issues, although they would be within the remit of the most senior members of an organization. *Sequential interdependency* takes a serial form. Direct interdependency between the parts can be identified and the order of the interdependency can be specified. For example, bills of quantities must be prepared before tenders can be invited (using this particular form of procedure). *Reciprocal interdependency* is when the outputs of each part become the inputs for the others and the process moves forward through a series of steps. Each step requires interaction between the parts and each part is penetrated by the others. This is seen, for example, when preparing an outline proposal for a building which must be functionally and aesthetically sound and also feasible from a structural and cost point of view.

The three types of interdependency can be arranged in ascending order of complexity as pooled, sequential and reciprocal. A more complex type also contains the less complex types. The order of complexity is also the order of most difficulty of integration. Integration can be defined in construction terms as the quality of the state of collaboration that exists among contributors to projects who offer specialist skills and who are required to achieve unity of effort by the environment (Lawrence and Lorsch, 1967). If, therefore, there are different types of interdependency, there would need to be different methods of integration. As reciprocal interdependency is the most difficult to integrate, and as this type of interdependency dominates in the construction process, the quality of integration and effort needs to be of a high order.

It has been found that the integration of pooled interdependency is best achieved through standardization and formal rules, and sequential interdependency through planning. That is, the tasks to be undertaken can be anticipated and their sequence planned so that sequential interdependency is identified and recognized at an early stage. Whilst much of the integration needed may be achieved by exchanging information without working closely in groups or teams, it is likely that on anything other than the simplest project, a team would need to be formed for this purpose. Reciprocal interdependency is integrated by mutual adjustment and feedback between those involved to ensure that the required input takes place at the appropriate time and that account is taken of the various inputs in the process. The management of reciprocal interdependencies requires that a balance be maintained between inputs in order to achieve clients' objectives. A clear perception of clients' objectives is required, together with the diplomacy and expertise necessary to integrate a group of highly skilled professionals. The iteration required to solve the complexities of developing and constructing projects can only be achieved through team work of the highest order in which reciprocally interdependent, highly skilled professionals meet face to face regularly in teams under sound leadership manifest in the abilities of the project manager.

However, the structure of the industry presents problems of integration to interdisciplinary project teams, the distinctive nature of which has been examined by Baiden et al. (2006). They named such a team an 'integrated construction project team' and proposed on the basis of the literature that an integrated project construction team can be described as 'fully integrated' when it:

- Has a single focus and objectives for the project
- Operates without boundaries among the various members
- Works towards mutually beneficial outcomes by ensuring that all the members support each other and achievements are shared throughout the team
- Is able to predict more accurately, time and cost estimates by fully utilising the collective skills and expertise of all parties

- Shares information freely among its members such that access is not restricted to specific professions and organizational units within the team
- Has a flexible member composition and therefore able to respond to change over the duration of the project
- Has a new identity and is co-located, usually in a common given space
- Offers its members equal opportunities to contribute to the delvery process
- Operates in an atmosphere where relationships are equitable and members are respected, and
- Has a 'no blame' culture.

Using successfully completed projects by award-winning construction managers as their sample, they examined whether they actually worked in an integrated manner. The broad finding was that 'construction project teams exist as individual competent units within their organizationally defined boundaries. They exhibit varying degrees of integration, which are determined by the team practices adopted and their congruence with the procurement approach'. None of the project teams exhibited the full range of the criteria proposed as signifying a truly integrated operation, in particular that a seamless operation is a fundamental requirement of integrated team performance. The authors conclude that 'this infers that fully integrated teams are not necessary for effective team operations within the industry, or that the sector must overcome significant organizational and behavioural barriers if the benefits of integration are to be fully realized in the future'. Whilst many of the criteria for integration were met in full or in part, those which were deficient were due to allegiances to firms and/or professions by members of the project teams: that is they did not 'belong body and soul' to the project team which is hardly surprising given the vast numbers of members of independent professional and contracting firms contributing to projects, each with their own culture, expectations and way of working, even in the case of the project managers of the high calibre considered here. The procurement approach was seen to have an effect on the quality of integration, demonstrated particularly by design and build. This is unsurprising because in its 'purest form' design and build brings all skills in-house to the integrated construction project team, thus removing many of the 'organizationally defined boundaries'. Even in this case: 'The team did not, however, provide a totally seamless project delivery operation with a complete absence of professional and organizational boundaries. Additional effort would have been required to fully break down professional and organizational barriers.' What the results reflect is simply the reality of the way in which the industry is structured with many professional and contracting firms of various sizes, some of which over the years have amalgamated to provide a wider range of services, to the extent that some can now offer a total design and build service all of which reduces the differentiation between specialists and makes integration so much easier to manage. But the problem in providing such a total service lies to a large extent in workload management, as larger, fully integrated organizations need a continuing demand for all their services which reduces flexibility and increases risk.

9.6 Structure and roles

9.6.1 Structure

Group or team structure refers to the way in which members relate to one another. People in groups are individuals and as such demonstrate different characteristics reflected in the way they behave in group situations. For example, some act confrontationally, others are conciliatory; some speak a lot, others little. So they bring more than their specific skills and knowledge to meetings and, mainly through their individual characteristics, form relationships with other team members. This pattern of relationships is called the group's structure and is of a distinctly different concept than organization structure which is the juxtaposition of formal roles in an organization. The term 'structure' implies permanence but group structure is fluid because it is modified by the interaction of members as they adjust to each other and as new members join the group and others leave. Hence structure constitutes a distinctly social aspect of group life and may act as an objective constraint on members' activity (Fincham and Rhodes, 2005).

Huczynski and Buchanan (2007) state that differentiation within a group occurs along several dimensions, the most important of which are power, status, liking, communication, role, and leadership which have a strong influence on members' relationships. Such differentiation is additional to that created by the task being undertaken as discussed in the previous section and does not involve interdependency. The power structure of groups will be strongly influenced by the authority/power each member brings from their formal position and reputation in the organization generally as well as the power they generate through their contribution to the group. Status is closely associated with the concepts of authority/power and is an active force within teams. Formal status is associated with a member's rank in an organization such that their contribution to group work is reinforced by being high in the company's hierarchy and therefore in all probability being senior to many of the members of the group. An example from construction would be if one of the professional firms sends along a director to a project team meeting which would normally be attended by a project level staff member for the purpose of obtaining a particular outcome. The intimidation of other members of the team meeting, not only the director's own staff but also those from other companies, by the presence of a director can be seen as the use of referent power. Status is also gained by members of a group due to their prestige power; that is, the respect in which they are held as a result of their superior knowledge of the topic under consideration or because through their personal characteristics they are major contributors to group cohesion and progress. They will have particularly high status if their status is founded on both these attributes, which could in certain circumstances transcend those members who have higher formal status.

The pattern and quantity of communication depend on the relation between members in carrying out the group's task and the characteristics of the individuals, for example are they talkers or listeners. The effect of roles is dealt with later in this chapter. In terms of leadership they believe that it '... seems more useful to view leadership as a set of behaviours that change their nature depending on circumstances, and which switch and rotate between group members as circumstances change, rather than a static status associated with a single individual'. So, group leadership is generally seen as leadership which emerges from a group as a function of its structure and not by external appointment; that is the leader is selected by the group. Group leadership has also been recognized as any member of a group performing acts of leadership. Whilst interesting and no doubt valuable in certain situations, this concept of group leadership is unlikely to occur in construction teams because of the focus and dynamics needed to develop projects which require leaders to be appointed. More conventional leadership theories are dealt with in Chapter 11. However, the concept of a leader emerging from a group could be found in informal groups in the construction milieu.

9.6.2 Roles

Individuals play different roles within groups and adopt specific behaviour, some of which is helpful to the group's development whilst some is not. Helpful roles are, of course, highly regarded and give rise to the earlier reference to power and status within groups. Benne and Sheats (1948) identified 27 different roles played by group members which McShane and Von Glinow (2003) condensed into the following 10 roles:

- *Initiater*: identifies goals for the meeting, including ways to work on those goals
- *Information seeker*: asks for clarification of ideas or further information to support an opinion
- *Information giver*: shares information and opinions about the team's task and goals
- *Coordinator*: coordinates sub-groups and pulls together ideas
- *Evaluator*: assesses the team's functioning against a standard
- *Summarizer*: acts as the team's memory
- *Orienter*: keeps the team focused on its goals
- *Harmonizer*: mediates intra-group conflicts and reduces tension
- *Gatekeeper*: encourages and facilitates participation of all team members
- *Encourager*: praises and supports the ideas of other team members, thereby showing solidarity of the group.

They go on to say that some team roles are formally assigned to specific people (functional roles) but that team members often take on various of the roles given above informally. Formally assigned functional roles could include, for example, team leaders, secretary, resource controller. This highlights the major difference between the groups and teams perceived by the mainstream OB literature and those which occur in construction. In construction teams, at whatever stage of project development, all members offer specific individual skills (functions) and are in formally assigned roles. For conventionally arranged projects individuals may also adopt one of the informal roles identified above as reflected by their personal characteristics but their dominant contribution is likely to be their professional skills. It is likely to be particularly the case that informal roles are not prominent when project teams consist of members from independent firms as members may be reluctant to usurp roles that others may wish to adopt. But for other relationship-type project forms such as partnering informal roles may be more prominent.

A popular approach to roles within groups is that of Belbin (1993) who developed the Belbin Team-Role Self Perception Inventory, a toolkit to assess the suitability of personalities in relation to team roles. He has described team roles as 'a pattern of behaviour that characterizes one person's behaviour in relationship to another in facilitating the progress of the team' (Belbin, 2000). He considers that there are only a limited number of ways in which members can contribute to team work from a relationship angle and identified nine (Belbin, 1993). He believes that someone who is strong in one role is often weak in another and is said to have an allowable weakness, as shown in Table 9.1.

Belbin sees members contributing to a team in more than one role so they can be described in terms of their team role contributing pattern, and also sees team members having complementary roles in order for teams to be effective and avoid members with similar profiles competing against each other. But, again, the opportunity to choose members with such attributes in many construction-related teams may not occur often. Rather, construction teams are selected on the basis of their specialisms and experience; only in the case of the client's representatives is there likely to be the possibility of selecting primarily on the basis of matching characteristics and role, and even the ability to do that is often constrained. Project team leaders should be conscious of other than specialist roles and seek to incorporate them whenever possible. Even though construction project teams are distinctive and often complex, Belbin Team-Role Theory has been used successfully by construction companies to help with personal development and review the behaviour of members of project and other teams.

Dainty et al. (2006) reviewed the application of the Belbin Team-Role Self Perception Inventory Toolkit to a construction, property development and consultancy company which aims to provide a 'one-stop-shop' for clients wishing to procure construction services. In this case to develop residential accommodation for patients with learning difficulties and behaviour disorders. All the staff involved together with the client, specialist professional firms and key

Table 9.1 Belbin's nine team roles (Belbin, 1996). Reproduced by kind permission of Belbin Associates, UK www.belbin.com – *Home to Team Role Testing.*

Team role	Contribution	Allowable weakness
Plant	Creative, imaginative, unorthodox. Solves difficult problems.	Ignores incidentals. Too preoccupied to communicate effectively.
Resource investigator	Extrovert, enthusiastic, communicative. Explores opportunities. Develops contacts.	Over-optimistic. Loses interest once initial enthusiasm has passed.
Coordinator	Mature, confident, a good chairperson. Clarifies goals, promotes decision-making, delegates well.	Can be seen as manipulative. Offloads personal work.
Shaper	Challenging, dynamic, thrives on pressure. Has the drive and courage to overcome obstacles.	Prone to provocation. Offends people's feelings.
Monitor evaluator	Sober, strategic and discerning. Sees all options. Judges accurately.	Lacks drive and ability to inspire others.
Teamworker	Cooperative, mild, perceptive and diplomatic. Listens, builds, averts friction.	Indecisive in crunch situations.
Implementer	Disciplined, reliable, conservative and efficient. Turns ideas into practical actions.	Somewhat inflexible. Slow to respond to new possibilities.
Completer/ finisher	Painstaking, conscientious, anxious. Searches out errors and omissions. Polishes and perfects.	Inclined to worry unduly. Reluctant to delegate.
Specialist	Single-minded, self-starting, dedicated. Provides knowledge and skills in rare supply.	Contributes on only a narrow front. Dwells on technicalities.

suppliers undertook the inventory. It was found that there was increased awareness of team roles and behaviour leading to more synergistic teams. This enabled the company to '… support the development of a corporate culture that values people by encouraging well motivated employees to commit themselves to company success and thus to a high level of quality and customer service' and that 'In some instances there was disagreement with the findings, but this in itself was positive in that it generated discussion around their strengths and contributions'. However, this case study is of an essentially in-house team developing a partnered project. Blockley and Godfrey (2000) recognize that generally it is unlikely that a construction team could be designed to incorporate all Belbin's profiles. Also Dainty et al. (2006), the reviewers of the case study, refer to Cornick and Mather (1999) who 'recognize it would be highly unlikely for a client's project manager to convince a client that a particular firm or individual(s) (who comply with all prequalification criteria) should not be appointed because of a risk of personality clash with another team member, firm or individual!'. This represents a common situation at all levels of management and development of construction projects due to the structure of the industry and the nature of most contractual arrangements.

9.7 Group development

The most established model of group development is that of Tuckman and Jenson (Tuckman, 1965; Tuckman and Jensen, 1977). Their model consists of five stages:

1. *Forming*: members get to know each other, ground rules are established; members are keen to make an impression; develop understanding of their task.
2. *Storming*: uncomfortable conflict stage, members state opinions and perspectives more assertively; negotiation stage as members put their point of view; resistance to being controlled by other group members.
3. *Norming*: cohesion stage; closer relations and norms of behaviour developed; sense of 'groupiness' develops; increased data flow.
4. *Performing*: effective structure has developed; working effectively towards accomplishing objectives; interdependence is a feature; problem solving is a major feature.
5. *Adjourning*: group disbands because objective has been achieved or members have left and so is unsustainable.

Jones (1973) complemented this model by describing the personal relation issues that affect group members and the task functions that are addressed as the stages develop. Personal relations progress from dependency at the beginning through conflict, cohesion to interdependency and the tasks from orientation through organization and data acquisition to problem solving.

Dainty et al. (2006) draw attention to the dynamic nature and characteristics of the construction industry and its projects resulting in group development being more problematic than in more static environments; to this can be added the diverse background, both professionally and organizationally, of group members. They continue that the temporary nature of construction project teams 'defines the interaction and communication within the construction project team environment and renders it amongst the most complex of all industries'.

9.7.1 Team building

Team building activities are seen by some as ways to improve the development and functioning of teams. Team building workshops may take the form of dialogue between team players or 'game'-orientated activities. The focus of workshops would be aimed at improving members' role definition, goal setting, problem solving, and interpersonal processes such as communication, negotiation and conflict management skills. 'Game-type' team building activities are more unconventional and include such activities as paintball wars, assault courses, survival tests and other off-beat activities which demand combined effort. Team building is seen to be particularly valuable in societies which traditionally favour individualistic effort, such as the UK and the USA, rather than collective effort. McShane and Von Glinow (2003) ask whether these team building programs, both workshops and unconventional, are effective and whether they are money well spent. They believe 'the answer is an equivocal maybe'. Problems are seen to be that team building activities are assumed to be general solutions to general team problems, yet specific teams have specific needs; team building is seen as a one-shot event when a new team is formed but should be an on-going process; and that team building takes place on the job not in artificial situations. Team building has not been a common feature of construction project management teams due perhaps to the absolute necessity to work in teams. This is not to say that team building would not improve team performance but rather to the reaction that 'we always develop projects through teams so we do not need to be trained how to work together'. However, the emergence of relationship-type arrangements such as partnering has shown that 'Team building consistently emerges as a desirable and often necessary way of helping to align teams behind project goals and objectives (even with long term partnering, since collaboration depends as much on individual behaviour). However one obvious limitation is the danger of not setting aside enough time or resources for effective team building because of the need to "get on with the real work". It is also important to realize that formal team building by no means guarantees collaboration, and that teams can suffer from the dysfunctional effects of over-cohesion' (Bresnen and Marshall, 2000a).

9.7.2 Managing groups

One role of managers in organizations is to manage groups under their jurisdiction. Whilst a group will have a leader, the hierarchy of the organization will wish to oversee a group's work to ensure that it is serving the purpose expected of it. This can result in a difficult situation if it is seen as interference in a group's work by a manager. It can de-motivate the group as it can create conflict between the organization and the group; on the other hand, such conflict can work as a grain of sand and produce creative results. Managers need sensitivity in overseeing the work of groups. They therefore need to understand the dynamics of groups and be positive in encouraging (without interfering) trust and open communications, stimulating discussion, providing information particularly on external factors and competition. A particularly delicate task for managers is rectifying poorly performing groups. Replacement of individuals who are not performing effectively may be necessary, leadership may have to be changed and communication between members may need attention, the correction of which may cause disruption to the group's activities. Consultants could be used to observe the work of a group and evaluate its performance as a basis for executive action.

Such management and maintenance of groups is possible in the construction industry for in-house teams in professional and construction firms but is considerably more difficult, if possible at all, for project management teams where members are drawn from a range of separate firms under the direction of a project manager. The project manager may be from a firm separate from the other members, or possibly from one of the professional firms also providing specialist skills, e.g., project management and design. In either case, the project manager's manager will have limited power to interfere with the project team, whose members will see themselves as having professional independence. If the project manager is from the client organization or a multi-disciplinary firm, management and maintenance of a project team should be easier to achieve.

9.7.3 *Size and composition*

There is a strong tendency in the mainstream OB literature to claim that small groups are more effective. Robbins and Judge (2008) believe that 'most effective teams should have fewer than 10 members and that experts suggest using the smallest number of people who can do the task'. Wilson (1999) agrees with fewer than 10 but adds that effective teams can range from 2 to 25. Essentially, size depends on the function of the group which in turn determines the range of skills and characteristics needed in its membership. For example, for a specific, well-defined task a small group would probably be better as it is faster at completing tasks and individuals are considered to perform better in small groups. For problem-solving requiring a diverse input and for fact-finding, larger groups are considered to be better.

However, Robbins and Judge (2008) find that managers err on the side of making teams too large, leading to communication and other problems, including more formalisation requiring protocols to control their business, inhibiting participation of some members and leading to absenteeism. Nevertheless, teams require the right composition in terms of both functional expertise and characteristics. In the construction industry, this can create very large teams, particularly on complex projects that entail many specialist consultants and subcontractors. Whilst it may be possible to keep the size of in-house groups relatively small, project development meetings (and particularly site meetings) involving all those making a design input have no choice but to have a large membership. Much of the business of such meetings is to share information and, most importantly, provide coordination which backstops misunderstanding and prevents loopholes. Most interestingly, Royer (2005) also argues for making room for 'the role of "exit champions": managers with the temperament and credibility to question the prevailing belief, demand hard data on the viability of the project, and, if necessary, forcefully make the case that it should be killed'. Her message is very relevant to the early stages of construction project development, and no matter how large the project team the inclusion of an 'exit champion' could produce the ultimate in effectiveness.

9.7.4 Norms

Norms refer to the standards of behaviour expected of team members by the other team members and so inform members what they ought and ought not to do in specific situations. Norms are the informal rules relating to behaviours important to a team that team members establish through their unwritten expectations by showing approval or disapproval of the actions of team members. Thus they are established by 'peer pressure'. Norms apply to behaviour only and not to the thoughts and feelings of members, hence members may do one thing whilst believing another. Examples of norms include: expectation to work long hours; reluctance to be absent from work; promptness; smart dress code; courtesy to other members; (or the opposite of these!) and can be seen to reflect aspects of culture. Most norms are established early in the life of a new team, often in the storming stage of group development. If they are felt to be inappropriate by managers they can be difficult to dislodge but may be achieved by the introduction of new or replacement members with more acceptable attitudes. Huczynski and Buchanan (2007) identify external and internal norms. External norms are seen to be 'imposed' by the group and are 'of such personal benefit to us that we are prepared to suppress any personal desires and are thus willing to limit our individual freedom and abide by them'. They see external norms as not being of equal importance and divide them into pivotal and peripheral norms. Pivotal norms are central to the group's work, for example output levels and pre-meeting preparation, whereas peripheral norms guide less essential behaviour such as mode of dress. There are also internal norms which emanate from inside the individual and include 'a desire for order and meaning in our lives' so we welcome external norms which are seen as reducing uncertainty and conform to them in order to feel that we know what is going on.

Emotional intelligence in general has been deal with in Chapter 3 but Drucksat and Wolff (2001) also see groups as having emotional intelligence (EI). They are not referring to the individual EI of members of groups collectively but to a group itself having EI. They see a group creating its EI through its norms or 'small acts that make a big difference'. However, EI of individuals is not readily accepted as a new phenomenon and is perhaps even less acceptable in groups. As with EI generally, the question remains of whether EI of groups is also a case of old wine in new bottles.

9.7.5 Leadership

Leadership of groups has already been referred to in this chapter and Chapter 11, devoted to leadership, appears later, so their content will not be repeated here; rather, leadership is referred to here only in relation to group development.

Leadership of groups is treated ambivalently in the OB literature in that it focuses on group leaders emerging naturally from the group and being made leader by the group members. This situation has the characteristics of informal groups rather than groups formally constituted by business organizations to undertake specific tasks. In such cases, one would expect that the hierarchy of the organization would appoint the leader of the group so that it could have confidence in the ability of the leader to direct the group and so produce results. Certainly in construction, at both the professional and construction firm level and at the project level, the hierarchy of the firms or the client would appoint the leaders of the teams charged with carrying out the development and construction of the project. Thus, in this context, the later chapter on leadership is relevant to leadership of teams in construction as well as to leadership of construction organizations generally.

9.8 Behaviour in teams

The social behaviour of employees has a great impact on an organization's effectiveness. Cooperation, purposefulness, flexibility are the products of individuals' personal characteristics which enhance organizations whereas deviant behaviour such as slacking, dishonesty and harassment lead organizations in the opposite direction. Many aspects of social behaviour manifest themselves in the interaction of team members. Working in teams magnifies and intensifies behavioural characteristics as a result of the close encounters that the members have with each other in both informal and formal settings where rapid responses are often required as teams move towards solutions to the problems and issues they are faced with. The intensity may not be so great in teams in which members work together on a continuous basis and/or on less critical activities. In construction project teams, however, consisting of professionals drawn from a wide range of independent firms and client representatives, it is frequently imperative that team members' behaviour aids effectiveness in order to move the project forward so that the professionals return to their firms able to continue their work of developing the project with confidence.

In such high-powered project meetings, and in the earlier subsidiary project meetings which will have taken place within the individual professional firms, the behaviour of members will have proved significant to the outcomes. The following content of other chapters (identified in italics) will have a direct influence on the performance of teams. The *perceptions* and *personalities* of the players, their *emotions and feelings* about the project and their *motivation* will have determined the way in which they have presented their contribution and, hence, the way in which it has been received by others, particularly its persuasiveness. In this respect a statement by McShane and Von Glinow (2003) points up the different perceptions of the place of teams in organizations between those in the construction industry and in the general OB literature: 'Effective teams require individual team members with the motivation and ability to work in a team environment. With respect to motivation, every member must have sufficient drive to perform the task in a team environment.' Here employees are seen to have special attributes in order to work in teams, but in construction working in teams is a way of life otherwise projects would never be designed and constructed effectively. It follows that members cannot be selected for team working primarily on the basis of their team-working abilities. Differences in *moral and ethical values* and lack of *trust* between team members can destroy the effectiveness of teams. There are frequently a number of ways to satisfy a project's objectives and these issues have a profound effect on the solution arrived at. *Communication* is central to a team's effectiveness through the way matters are presented as referred to above, including non-verbal communication and formal communications which identify that which was agreed by a team. This aspect includes the *power* structure and the status of members that controls the issue of minutes and formal instructions to team members which leads into the important matters of the *leadership* of teams and *politics*. The operation of teams and their formal meetings are fertile ground for political games which can work to the detriment of goal achievement if not controlled by team leaders. Underpinning all these forces at play in teams is *decision making* which is absolutely necessary if teams are to make progress in achieving their task. *Culture* is a particularly interesting aspect of team interaction. Organizational culture has been seen to be an ephemeral and complex issue and is further complicated by the idea of a team having a culture. This raises the issue whether, if this is the case, a team culture has to be consistent with the culture of the organization in which it is set. If not, would conflict between the team's culture and the organization's culture be disadvantageous and something to be 'corrected' or would it be a positive force for innovation? Also, what would be the relationship between the culture of a team formed of members from independent firms, as in construction, and the organizational cultures of the firms providing the members of the team?

Other features identified in the OB literature that specifically affect social interaction in teams are outlined below.

Synergy which contradicts the perceived wisdom that decision making by groups or committees, etc., is inferior, in terms of decision quality and time taken, than for individuals. Much of the research on groups reveals the opposite. In most conditions, groups outperform their best members because discussion within groups produces more alternatives, tends to eliminate inferior contributions, averages out errors and supports creative thinking (Fincham and Rhodes, 2005). Or as Kakabadse et al. (2004) put it: 'Synergy is the aim of all teamworking – to produce an output where the whole is greater than the sum of its parts.' If inappropriate tension, conflict and problems arise in teams and are not managed properly, groups develop negative synergy when process losses emanating from a team are greater than the benefits which arise from working in a team.

Social loafing is an aspect of negative synergy. It arises when people perform at a lower level than they are capable of by exerting less effort. Social loafing is more likely to occur in large teams than in small ones, where members perform similar tasks and their output is pooled, and particularly where individuals' output is difficult to identify and has low task interdependency. Therefore its incidence is likely to be very low in construction-related teams. Social loafing has been seen to be a result of the following attitudes (Huczynski and Buchanan, 2007):

- Others are not contributing, why should I?
- I'm hidden in the crowd, no one will notice me.
- Everyone will get the same reward, why should I work harder?
- People are getting in each others' way.

Cohesiveness of teams is the extent to which a team is committed to remaining together. All teams must have a minimal level of cohesiveness. Members will be committed when they feel the team will be useful to them in achieving their goals, satisfying their need for affiliation and status or provide social support. Sufficient members need such motivation for the team as a whole to wish to stay together. The most important effect of cohesiveness is on the potency of group norms (Fincham and Rhodes, 2005), resulting in the reinforcement of the extent to which norms determine the behaviour of the team. This can create major problems for organizational management by strengthening conformity to low productivity norms such as absenteeism. On the other hand, given norms which promote pro-social behaviours and organizational citizenship, cohesiveness can be beneficial to the organization. Factors which increase the strength of cohesiveness include homogeneousness of members, team size (smaller the stronger), regular interaction, eliteness, team success, external competition and challenges (McShane and Von Glinow, 2003). Benefits of highly cohesive teams are seen to be high motivation, cooperation, good interpersonal relations and low conflict. However, they can lack creativity and innovation through introspection and can develop groupthink. Elements which generate high cohesiveness are not generally present in any force in interdisciplinary construction project teams as they are temporary and regularly form and disband as projects come and go. There may be more of a tendency for groups within the separate firms to develop this characteristic, particularly if they have contributed to a series of successful projects. It is also possible that interdisciplinary project teams may attempt to stay together following a successful project as it may be possible for them to market their collective abilities.

De-individualisation refers to an individual's loss of self awareness and self monitoring in a team situation (Huczynski and Buchanan, 2007) which involves loss of personal identity and greater identification with the group. They quote Hampton's (1999) description of de-individualisation:

> There are moments when we can observe ourselves behaving irrationally as members of crowds or audiences, yet we are swept by the emotion, unable to check it. In smaller groups too, like committees or teams, we may experience

powerful feelings of loyalty, anxiety or anger. The moods and emotions of those around us seem to have an exaggerated effect on our own moods and emotions.

Interactions in teams can take many forms depending on the purpose and objectives of the team, the organizational setting, the environmental forces and, most significantly, the characteristics and idiosyncrasies of members. The mechanics of face-to-face interaction includes initiating topics and issues, responding, taking turns to speak (either by introduction or by interference), interrupting, not allowing interruption, and the intensity and emotion of exchanges. Emmitt and Gorse (2007a) refer to the possibility of uneven participation by team members leading to their work being dominated by relatively few members, particularly when teams are large. They point to Yoshida et al. (1978) who found that 'the stronger combined group forces often overruled individual expertise and experience'. Bell (2001) found that in multi-disciplinary teams limited contribution by some specialists prevented a truly multi-disciplinary view being presented (although this finding was in the context of child protection teams comprising mainly female participants). Emmitt and Gorse also point to the contrary view; that Littlepage and Silbiger (1992) 'found that, regardless of uneven and skewed participation rates, groups were able to recognize and use individual expertise confidently' and that skewed and uneven participation 'does not necessarily hinder group performance'. So the picture is far from clear. Such situations are common for multi-discipline construction project management meetings which usually have a large membership. This situation is also brought about to some extent by much of the business of such meetings being taken up by information exchange which may rely on only a small number of members. Nevertheless, much business requires interaction by the different disciplines and the key to making progress is an effective chairperson (project manager) who ensures the appropriate balance of contributions by members, as even in the case of highly-skilled professionals there may be some who are reluctant communicators. Teams in-house to construction and professional firms could be expected to operate in a similar manner to groups in other types of organization, but because these groups comprise highly-skilled professionals it is important that each member is given a chance to contribute. It is the responsibility of the leader to ensure that everyone is given that opportunity.

A technique for categorising the content of team members' face-to-face exchanges called 'interaction process analysis' (IPA) was devised by Bales (1950, 1955). Each person's spoken statements are classified into the following 12 different categories:

Shows solidarity	Shows tension release
Shows agreement	Gives suggestion
Gives opinion	Gives information
Asks for information	Asks for opinion
Asks for suggestion	Shows disagreement
Shows tension	Shows antagonism

The percentage of each type is mapped to show an individual's profile, thus the type of contribution made by each member in face-to-face interaction can be identified and studied.

We can see that interaction between group members is controlled to a large extent by group norms, which can lead to the style and content of discussion being constrained to the disadvantage of the quality of a group's output. Research by Asch (1951) showed how difficult it is for individual members to express an opinion contrary to the overarching view of the members of a group. This illustrates the amount of pressure a group can exert on its individual members. Inhibition of contrary ideas can work strongly against creative and innovative objectives. Leaders of groups with creativity as an objective must ensure that norms are not developed that work against such objectives and should ensure that during interaction every member is able to contribute without censure. The extent to which a member conforms to norms when interacting within a group is conditioned by:

- the characteristics of the individual
- uncertainty (conformity increases with uncertainty)
- the size of the group (conformity increases with an increase in group size)
- unanimity of the majority, and
- the group's structure.

In spite of inhibition by norms, research (and history) has shown that minorities can persuade majorities in a group (Moscovici, 1980; Nemeth, 1986). Huczynski (2004) summarized that, to motivate the majority to listen, the minority have to be heard, be noticed, create tension, be consistent, be persistent, be unyielding, be self-confident and seek defectors from the majority.

An influence on the quality of interaction in teams which can have a profound effect is the 'liking structure' which is the combined feelings of members towards each other; that is, whether they like, dislike or are indifferent to each other. Their feelings toward each other will affect the way in which they interact as individuals and the extent to which they give weight or not to contributions from other group members. The liking structure can be studied using the technique 'sociometry' (Moreno, 1953) which diagrammatically maps the emotional relationships between individual members of groups. It is now known as 'social network analysis' and has been used for mapping the informal organization, selecting group members, revealing feelings and modifying the group structure (Huczynski and Buchanan, 2007).

Emmitt and Gorse (2007b) identify that the field of group interaction and communication behaviour of construction professionals has received little attention from researchers (Emmitt and Gorse, 2003, 2007a; Dainty et al., 2006). They continue by using Bales IPA to model both task and relational interaction which, they state, 'represents the first attempt to model the construction team's interaction in live project meetings'. The results were contrasted with other studies of work, social and academic groups. In comparison with social and academic groups the amount of positive socio-emotional interaction was considerably less for construction project teams as was the amount of negative socio-emotional communication. They state that for construction project teams, 'A combination of the members' temporal relationships and the commercial context appears to restrain interaction, thus restricting socio-emotional development, as achieved in more stable groups.' In contrast, participants in construction meetings use high levels of task-based interaction in common with other work groups of professionals. In construction teams the level of 'giving information' was higher than for 'giving opinions, analysis etc'; Bales (1950) suggests this pattern arises from a type of communication efficiency which Emmit and Gorse construe as construction professionals using construction terms that have implied meanings without the need to fully explain them. It is claimed that 'The high level of task-based interaction resulted from the low level of socio-emotional interaction, owing to the group's restricted development ... Restricted development of the group is common to construction; indeed given the temporary nature of construction projects it is highly unlikely that many groups ever reach maturity', which has implications for teamwork and relational forms of contracting. A somewhat surprising finding is that 'The high levels of conflict reported in construction literature were not evident in the data; indeed if the levels of conflict reduced much further the groups may be subject to groupthink'.

Conflict is an unavoidable aspect of organizational life and is frequently exposed in teams. It is essentially disagreement between parties over significant (or what the parties perceive to be significant) issues for which the term 'conflict' gives a dramatic image. Disagreements can vary between a number of valid points of view that need resolving in order to make progress and which can be discussed rationally to reach a conclusion to, at the other extreme, political battles over such matters as position, status or resources which are more in the nature of a 'dog-fight' and likely to be damaging to the organization. All disagreements involve interactions between protagonists which can vary from rational to vindictive and ultimately leading to no communication whatsoever. Team meetings are often considered to be fertile ground for airing such disputes, either legitimately if the issue is within the remit of the team or by using team meetings for fighting personal or organizational disputes. However, it should be recognized that in certain situations disagreements can be beneficial by stimulating new ideas, promoting competition and producing well-thought-out solutions.

The perceived importance of the topic of a dispute has a large effect on how the dispute is handled relative to the importance of maintaining good relationships with other parties. At one end of the scale, insignificant disagreements will be avoided if maintaining good relationships are important and, at the other end, if the topic in dispute includes a matter of principle, a battle may be fought to the end and important relationships sacrificed in order to win the dispute, with a range of dispute/relationship balances in between which business organizations generally resolve by compromise. In groups, a third party is often used to reach a compromise, usually a manager with appropriate status and negotiating skills. In teams associated with the construction industry, disputes that occur within professional firms will mainly be project based and involve professional opinions. They will have characteristics similar to those described here but with the benefit of a common professional grounding so that they will be infrequent and based on rational argument. A level up from this are interdisciplinary project teams, generally of professionals from different firms. A balance between prosecuting a dispute and maintaining relationships becomes more acute in these teams as relationships tend to be more formal and future work may depend on maintaining good relationships. But discussion needs to be rigorous and open in order to produce an appropriate solution whilst still maintaining relationships; a balance which is usually achieved due to this being the *modus operandi* of the construction industry. The level at which disputes become highly significant, often to the extent that relationships are sacrificed, is at the contractual level between clients and contractors in which large sums of money are often involved.

9.9 Teams in general vs. project teams in construction

We have seen that there are major distinctions between the nature of teams used in the construction industry, particularly interdisciplinary project teams, and teams used in organizations generally. Teams in organizations generally are driven to a large extent by the expectation that they will perform better than individuals but also by social agendas to fulfil psychological needs of individuals. In construction, teams are used simply because there is no other way of developing and constructing projects. A significant difference is illustrated by an example used by Moorhead and Griffin (2001) which describes planning for a team start-up: 'The actual planning took eight and a half months. It often takes a year or more before performance levels return to at least their before-team levels. If teams are implemented without proper planning, their performance may never return to prior levels. The long lead times for improving performance can be discouraging for managers who reacted to the fad for teams and expected immediate returns.' Construction project management cannot afford the luxury of such timescales, confirmed by Emmitt and Gorse's (2007a) statement: 'Problem solving during the construction process is subject to time pressures, and these problems need to be resolved or 'closed' if the programme is to be maintained', and reflected by De Grada et al. (1999) who found that 'time pressure prevented groups [in construction] from engaging in 'social niceties' hence resulting in groups emitting a lower proportion of positive socio-emotional acts'.

Reinforcing these differences is mainstream OB's view of the factors which create effective teams (and by implication, high performance). Robbins and Judge (2008) see the key components as follows (in italics followed by this author's brief and basic comments on their relevance to construction):

Adequate resources: Should not usually be an issue for construction projects.
Effective leadership and structure: Need for leadership readily recognized, structure not usually considered.
Climate of trust: Generally high amongst professionals.
Performance evaluation and reward systems: Not needed, all part of the job.
Abilities of members: Generally a given for construction.
Personality: Embedded in team members appointed for their professional skills.
Allocating roles: Not specifically done.
Diversity: A result of the specialist skills needed.
Size of teams: Determined by the skills needed by the nature of the project.
Flexibility (members' ability to fulfil other members' skills): Not necessary.
Members' desire to be part of a team: No choice.
Work design: A given.
Common purpose: Clear.
Specific goals: Clear.
Team efficacy (team believe they can succeed): Generally a given as members are specifically trained for their tasks.
Conflict levels: Usual, controllable, procedures in place for resolution of some.
Social loafing: Not usually an issue.

Whilst this list helps to demonstrate the distinctiveness of construction project teams, there is much in the OB literature on groups and teams which is useful in aiding understanding of how members of construction project teams behave even though the context of the team may be distinctive. Issues such as structure and roles, group development (team building, composition, norms) and individual behaviour are important issues for construction project teams but which are not usually considered explicitly. The area in which the whole gamut of the OB literature on teams is most valuable is in aiding understanding of the requirements of construction project clients. It is very rare that the definition of construction clients' requirements for their project is decided by one person from the client organization. Even for a simple single dwelling there are usually at least two people involved! Anything more complex produces many more contributors. Frequently a client's employees who are required to contribute to defining the client's requirements are formed into a team and the team or its leader will instruct the construction project team. The team defining the client's requirements is embedded in the organization structure of the client organization which can be for the purpose of any commercial or social function. In such circumstances the full range of factors involved in the activities of teams as identified in OB literature come into play in the client's team. For this reason, members of construction project teams need to understand how clients' teams function and the forces that come into play in their decision making which will help the construction project team to understand the forces which have generated the clients requirements and to anticipate how the client's team may react to their proposals.

Construction project teams find it difficult to deal with clients' teams in the project briefing process. This is due partly to them not understanding the dynamics of clients' teams, which is often manifested by frequent changes in direction leading to abortive work that can lead to often unrecoverable costs and create frustration for project team members. Construction project teams would rather deal with one point of contact in a client's organization, which can be achieved by the client organization appointing a client's project manger from amongst their employees, but even then there will usually be a team behind the client's project manager. Whilst this arrangement may simplify communications, it is unlikely to reduce changes of mind by the client's organization as it seeks to define its needs and accommodate the political forces at play during the project definition process. Cherns and Bryant's (1984) well-known account illustrates examples of such forces which may well underlie the activities of clients' teams charged with defining their projects:

> Consider the case of the corporate client that always has more projects competing for finance than it has capital available. Thus the agreement to invest in X (say, building a new warehouse) is a decision which not only pre-empts X^1 (extending an existing one) but also gives priority over Y (building a new laboratory) or Z (installing a computerized production system). Since Y, Z and X^1 are projects in competition with X, and all have their organizational supporters, the decision for X is a victory for the supporters of X – and a defeat for the supporters of the other competing projects. (Examples of client interest groups supporting competing projects include operating divisions, specialist departments, political factions, professional groups, etc.) The victory for X then marks a shift in the balance of power within the client organization (if only to confirm the dominance of one interest group over another). But the victor has now given a hostage to fortune. In fighting the battles, he has probably shaded the risks and been optimistic in his estimate of cost in order to get his budget just below the organization's threshold of approval for capital expenditure. He has promised a return and now has to deliver – or lose credibility.
>
> The study has provided us with vivid glimpses of the various "hostages to fortune" that the different interest groups within the client complex may have offered in promoting a particular project in competition with others. Each participant can be seen as bringing to the table his own sense of what is at risk personally, as well as what is at stake professionally or departmentally, in the forthcoming project experience. Some client participants have a high stake in meeting the target completion date; others in working within the promised budgetary limit; others in the operating performance of cherished design concepts. Many of the stakes are reputational (e.g., the operational manager's reputation as a skilled negotiator in the best interests of his own department; the project manager's reputation as a hard-headed realist who gets results in the face of whatever obstacles).
>
> In considering the role of the client in construction, then, we cannot treat the client as unitary; nor we can ignore the events which preceded the decision to build. The progress of a construction project involves various groups within the client organization whose interests differ and may be in conflict, and whose observed behaviour cannot adequately be explained without reference to the past.

General opinion of the value of groups and teams in organizations seems to be ambivalent at best compared to construction (and no doubt some other industries) in which they are essential. Fincham and Rhodes (2005) summarize the conflicting views, observing that 'Teams and team working are still seen by management consultancies as the best way of marrying the fulfilment of fundamental individual psychological needs with senior management demands for more flexibility, less "down time" as members cover for each other's absences or variable work rates and more self-regulation by team members, reducing the cost of supervision'. Yet they also comment that:

- Team working can have a profound and complex effects on individuals' behaviour
- Group membership can cause distorted perceptions of reality
- Social identity in groups can cause stereotyping and discrimination against non-group members
- Competition with other groups in the same organization can be dysfunctional
- Dependency and conformity in groups can stifle innovation and originality
- Cohesiveness may inhibit contrarian thinking
- Norms may work against the interests of senior management.

10 Decision-making

10.1 Introduction

Sound decision-making is the bedrock of effective management. Decisions are made at all levels of an organization and vary in significance from those crucial to the survival of the organization to those which can be seen to be trivial but nevertheless necessary. Since early research and writing on management issues, decision-making has been recognized as central to management. Key publications in the development of management thinking (Barnard, 1938; Simon, 1957) have placed decision-making at the centre organizational life. Mintzberg (1989) believed that 'decision making is one of the most important, if not the most important, of all managerial activities …'. Understanding how decisions are made at the different levels of business (individual, group and organizational) is vital to improving decision-making and organizational performance. Business decisions taken by construction organizations are as significant as for any type of organization but of particular significance are decisions relating to progressing projects; the lifeblood of their business. Points at which major decisions (primary, key and operational) need to be made define the organization structure for developing a construction project (Walker, 2007).

To assist decision-makers a wide range of quantitative decision methods is available arising from management science, such as decision trees, break-even analysis, queuing theory, game theory and CPM and PERT (which will be familiar to those in construction), many of which use probability theory in simulations of decision situations. However, such techniques, whilst valuable in analysing complex decision situations, invariably require assumptions to be made regarding inputs and judgement to be exercised in making final decisions, and many business decisions are made without the aid of such analysis. The majority of decisions are made on the basis of the knowledge and experience of managers and other individuals and groups in organizations and are strongly influenced by the behavioural characteristics of owners and employees acting alone or in groups. Thus, in tune with the topic of this book, the focus of this chapter is on the behavioural aspects of decision-making. Quantitative

Organizational Behaviour in Construction, First Edition. Anthony Walker.

decision-making is a field in its own right on which there is a considerable literature. The initial focus of the chapter is on decision-making in general relating to individuals followed by the particular aspects of decision-making by groups.

10.2 The rational model

The traditional model of decision-making is based on the rational approach arising from economic theory and the idea of 'rational man.' However, the assumption that decisions are made rationally does not correspond to how they are actually made, as, in reality, decision-making incorporates the behavioural characteristics of those making decisions, which have been discounted by economic theory as irrational. The rational view saw behavioural elements as playing no part in the decision-making process. In the real world these characteristics have a fundamental effect on decision-making, an understanding of which is vital to knowing how and why decisions are made. Even economists now accept their traditional position is unrealiztic and recognise that decision-making includes aspects of human influence. Nevertheless, the basic rational model, which comprises the following stages, remains useful as a starting point for considering the decision-making process and its behavioural elements:

1. *Define the problem.* This is the most important step. The rational model assumes the problem is clear and unambiguous. If the problem is not clear or the wrong problem is identified there is no chance of making a correct decision. Often organizational goals are ambiguous or in conflict with each other. When Einstein was asked how he would save the world in one hour, he said that he would spend the first 55 minutes defining the problem and the last 5 minutes solving it.
2. *Search for alternative courses of action.* The rational model assumes all possible courses can be clearly identified.
3. *Gather and analyse data about the alternatives.* Assumes all relevant data can be identified and obtained.
4. *Evaluation of alternatives.* Assumes appropriate techniques for evaluations are available and are used so that weights reflecting the importance of factors can be correctly assigned and each alternative accurately rated to identify the alternative that yields the highest value.
5. *Selection of the best alternative and its implementation.* Assumes that the best alternative from the rational process is accepted and is implemented exactly as chosen.

In addition to the assumptions referred to above, another major assumption of the rational model is that there is no time or cost constraints affecting the decision process.

As stated previously, the idea that decision-making is based on an absolute version of rationality is not now generally accepted; rather, decision-making is viewed as a socially constructed phenomenon constrained by the real-life limits of social groups, social structures and human factors, which have been the inspiration behind managerial theories of decision-making. Decision-making is a human process constrained by the limits on ability and by the conflicts of social and economic life (Fincham and Rhodes, 2005). As an example, they point out that the idea of organizational goals suggests an ideal state objectively identified and agreed. Although they accept that this must be true to an extent otherwise nothing would get done, this view is too simplistic. They believe that the real goals which employees pursue may be very different from formal organizational goals. Personal goals (promotion, power, income, etc.) may conflict with official goals. In pointing to the assumption of the rational approach being unreasonable, they say, 'if we always waited for complete information before reaching a decision we would never take a decision at all.' In taking this view a step further by identifying organizational irrationality they quote Jackall (1988) as saying 'that the realities of decision-making, as managers saw it, took into account the likelihood of failure right from the start – quite different from conventional theories' and:

> 'The actual rules for making decisions are quite different from managerial theories about decision making, and … managers themselves were well aware of the rational approach but this formed a kind of rhetoric or way of legitimizing courses of action. Managers knew they had to repeat the rhetoric – while the reality was not only different from this ideal but bore no relationship to it at all. As one manager put it, 'the basic principles of decision making in this organization and probably any organization are: (1) avoid making decisions if at all possible, (2) if a decision has to be made, involve as many people as you can so that, if things go wrong, you're able to point in as many directions as possible.'

This rather jaundiced view is hardly a sustainable approach to construction projects, otherwise nothing would be done and projects would never be completed, but it may well have resonance for the more general management of construction and professional organizations.

The rational model is classified as a prescriptive model of decision-making. Prescriptive models categorically specify how one should approach a problem or which styles work in which situations and hence how one should behave to achieve a desired outcome. They are promoted by management consultants and often contain specific techniques, procedures and processes which are claimed to lead to more accurate and efficient decision-making. Common features of prescriptive models have an emphasis on rationality – a logical framework including a list of stages to be carried out – but we have seen that rationality in decision-making does not represent the reality of decision-making in organizations and that generally flexible approaches to problem solving and decision-making are necessary.

10.3 The types and nature of decisions

At a basic level decisions can be categorized as *programmed* and *unprogrammed* types. Programmed decisions are routine decisions which recur sufficiently often for a decision rule to be created. A decision rule informs decision makers which alternative to choose once they have information about the decision situation. For programmed decisions the goals are clear, the procedure established and information needed is defined and obtainable. In construction, procedures under construction contracts may fall into this category of decisions, for example, preparing valuations for certificates of payment, as will many structural calculations. The mainstream OB literature suggests that such decisions are capable of being made by low-level employees but this is not usually the case in construction and other professional activities as, although programmable, judgement is often needed to determine the actual input to a decision. Unprogrammed decisions have to be made when a problem which has not occurred before is encountered. In such unique situations the problem needs to be clearly defined as it is often vague and alternative solutions have to be generated and evaluated in the context of ambiguity and uncertainty which will require judgements to be made in arriving at a decision. Many such decision situations occur in construction organizations, particularly in the actual management of projects as new projects are originated and developed. However, this classification of programmability of decisions is perhaps simplistic in the context of construction as, although the majority of construction projects are original and on the face of it require essentially unprogrammed decisions, they can bear similarity to previous projects and previous decisions and fall somewhere between the categories of programmed and unprogrammed.

Adaptive decisions require human judgements of a kind that computer programs cannot make. Once judgements are made adaptive decisions may be made using quantitative decision tools on the basis of the judgements. The judgements then become the assumptions on which the output of the decision tool is based.

Innovative decisions are made in novel situations for which there is no precedent. They occur rarely and in areas of the unknown and are made at the top echelons of companies, usually on matters of future strategies, for example, embarking on a new field of business and forming new business alliances. Both adaptive and innovative decisions can be envisaged in conceiving, designing and developing innovative buildings such as the Sydney Opera House, the Pompidou Centre, Paris and the HSBC Building, Hong Kong.

Probably the most useful perspective on types of decision is that of ***risk***. Every decision is made under condition of *certainty, risk* or *uncertainty*. This concept is central to probability-based decision tools but is equally important to thinking about decisions made without formally using such tools. With *certainty* all outcomes are known in advance so all that has to be done is to select the outcome with the greatest pay-off. It is generally accepted that total certainty is so rare, except in trivial circumstances, as to be non-existent. The reason for considering it is to help define the other two more common states. Many decisions in business and life in general can be seen to be made under conditions of risk. One of the reasons for having managers is for them to exercise judgement when there is some risk involved in decisions. If all conditions were certain, managers would not be needed and computers could be programmed to make all the decisions. In ideal circumstances, in conditions of risk managers would be able to obtain accurate data in order to calculate the probability of each possible outcome occurring. The probabilities can then be input to the well-established decision tools for conditions of risk that require managers to specify the actual numerical probability they ascribe to each alternative outcome in order for the outcome with greatest value to be calculated. However, in conditions of risk such data are rarely readily available in the form required so, on the basis of their experience, knowledge and research, managers need to be able to assess the probability of alternative outcomes occurring for inputting to decision tools. But in reality in their general management activities, rather than using decision tools, decision-makers may not formally determine the probabilities associated with each alternative outcome but, in the process of deciding which alternative to choose, they may act more informally by relying on their sense of the probabilities associated with each outcome.

Whilst many decisions fall under conditions of risk, increasingly in the modern world where decision-makers work in rapidly changing and uncertain markets, decisions may have to be made under conditions of *uncertainty*. Powerful examples would be decision-making following major natural and man-made catastrophes, e.g., volcanic eruptions, 9/11 and the many financial crises. In conditions of uncertainty, the likelihood of outcomes and their payoffs is unknown so probabilities cannot be assigned. In uncertainty there is no direct relevant data and, on the face of it, the situation is so unique that there is no basis for judgements. However, in business situations, there are mathematical models available, such as Monte Carlo simulation and game theory, but they are rarely used in practice. Paradoxically, they are often seen as too complex yet still cannot capture the complexity of the most difficult business decisions. Even if they could, the data collection and analysis would, in most cases, be too great to make collection and implementation practicable. As a result, heuristics and intuition are used by managers in uncertain market environments. However, it is claimed that the systematic transparent frameworks which models provide for fact gathering and discussion of key complex decisions constitute their major benefit and are valuable components of any decision-maker's armoury.

10.4 Behaviour in decision-making

Rational decisions equate with scientific reasoning, empiricism and positivism using decision criteria of evidence, logical argument and reasoning (Huczynski and Buchanan, 2007). But decision-making in practice is not a wholly rational process and the behavioural approach to it attempts to account for the limits on rationality in the decision-making process. As a result, actual real world decision-making is seen to incorporate features of both the rational and behavioural approaches. A major challenge to the rationalist approach came from Mintzberg (1973) and others who asked questions like, 'What do managers actually do?' and, using observational techniques, found little evidence of rational decision-making. Managers were found to engage in a great deal of reactive decision-making, making snap decisions and having short interactions. The rationalist image of managers decisively planning, controlling, leading and devising strategies was contradicted; rather they were under great pressure from an unrelenting pace of work and their activities were defined by brevity, variety and fragmentation, causing Mintzberg to say, 'In order to succeed, the manager must, presumably, become proficient in his superficiality.' However, Fincham and Rhodes (2005) point out that Kotter (1982), whilst finding similar behaviour, also found that an element of rationality was provided by managers working to flexible internalized agendas of interlinking goals which did contain the elements of planned and prioritized activities, concealing a more rational pursuit of objectives. They continue by saying, 'In this way, behavioural theorists have shown that rationality does not have to be applied in an all-or-nothing sense. Our understanding of management rationality can be "softened" and distanced from the absolute rationality of much conventional management theory in order to bring it closer to the actual behaviour of managers. But at the same time, there may still be an underlying sense of purpose or a guiding rationality to what managers do.'

The fundamental feature of the behavioural approach is *bounded rationality* which recognizes that, in the real world, decision-makers are not capable of identifying and considering all possible alternatives before making a decision. This is due to decision-makers' inability to:

- Completely define the decision situation
- Obtain complete information
- Generate all possible alternatives
- Predict all the consequences of each alternative
- Avoid personal and political influences
- Avoid the constraints of time and cost.

As decision-makers are restricted in their choices, they have to settle for less than a fully researched solution. That is, most people who make decisions are limited in their decision processing abilities so react to a complex problem by reducing it to a level they can deal with and have to be prepared to settle for what may be a less-than- perfect solution.

Thus, whilst the rational approach aims to optimize (or maximize), in reality sub-optimization actually takes place. Optimizing requires all possible solutions to be considered simultaneously and the best one selected. However, what usually takes place is known as *satisficing*; that is, decision-makers will consider each alternative sequentially as it is identified until they find an acceptable one, which may or may not be the optimal solution, but one that 'will do' and they will then decide to accept it and consider no more possible options.

Robbins and Judge (2008) point out that, surprisingly, little research has been done on the relationship of *personality* and decision-making but that what has been done suggests, as would be expected, that personality does influence decision-making. An obvious example would seem to be a decision-maker's attitude to risk, an aspect of personality that will lie on a spectrum from risk-averse to risk-taker. They also make reference to gender differences in decision-making styles. They say that it has been found that '[Women are] more likely to over-analyse problems before making a decision and rehash the decision once it has been made. Thus women in general are more likely than men to engage in rumination. On the positive side, this is likely to lead to more careful consideration of problems and choices' but also that 'gender differences seem to lessen with age'.

Moorhead and Griffin (2001) suggest *the practical approach* which is aimed at combining the rational and behavioural approaches in which 'rather than generating all alternatives, the decision maker should try to go beyond rules of thumb and satisficing limitations and generate as many alternatives as time, money and other practicalities of the situation allow. In this synthesis of the two approaches, the rational approach provides an analytical framework for making decisions, whereas the behavioural approach provides a moderating influence'.

Political factors have been referred to earlier and Fincham and Rhodes (2005) point out that power and politics (see Chapter 7) challenge rationalistic ideas of decision-making in organizations, an idea originated by Simon (1960). The rational model was conceived in a vacuum without recognizing the influence of power and political forces but they cannot be ignored in the real world. Hence political limits were linked with cognitive limits to rebut the rational approach to decision-making in favour of the behavioural approach. Decisions at all levels in organizations involve power and conflict as individuals and groups jockey for position. This is often to the detriment of the quality of decisions' contributions to the wellbeing of organizations. Rather it is to the benefit to individuals' and groups' ability to impose themselves on other individuals and groups for their own advantage. Such hidden agendas and back-door dealing distort rational decision-making as they force decisions to be made in the light of likely consequences which aid the power and political objectives of individuals and groups and are not in the best interests of solving the problem under consideration.

Political games thrive in conditions of uncertainty. Political forces may well come into play in professional construction organizations and result in distorted decisions that work against the interests of the firm and its projects. This is especially so in those firms which are multi-disciplinary if the various specialties compete for power and resources, but can also come into play in single discipline practices and contracting organizations if battles are joined for directorships and partnerships. But the greatest impact on projects is likely to arise from power struggles in client organizations resulting in inappropriate decisions about the objectives, brief and 'ownership' of projects leading to uncertainty on the part of project teams.

Intuition is the ability to know when a problem or opportunity exists and to select the best course of action without conscious reasoning; is a factor which is recognized in decision-making but which remains controversial (Behling and Eckel, 1991). Managers tend to believe in the value of intuition but use it in connection with more conventional decision-making approaches to come to a final conclusion. It is likely to be most valuable when used by experienced managers with well developed knowledge of their industry and organization (tacit knowledge). People are most likely to use intuition when the following conditions exist (Robbins and Judge, 2008):

- High level of uncertainty
- Little precedent
- Low predictability of variables
- Useful information is limited, particularly for analysis
- No clear sense of direction given by the facts
- Several plausible options all with good supporting arguments
- Limited time to come up with a decision.

Although it appears obvious to state that identifying a problem is critical to its solution and that a decision which solves the wrong problem is worthless, nevertheless problems are frequently misdiagnosed because decision-makers are not objective in their thinking. They lack diagnostic skills and/or have perspective biases which distort the diagnosis of problems. Peoples' biases are created by their perceptions, values and assumptions. Such biases are present throughout the decision process. Hence they misinterpret information, make invalid assumptions and can be influenced by others in positions of power. People often very quickly come to a conclusion of what a problem constitutes based on stereotypes, previous problems and false information. They also frequently come up with a solution quickly, usually based on the solution to a previous problem which was probably not the same as the one currently being considered. All of this is compounded by the problem not having been clearly defined in the first place. The tendency to do this is not uncommon on construction projects, for example, choosing contractual arrangements because 'they worked last time' even though conditions differ this time; another example would be choosing a particular cladding type simply because it had been used before. Nevertheless, those with good intuition, which includes 'reading between the lines', are valuable to organizations for their ability to make decisions that, if their intuitive solution is accepted, move things forward but they may have difficulty in arguing their case.

In this process managers rely on judgement shortcuts called *heuristics* to speed up their decision-making which, although faster and simpler, expose users to biases which are inherent in human intuition (Huczynski and Buchanan, 2007). The three most common are the representative, the anchor and adjustment, and the availability heuristic. They are subconscious and have an instant effect on a person's judgement.

The *representative heuristic* is a situation in which people judge on the basis of how well things match so that the similarity of one object to another infers that the first object acts like the other, for example using packaging to infer the quality of a product.

The *anchor and adjustment heuristic* states that starting from somewhere is easier than starting from nowhere. Both rational theories of decision-making and also more realistic approaches imply that decisions are seen to be taken to totally solve a problem at the point of decision but, in reality, decision-making moves forward incrementally in seeking to achieve its ultimate goal. In this incremental process, decisions tend to build on previous decisions, which inhibits rationality as the decision-maker does not start again at the beginning of the problem. In construction, this process can be envisaged as using the specification for a previous project as the basis for drafting the specification for a similar new project. It can also be envisaged in preparing a budget for a new project based on the cost of a previous project. This means that estimated figures for a new project based on the actual figures from a previous similar project are biased towards it, including any distortions it may contain for which adjustments should be made.

The *availability heuristic* is used to estimate the probability of an event occurring by assessing how readily instances of it come to mind. More vivid events come to mind more often than vague events but in actual fact may not occur more frequently.

Huczynski and Buchanan (2007) state: 'While helpful in many situations, heuristics can lead to errors and systematically biased judgements. Although the three main biases have been discussed, many other errors, fallacies and biases exist. People have ideas about order, randomness, chance and so on. Studies have shown how people's judgements become biased and hence less rational.' Some of the biases they identify include the following:

Contrast bias: For different items that are presented one after the other, the tendency is to see the second as more different than it really is.

Commitment and consistency bias: Individuals become bound to their actions and through these to their beliefs. Indicates how we will behave. After making a decision, people will adjust their attitude to make it consistent with their action, and become committed to it.

Social proof bias: People decide what to believe or how to act in a situation by looking at what others believe or do and copy them in conditions of uncertainty, especially others they perceive as being similar to themselves.

Liking bias: Put simply, we enjoy doing things for people we like.

Authority bias: We are trained from birth to believe that that to obey legitimate authority is right and comes from systematic socialisation which has led to the perception that this is correct conduct.

Scarcity bias: We use an item's availability to decide quickly on its value. Hence as things become less available we want these things more.

Also Robbins and Judge (2008) identify:

Overconfidence bias: Overconfidence has been said to be the most prevalent bias and potentially the most catastrophic (Plous, 1993).

Confirmation bias: The acceptance of information without question when it confirms our preconceived views whilst rejecting information that challenges them.

The generation of a limited range of alternatives from which to choose in making a decision (bounded rationality) inhibits successful decision-making. Generating alternatives requires *creativity* and divergent approaches to problems which involves reframing problems in ways other than the conventional. Creative people will frequently come up with insights that have not been seen by more conventional thinkers. Four characteristics are said to typify creative thinkers: intellectual abilities; relevant experience; strong motivation and an inventive thinking style (McShane and Von Glinow, 2003). Intellectual abilities allow people to analyse data, synthesize information by connecting facts in ways that have not occurred to others and to see how their ideas can be applied to problems in the real world. A good foundation of knowledge and experience provides a basis for understanding and hence is fertile ground for creativity. Motivation and perseverance are needed to ensure that the results of creativity are realized. Thomas Edison remarked that genius is one per cent inspiration and ninety-nine percent perspiration and the same is still said today about achieving a PhD. An inventive thinking style can be seen as a gift but one which will not be used effectively without the other characteristics referred to here. Creativity in the construction field is invariably seen as present only in design and, although visible predominantly in this field, it is also present in a less transparent way in engineering, cost control and management when inventive solutions to problems are found.

Robbins and Judge (2008) identify *organizational constraints* which directly affect the decision-making behaviour of employees. Performance evaluation of employees is frequently used to assess individuals and the criteria used for this purpose is usually published. Hence, employees will focus on meeting such criteria and in doing so may distort their decisions to satisfy the criteria they are expected to meet rather than make what they believe to be the best decision. The tendency is to render decision-makers risk averse, or the opposite, depending on the criteria and so remove objectivity. In a similar manner, an organization's reward system may bias decision-making as employees seek to maximize their income by manipulating the reward system. Also, formal company regulations constrain employees' decisions but in many cases these are seen to act in the best interests of the organization. Similarly, organizationally imposed time constraints can mean that problems are not fully explored before a decision has to be made but this situation is a trade-off, usually in favour of making timely progress.

Topics of some previous chapters are particularly significant for behaviour when making decisions. *Ethics*, both individual and organizational, underlie all decisions other than the most trivial. Ethical standards range from the very highest in which no compromise is tolerated to the lowest for which anything goes. Although knowledge of an individual's or an organization's ethics is difficult to identify unequivocally, a sense of their ethics does tend to emerge over time and needs to be taken into account in assessing the quality of a decision. In practice, the significance of ethics in business is essentially about making decisions that are in the interests of the organization as opposed to the best interests of the decision-maker but this simple statement cannot hide the fact that ethical dilemmas can frequently face decision-makers from both a business, personal and social perspective. In construction, ethics arise particularly in decision-making based on advice given to the client by professionals. Ethically, professionals should give advice that leads to decisions which are in the best interests of the client but the opportunity exists for them to act unethically by making recommendations which are in their own interests. However, as pointed out in Chapter 4, codes of conduct of professional institutes are designed to guard against abuse.

Culture is covered in Chapter 8, the content of which can impact significantly on decision-making. Differences are particularly marked between the decision-making behaviour of nations and hence their individual people. For example, in Japan the system of *ringisei* ensures that subordinates have an opportunity to voice their views and possibly influence final decisions; in Korea groups are seldom used, decisions being made by family members who own most businesses; in Israel decision-making is characterized by collectivism but there is a move towards more individual freedoms; and South Africa is still making progress with novel decision-making arrangements in multicultural groups which reflect the new order (Francesco and Gold, 1998). Robbins and Judge (2008) point out differences in time scales for decisions, citing the USA for speedy decision-making in contrast to decision-makers Egypt who move much more slowly, and the USA's emphasis on problem solving in contrast to Thailand's and Indonesia's tendency to accept situations as they are. Organizations themselves also exhibit different decision-making cultures, for example: autocratic or collective decision-making; degree of reliance on rationality; experience or intuition; degree of urgency; and confronting problems or ignoring them, all of which reflect a company's organizational culture which is then overlaid by its national culture. Hence decision-making in multinational, multicultural organizations can be extremely difficult.

10.5 Group decision-making

Groups and teams are common organizational components and the source of many decisions. The manner in which they make decisions differs in significant ways from individuals' decision-making due to the specific factors that arise in group situations. These factors, if not controlled by group leaders, can result in poor decisions; nevertheless, group decisions are generally assumed to be better than individual ones. This assumption is based on the common belief that 'two brains are better than one' as a result of groups having a range of contributions on which to base decisions and the advantages of generating a greater pool of knowledge, having a variety of perspectives leading to greater understanding of the problem and wide acceptance of a decision. However, it is generally accepted that, whilst the average quality of a decision made by a group is greater than the average quality of a decision made by an individual, the quality of decisions made by groups are below that made by the most capable individual.

A high level of cohesiveness is generally seen to be beneficial to group performance but can lead to negative consequences. These are usually associated with a wish not to 'rock the boat', leading to reluctance to question a group's consensus view or to be overly influenced by the direction the group is taking, resulting in poor decisions. The specific factors that come into play in these and similar respects in group decision-making, many of which are interrelated, include the following.

Time constraints are usually present in most decision-making as organizations have to respond to pressures over which they do not have total control. They are highlighted in construction by Emmitt and Gorse (2003) and are very prevalent on construction projects which are invariably tightly programmed to meet the needs of users and/or investors but perhaps less so for general business decisions by construction organizations. Time pressure is exerted on individuals to make their technical decisions relating to their speciality to feed into the generally more complex decisions to be made by the project team, which are themselves subject to the more challenging group decision-making phenomena. Time pressure has been found to work against the cohesiveness of groups as there is insufficient time and opportunity for members to engage in team-building activities. Time constraints were also found to lead to group members with inherent problem-solving skills being ever more dominant as pressure for a decision meant that less active members were prepared to accept the decisions of others (De Grada et al., 1999). Time constrained decision-making has an impact on relationships within groups, as trade-offs have to be made between providing time necessary to develop relationships and the urgent need to come to a decision, which could mean riding roughshod over others' opinions. The risk for construction projects is that technical points may not be given sufficient weight in the rush to decide, which could be to the detriment of the project, particularly in cases where some professionals, although technically and professionally competent, are not good communicators in the rough and tumble of debate on urgent issues. Such situations can be intensified by the temporary nature of construction project teams in which the lack of familiarity of members may allow the more confident members and those in positions of power to dominate discussion and decisions.

Group polarization (also referred to as *Groupshift)*, which refers to the intensification of the position held by a majority of group members as a result of discussion (Lamm, 1988), usually towards a more risky position (risky shift), although it may also occur towards more caution (caution shift). Group polarization is more likely to occur when members find that many other members share their opinion, thus reinforcing their convictions and hence moving the group to a more extreme view. Also, articulate expression of views, particularly of the same view by influential members, is likely to persuade others who then reinforce the direction being taken, again leading to a more extreme position being adopted. A perceived lack of responsibility for decisions on the part of group members can also lead to extreme positions being adopted. As one of a group making a decision, individual members may not feel personal responsibility for a decision which they would find extreme if they were making it as an individual, hence with individual responsibility.

Groupthink originated from Janis's (1972) work on United States' foreign policy disasters, particularly through a detailed analysis of the Bay of Pigs fiasco in 1961, and was summarized by Fincham and Rhodes (2005). Groupthink was also identified as being a cause of some other catastrophic decisions in US foreign policies, such as being unprepared for the Japanese attack on Pearl Harbour and the escalation of the Vietnam War between 1964 and 1967. Groupthink is, in Janis's words, 'a mode of thinking that people engage in when they are deeply involved in a cohesive in-group, when the members' striving for unanimity override their motivation to realistically appraise alternative courses of action' (Janis, 1982). Many seriously bad decisions have been put down to groupthink, including the decision by the operators responsible for the catastrophic accident at Chernobyl to experiment with one of the reactors (Reason, 1987). When groupthink is present, a group is aiming for unanimity, not the best decision. Conditions which create groupthink include:

- Cohesiveness
- Members hold a positive image of their group
- Group sees a threat to its positive image
- Leaders have a preferred and promoted solution
- Insulation of group from experts' opinions.

Groupthink has eight characteristics (Janis, 1972):

An illusion of invulnerability. Excessive optimism that past successes will continue, leading to extreme risk taking.

Collective rationalization. Rationalizes away data and information which contradicts assumptions and beliefs on which a desired decision is based.

Illusion of morality. Members are inclined to ignore moral and ethical issues which contradict their own.

Stereotyping of out-groups. Over-simplifies external threats to the team's decisions. 'Enemies' are dismissed as evil or moronic.

Pressure on dissenters. Members are pressured to fall into line if they express disagreement with decisions of the group and are required to be more loyal.

Self-censorship. Team members suppress their doubts to maintain harmony.

Illusion of unanimity. Self-censorship leads to a sense of unanimity as doubts are suppressed and silence of team members is perceived as evidence of consensus.

'Mindguards'. Some team members are self-appointed guardians of the team's perceived beliefs and values and screen out information that might endanger the team's complacency about the effectiveness and morality of their decisions.

Positive action is needed to overcome groupthink and avoid inappropriate decisions. Moorhead and Griffin (2001) summarize the actions that can be taken to facilitate the critical evaluation of alternatives and discourage the single-minded pursuit of unanimity:

(a) *Leader prescriptions*
- Assign everyone the role of critical evaluator
- Be impartial; do not state preferences
- Assign the devil's advocate role to at least one group member
- Use outside experts to challenge the group
- Be open to dissenting points of view.

(b) *Organizational prescriptions*
- Set up several independent groups to study the same issue
- Train mangers and group leaders in groupthink prevention techniques.

(c) *Individual prescriptions*
- Be a critical thinker
- Discuss group deliberations with a trusted outsider; report back to the group.

(d) *Process prescriptions*
- Periodically break the group into sub-groups to discuss the issues
- Take time to study external factors
- Hold second-chance meetings to rethink issues before making a commitment.

However, Moorhead and Griffin (2001) state that, although the arguments for the existence of groupthink are convincing, the hypothesis has not been subject to rigorous empirical examination and that research supports part of the model but leaves some questions unanswered. Nevertheless, the ideas underpinning groupthink are evident, if not always in their full-blown form, in decisions made by groups in business and other organizations, including those in construction.

In the construction milieu, the most fertile ground for groupthink to occur is the conception stage of a project when the will to establish the project is greatest, but opportunity also occurs for groupthink to produce sub-optimal decisions at all stages of a project. Major decisions of outline design and detail design can be subject to groupthink forces, particularly within successful design offices in which architectural teams have worked together on a number of successful projects and also in conflict situations between clients and contractors when both sides are likely to fall into the groupthink trap as they convince themselves that they are pursuing a just cause. A major factor which may inhibit groupthink occurring in construction is that many major decisions are made by multidisciplinary teams drawn from a range of different professional firms and contractors. Their allegiance to their profession and firm gives them an independence which may resist many of the forces that induce groupthink; rather the problem becomes 'how does the project manager (or other leader) obtain sufficient unanimity to move the project forward?'

Ego-driven decisions can be evident in construction and may involve elements of groupthink, particularly for very high profile projects in the public sector which are driven by the egos of prominent politicians and other public figures. Such projects are often labelled 'white elephant' projects, defined by Greenacre (2008) as 'a rare and/or expensive possession whose cost of upkeep exceeds its usefulness'. Examples over recent years could include a number of Olympic projects, Shanghai's World Finance Centre (Moore, 2008) and Birmingham's 'planned multi-purpose domed facility' (Abbott, 2009) but counter arguments regarding their benefits have been made by Greenacre (2008) who believes that 'More often than not, they are iconic structures that follow in the vein of the regeneration theme and make a positive aesthetic impact on their surrounding area'. Using the Millennium Dome as an example, he says that it 'regenerated an old gasworks area of London bare to the elements, and in turn created about 2,000 construction jobs plus thousands more to maintain it. It is instantly noticeable outside and features very prominently on the riverside'. Examples of white elephants have been analysed by economists, including Robinson and Torvik (2005) who saw them as misallocations of investment. They argue that 'they are a particular type of inefficient redistribution, which are politically attractive when politicians find it difficult to make credible promises to supporters' but they do not make reference to groupthink in this process. Ego-driven decisions also occur in the private sector and may include groupthink, but usually without such public exposure. For example, they could result in a decision to construct a new building (often to satisfy the egos of the businesses leaders) when other more viable alternative solutions are available such as leasing or refurbishing, or possibly a solution that does not require construction work, for instance a takeover of another business.

Advocacy versus enquiry are different modes of operation employed by groups to arrive at a decision, as identified by Huczynski and Buchanan (2007) drawing on Garvin and Roberto (2001), which relate to groupthink. An advocacy approach is about persuading a group to come over to the advocates' view using spokespeople who aim to persuade, defend their position, downplay weaknesses and discourage or dismiss minority views and is seen as a contest with winners and losers. This approach is akin to inducing groupthink. The enquiry approach requires critical thinkers to collaborate in problem solving using balanced arguments and constructive criticism to arrive at alternatives, taking into account minority views to arrive at a collectively owned decision; the opposite of groupthink.

Participation, in the mainstream OB literature, is about the participation of employees in group decision-making within the context of managing the organization and is associated with the agenda of empowerment. Whilst this movement is relevant to managing construction and professional firms, it is also subject to the finding pointed out by Moorhead and Griffin (2001) that 'Numerous research studies have shown that whereas employees who seek responsibility and challenge on the job may find participation in the decision-making process both motivating and enriching, other employees may regard such participation as a waste of time and a management imposition', which draws on very early work by Coch and French (1948) and Morse and Reimer (1956). It is believed that employees' involvement in decision-making increases their ego, motivation and satisfaction and for those that wish to be involved this is likely to be the case. The drive for more involvement may not be as high in construction, as the nature of the business is such that professionals from consultancy and contracting firms are involved in group decision-making in connection with their projects on a frequent basis and are axiomatically empowered.

Escalation of commitment is generally seen to be a decision-making phenomenon relating to groups but originally related to individuals (Staw, 1981). It is an increasing commitment to continuing with a course of action which has been shown to be unlikely to succeed due to:

- The identification of insurmountable problems, or
- Evidence of negative outcomes in the past, or
- Being a repetition of known bad decisions in the past.

Huczynski and Buchanan (2007) point out that escalation of commitment occurs not only in business situations but also in other areas such as interpersonal relations, gambling, economic investment and policy-making. They also point out that the nature of the task which is the subject of a decision to continue has a major influence on the severity of the escalation of commitment. Tasks that have high upfront investment in financial and emotional terms can soon lead to situations where to abort would incur large losses so the tendency is to increase commitment in the hope that things will improve. In less demanding decision situations the seriousness of escalating commitment may not be so severe but, nevertheless, if they had to abandon their task, the effect on team members' reputations could affect their careers so the tendency is also to escalate commitment. Huczynski and Buchanan also point out that escalation is linked to groupthink: 'Risky shift findings shows us that groups make riskier decisions than individuals, and the decision to escalate by groups in the light of past failings can be viewed as risk-seeking. From a groupthink perspective, since a majority view is sufficient to induce dissenters to conform to a decision to escalate, reliance on a group rather than on an individual to resolve an escalation dilemma is likely to increase the frequency with which an escalation occurs.' The mainstream OB literature abounds with examples of escalation, including EXPO '86 in Vancouver, Tokyo's subway loop, Denver Airport's baggage handling system and Concorde. Groupthink is likely to be the seedbed for escalation of commitment.

Many of the publicly-identified projects which demonstrate escalation of commitment are extremely large and frequently incorporate experimental technology. The launch and the process of developing such high-profile projects is often accompanied by the commitment of high-level public figures, frequently politicians who 'nail their flag to the mast' by promising success and large benefits. The development of such projects is strewn with potential problems as the process enters the realms of the unknown. Invariably estimates and timescales are shaved in order to have the project accepted and costs and timescales begin to increase. Overconfidence at the outset, driven by the egos of the main participants who will see failure of the project affecting their self-esteem and their belief that they can control the unknown, leads to underestimation of the risks. Pride often presents an insurmountable barrier to admitting the scale of the mistake of embarking on the project and reinforces the commitment of the group responsible for the project in the belief that it will work out in the end (or they will have moved on to some other venture leaving behind responsibility to someone else long before the final outcome).

The mega projects referred to above were in the public domain and treated as scandals by the press. Many similar instances also occur relatively regularly on a smaller scale in business and public organizations which, whilst not on the same scale, threaten the viability of companies, public organizations and the reputations of the people associated with them. A general pattern is that expenditure is committed to research and development of a new product which does not go as well as expected. The research group make a strong plea for more funding without which they say there is little hope of a return on the initial investment. Does the organization agree to the additional funding or does it cancel the product's development and accept the loss? If the funding is granted, the research group may continue their work, become more committed to the product and come back for further funding with some hope of final success but without being able to show that it is guaranteed. Does the company commit more funds? And so the dilemma continues as the team becomes more committed and the downside of cancelling the product becomes greater. Escalation of commitment occurs not only in the research group but also in the manager or group overseeing the research group but in the latter case perhaps reluctantly. Self-justification is a large motivator in this process as is the reluctance to abort the development of a product which has characteristics similar to the development of previously successful products or in situations in which a product's development is continued in order to try to recover resources already invested. The tendency to subconsciously distort information also leads to the illusion that the product can be successfully developed.

Construction projects are strong contenders for attracting escalation of commitment, particularly within client organizations, as they are often championed by units within businesses and public organizations intent on strengthening their position and power within their organization. Once they have obtained permission to proceed, they will be reluctant to admit they were wrong should it seem that what they want is not achievable in the light of increasing costs and timeframes. However, if appropriate feasibility techniques are used by competent construction professionals who are listened to by the client organization, then damage limitation is possible. Potential cost overruns from the initial budget should be exposed during design development so that the project can be aborted with only design costs being incurred. Although this would be wasted expenditure, it would be small relative to the costs of construction. However, it is not beyond belief that a client organization may continue to proceed in spite of professional advice, leading to an outcome which has a severe effect on an organization's viability.

The scenarios outlined above are characteristic of escalation of commitment and are also often underpinned by the egos of investors, sponsors, project champions and project managers. They can become so immersed in a project that their reputations become synonymous with the project to the extent that either cancellation or a scandalously expensive and/or delayed outcome is unacceptable but one of them cannot be avoided. The tendency is likely to be to 'push on'. At a level lower than the ego of senior people are the managers closely involved with projects who do not wish to expose their mistakes and appear incompetent. Whilst proposing cancellation of a project may be the mature and proper recommendation, the fear of 'loss of face' may persuade a manager to soldier on and act defensively if asked to justify the project before an audience, whether hostile or friendly. Organizational inertia has also been cited as a reason for not aborting a project; that is, the organization moves too slowly for timely action to be taken and, by the time the organization realizes, the situation is out of control and it is too late to redeem it. Also the degree to which a project is central to an organization's activities can make it more difficult to abort, particularly if other business activities have been sacrificed to support the project. The impact of technological, economic, legal and social environmental forces on an organization requires it to react and adjust to them. Reaction may mean that a project should be cancelled as it has become incompatible with some aspect of environmental change but an organization's commitment to the project may prevent it from recognizing and accepting environmental change so the project proceeds without a chance of ultimate success.

A project that demonstrated many of the characteristics of escalation of commitment but which, in many ways, contradicted an expected unsuccessful outcome was the development of the Hong Kong University of Science and Technology (Walker, 1994). It was conceived in the years approaching the handover of Hong Kong to China in order to expand the provision of higher education and give confidence to the people of Hong Kong. The first Planning Committee met in 1986 and a first estimate of about US$200 million (excluding inflation) was established. As the project developed, the estimated cost escalated until on completion the cost (without inflation) was approximately twice the original estimate. When the project was seen to be in trouble The Royal Hong Kong Jockey Club agreed to project manage the project and provide the shortfall in expenditure but with the Government playing an associated role. The Jockey Club, a prestigious organization in Hong Kong, had project managed many successful projects previously. The commitment to the project was enormous as it was seen as a symbol of Hong Kong's future and the Jockey Club was not prepared to lose face by not completing it. The escalation in cost was due to an initial lack of understanding of the characteristics of a 'world class' university of science and technology but as the reality of what was required emerged the commitment to the project remained exceedingly strong. Common sense said the Government should abandon the project but the Jockey Club's commitment was so great that it continued with the project. Whilst apparently irrational, the final outcome was spectacularly successful. As a result of the drive of the Jockey Club, Hong Kong was provided with a world class state-of-the-art university which had been developed from scratch to occupation in an extraordinarily short period of five years. Not only that, but the project was shown to represent value for money for the type and quality of university which was finally produced even though it cost about double what had been originally expected. If commitment had not escalated Hong Kong would not now have a world class university of science and technology. This was an extremely rare outcome for such circumstances; escalation commitment can frequently lead to financial disaster but fortunately in this case escalation was created by an organization that could afford the cost.

Whilst individuals can escalate commitment, groups are particularly vulnerable to this phenomenon and managers of groups need to take steps to prevent them acting in this way. Staw and Ross (2005) suggest the following steps (modified here a little to reflect construction projects) that can be taken by individual managers and organizations (which can also be taken by individual members of groups and particularly group leaders; and in construction by project managers) to guard against or deal with escalation of commitment:

Recognize over-commitment. The line between an optimistic, can-do-attitude and over-commitment is very thin and often difficult to distinguish.

See escalation for what it is. Do I know what constitutes failure? Have I staked my career on the project's success? Do I hear and consider seriously other's concerns about the project? Do I consider how various events affect the project before I think about their effect on the organization as a whole? Do I feel that if the project ends, there will be no tomorrow? If the answer is yes to one or more of these questions, a person is probably over-committed to the project.

Back off. Stand back from the project and try to look at the project from an outsider's perspective (not easy). Ask oneself, 'If I took over this project today would I support it or terminate it?'

Organization replaces those associated with the original project. Changing project managers can be disruptive and costly, whether project team manager or the client team manager. Not recommended by Staw and Ross.

Separate decision makers. Shift the issue from one group of decision-makers to another group at a time convenient to the process. Not really relevant to construction.

Reduce the risk of failure. This refers to reducing the fear of failure from managers and teams so that they will not continue to be committed to failing projects. Organizations can provide rationalizations for changing courses of action and excuses for their managers and group members so they do not appear stigmatized if their project is unsuccessful. This makes it easier for them to 'pull the plug on a project'.

Improve the information system. Improve the honesty of the information on the project as it is passed up the organization by not filtering it in favour of continuing the project when this is unwise. Reward managers for the rigour of this process. Such a reward system is different from the usual one which rewards success on projects and punishes failure. It is a system that should reduce many of the forces for escalation.

Further suggestions are (Royer, 2005):

Assemble project teams not entirely composed of like-minded people. Frequently project teams are formed with people who share enthusiasm for the project. Although this may not normally be the case for construction projects, members of construction teams may have worked together on successful projects in the past which can produce a lack of challenge to decisions. It is recommended that some sceptics should be included in teams from the outset who should be directly involved in decision-making. It is further recommended that over the course of the project some decision-makers should be replaced with others who will look at the project with fresh eyes.

Establish an early warning system. Ensure that control procedures and criteria for evaluating project viability are clearly defined, rigorous and actually met.

Use an exit champion. An exit champion needs the temperament and credibility to question the prevailing belief, demand hard data on the viability of the project and, if necessary, forcefully make the case that it should be aborted. Exit champions need to be directly involved in the project, fearless and prepared to put their reputations on the line as they are likely to lose friends in the company.

Escalation of commitment can occur in construction project teams when, for example, early estimates are increasing and promise to continue to increase (probably due to technical reasons) as more details of the project emerge, requiring a review of the project and someone from the project team (an exit champion) is needed to call for the review. The major source of escalation of commitment is likely to arise within the client organization (probably due to the scope of the project escalating) and, having been allowed to go unchecked, requires an exit champion within the client organization to call for a review. Attempting to deal with such a situation will have repercussions in the client organization if feedback from the construction project team suggests that the project is not viable due to cost and time issues. Correspondingly, if escalation has led to concerns and uncertainty within the client organization there is likely to be significant difficulties in obtaining a stable brief from the client with frequent changes in direction and much abortive work by the design team.

10.6 Group decision-making techniques

Reference has been made earlier to quantitative decision-making techniques. These are not the subject of this book but some techniques are behaviourally orientated.

Brainstorming is perhaps the most well-known. The term has found its way into common usage but is usually used inappropriately and its benefits generally overstated. It relies on the assumption that group decision-making is superior to that of individuals and aims to enable groups to produce new creative ideas. Basically this is achieved by forming brainstorming groups who introduce new ideas on the problem/issues placed before it. The idea is that the group members are uninhibited in their contributions, and to achieve this four procedural rules are used:

- Ideas should be put forward no matter how outlandish
- Ideas put forward should not be criticized
- Positive comments should be made about the ideas put forward
- Comments and suggestions should build on and develop the ideas put forward through free association.

Since the earlier research on brainstorming (Taylor et al., 1958), the technique has been shown not to be superior to individuals working independently. In fact, brainstorming has been shown to inhibit creative thinking. Huczynski and Buchanan (2007) believe that it is based on the following questionable assumptions:

- That amongst the torrent of ideas (actual associations) there are bound to be some good ones
- That solving problems is a matter of giving one's thoughts free reign
- Assumes the more ideas the higher their quality
- Inhibits creative thinking by not allowing more than one person to speak at a time.

The last of these is counter-intuitive and known as 'production blocking'. Interestingly, it has been shown that brainstorming by electronic means is superior or equal to face-to-face brainstorming because members do not have to wait to speak and can refer to others' contributions at any time. This therefore bypasses the production blocking problem (also techniques are available which can provide anonymity).

Nominal group technique aims to improve decision-making but not only in the creative phase. In the first part of this technique no discussion is allowed; rather, each member of the group writes a list of ideas. Individual members then present their ideas one at a time. The ideas are displayed for colleagues to refer to and they are asked to contribute to the ideas by association. After all the ideas have been presented, members continue to develop them. This part does not require face-to-face meetings but can be done by video conferencing, computer or other electronic means, but nevertheless benefits from face-to-face meetings. On conclusion of discussions, in a secret ballot to avoid intimidation, members rank the ideas that have emerged. Further discussion may take place to confirm agreement and the cycle can continue for as long as necessary. The benefits claimed for nominal group technique are that it:

- Allows groups to meet formally but does not restrict individual thinking
- Limits, to an extent, the effects of power and status
- Is useful in a range of decision situations
- Has low potential for conflict within the group.

Groups using the nominal group technique are generally not as cohesive as traditional interacting groups as the structure limits interaction and production blocking can also occur to some extent. Nominal group technique has been shown to provide superior performance to brainstorming and to traditional interacting groups (Faure, 2004; Hegedus and Rasmussen, 1986). Robbins and Judge (2008) highlight the *computer-assisted group* or *electronic meeting* which applies the nominal group technique through computer technology in which individual contributions are fed into members' computers and displayed anonymously onto a projection screen. The suggested benefits are anonymity, honesty and speed and that it allows contributors to be 'brutally honest without penalty'. It is also supposed to be fast as it reduces digression and many members can contribute at the same time. However, it appears that electronic meetings do not meet most of their expectations. It has been found that they have led to decreased group effectiveness, are slower and give less user satisfaction than face-to-face groups (Baltes et al., 2002) but, with applications of computing continuing to develop apace, they are expected to increase in popularity.

11 Leadership, Learning and Change

11.1 Introduction

It is generally accepted that leadership is necessary for management to be effective and this tends to create a general perception that management and leadership are somehow separate. Management is often seen as a mechanical process using techniques, responding to directives from elsewhere and controlling those being managed. In the public psyche, leadership is perceived as charismatic, inspirational and forward thinking. However, Cleland and King's (1972) operational definition of management (which is one that identifies a number of observable criteria) remains useful and identifies the observable criteria as 'organized acivity, objectives, relationship among resources, working through others and decisions'. The public perception of management as a series of mechanistic processes is better described as administration. Although there is no commonly accepted definition of leadership, most definitions refer to it being about influencing people. Hence Cleland and King see it as intrinsic to management by their inclusion of the phrase 'working through others' in their definition. Townsend (2007) believes that 'Most people in big companies today are administered not led. They are treated as personnel, not people' and he quotes *The Peter Principle* (Peter and Hull, 1969) in support: 'Most hierarchies are nowadays so cumbered with rules and traditions and so bound in by public laws, that even high employees do not have to lead anyone anywhere, in the sense of pointing out the direction and setting the pace. They simply follow precedents, obey regulations and move at the head of the crowd. Such employees lead only in the sense that the carved wooden figurehead leads the ship.' Whilst these references originate from some years ago, their basic premise is still relevant today and, although much has been written and many major research studies have been undertaken that have contributed to our understanding of leadership, there has not been a convincing and accepted definition and theory of leadership, nor is there likely to be due to the variety and complexity of organizations and their environments leading to rapidly evolving organizational structures. The changing world created these changes in organizations and hence changes in our concepts of effective leadership.

Organizational Behaviour in Construction, First Edition. Anthony Walker.

Nevertheless, significant perceptions of leadership have emerged to reflect the needs of modern organizations. Leaders are now expected to develop and transmit visions of the future of organizations, initiate and drive new ventures and, particularly, identify, initiate and create change in organizations which have to adapt to turbulent environments. Essential to these objectives are leaders' abilities to get others to sign up to their visions, inspire them to make them happen and live with and continue to develop change. Complementary ideas include increasing participation and leaders as nurturers of leadership in others. However, whilst these views of leadership, along with many others, are portrayed in the literature, academic, professional and 'management-speak', there is a sense that each school of thought is often overstated as the best and only way to lead, whereas the needs of organizations may not necesssarily fit the latest, or indeed any, existing model.

Leadership is a complex and controversial topic with many paradoxes. Traditionally it is seen as a function of hierarchical positions holding status and power but, with the emergence of broad organizational structures and team working, increasingly it is seen as attaching itself to a wider range of individuals lower down an organization's hierarchy and to have led to traditional leadership structures being challenged. Nevertheless, the concept of authority/power as described and developed in Chapter 7 is fundamental to leadership at all levels. Authority and all classes of power contribute to leaders' influence and form a backcloth to this chapter, and authority continues to carry great influence because of our acceptance of authority from a very young age, as does referent power, often referred to as charisma.

The common cry when problems arise of 'we need more leadership' finds difficulty in identifying a target with the emergence of new style organizational structures. Such developments have led to new perceptions of leadership and contributed further to the difficulty of devising a commonly accepted definition. Examples of definitions include one from many years ago by Stogdill (1950) which is still relevant: 'Leadership is an influencing process aimed at goal achievement.' Huczynski and Buchanan (2007) explain it thus: 'Stogdill's definition has three key components. First, it defines leadership as an interpersonal process in which one individual seeks to shape and direct the behaviour of others. Second, it sets leadership in a social context, in which the members of the group to be influenced are subordinates or followers. Third, it establishes a criterion for effective leadership in goal achievement, which is one practical objective of leadership theory and research.' They continue that most definitions share these processual, contextual and evaluative components. McShane and Von Glinow (2003) point out that scholars do not agree sufficiently on the definition of leadership and they 'cautiously' define leadership as 'the process of influencing people and providing an environment for them to achieve team or organizational objectives'. So, although writers seem to want to define leadership in their own terms there has been little, if any, meaningful change from early definitions.

At this early stage in the chapter it is necessary to point out the current state of research into leadership. Fincham and Rhodes (2005) believe that 'the attempt to build a broad theory of leadership, which captures both leadership style and the immediate organizational context it operates in, has largely failed'. Grint (2005) agrees, saying 'despite over half a century of research into leadership, we appear to be no nearer a consensus as to its basic meaning, let alone whether it can be taught or its moral effects measured and predicted.' He cites the staggering statistic that in 2003 there were 14,139 items relating to leadership for sale on Amazon.co.uk and that it would take 39 years, reading one item a day, just to read the material. Fincham and Rhodes (2005) find that the lack of the development of a leadership theory by the academic community has resulted in management consultants now dominating the leadership literature, leading to a shift from psychogical and sociological academic approaches to prescriptive managerial perspectives which are little different from self-improvement texts by self-styled 'gurus'. The major problem in achieving a leadership theory is that the field of interest is a moving target. Not only is the context of leadership changing rapidly as a result of unpredictable economic conditions, rapidly advancing technology and unstable political environments but the vast range of the scale of organizations, from the NHS at one end to about 80% of businesses employing five people or less at the other, generates what seems like an infinite range of complexity. Nevertheless, the leadership theories and approaches which have been developed, whilst not representing the holy grail of leadership, do have significant value in aiding understanding of how leaders may improve their performance and how followers may better understand their leaders' approaches.

More or less chronologically, the progression of leadership theories has been: trait approaches, behavioural aproaches, contingency approaches and inspirational approaches.

11.2 Traits approach

There exists a populist commonsense idea of what constitutes a leader, referred to as an 'implicit theory of leadership'. This view believes that leaders are 'born not made'. That is, some people are born with a set of characteristics (or traits) that mark them out as leaders, also referred to as the 'great man theory'. Trait approaches go back to the beginning of leadership research when leaders were examined to attempt to identify their distinguishing characteristics. 'Great men', such as Napoleon, Churchill, Kennedy and Mandela, were characterized and terms such as charisma, courage, integrity, tenacity, enthusiasm, judgement, decisiveness and many others were used to describe them. But trying to identify common traits led to dead-ends as the list became so long that it was of no practical value. Robbins and Judge (2008) refer to a review published in the late 1960s of 20 different studies which identified nearly 80 leadership traits of which only five were common to four or more of the investigations (Geier, 1967), so not much sense could be made of it all.

They also point out that the Big Five personality framework (see Chapter 2) has subsequently been used to rationalize the wide range of traits originally identified using the five components: extroversion, agreeableness, conscientiousness, emotional stability, and openness to experience. Extroversion has been shown to be related to leadership emergence rather than effectiveness; conscientiousness and openness to experience were shown to be strongly related to leadership; and agreeableness and emotional stability not so strongly related. Fincham and Rhodes (2005) point to conscientiousness as predicting effectiveness. Conscientiousness was defined by reliability, hard work, dependability, achievement orientation and concern for quality and standards producing a strong work ethic arising from childhood, together with very high self-expectations and lack of tolerance of laxness in others. Associated with these characteristics is a strong sense of right and wrong. However, they also point out that leadership effectiveness is more complex than indicated here. Whilst these characteristics are seen to lead to effective leadership, some effective leaders are quite maverick and individualistic. However, there is evidence of such leaders being 'high risk', leading to stretching of legal responsibilities and their downfall. They describe the talents of extrovert leaders as enthusiasm and zest for their leadership role and seeing opportunities rather than threats. Extrovert leaders identify closely with employees and groups, work hard at developing relationships, are open and direct and so can be seen as relationship-oriented. However, whilst they can make a good first impression and quickly develop rapport with others, this can be superficial and their charm self-serving and self-centered.

Robbins and Judge (2008) believe that, overall, it does appear that the traits approach does have something to offer. Nevertheless, it does seem that amalgamating the large number of traits into five categories removes the finesse for identifying leaders, as many people are likely to comply with an appropriate Big Five category but may not make effective leaders due to the breadth of categories seen as appropriate to leaders. It is also argued by proponents of emotional intelligence (see Chapter 3) that even with other relevent traits people will not make leaders unless they have sufficient emotional intellegence. However, the connection between emotional intellegence and leadership effectiveness has not been accepted, so perhaps it is no more than another trait.

In spite of the lack of conviction in the value of traits as determinants of leadership effectiveness, the belief that there is some merit in the approach seems to be ingrained in the human psyche to the extent that it appears to be being resurrected in contemporary approaches which are reviewed later in the chapter.

11.3 Behavioural approach

The behavioural approach emerged in the 1940s as the traits approach was shown to be unproductive. Focus shifted to how specific leaders behaved. It was felt that if the behavioural approach to leadership was shown to be valid it would be possible to identify the behaviours associated with effective leadership and hence to train people as leaders rather than attempting to find people with the 'right' traits. The underlying assumption was that the behaviours of effective leaders was the same in all situations and differed from the behaviours of ineffective leaders. The work was led by research studies at both the Ohio State and Michigan Universities which took place at about the same time.

The Michigan leadership studies aimed to determine the pattern of leadership behaviours which result in effective group performance. Two forms of leader behaviour emerged: 'job-centered leader behaviour' and 'employee-centered leader behaviour'. The former pays attention primarily to the task through organization, work scheduling and processes and hence performance; that is, on getting the job done. The latter focuses on relationships and employee needs and building effective groups through concern for human aspects of group performance. The Ohio studies used exhaustive dimensions to identify independent dimensions of leader behaviour which they narrowed down to two categories called 'initiating structure behaviour' and 'consideration behaviour'. The former is about organizing work by assigning group tasks, maintaining standards, meeting deadlines and expecting instructions to be followed, whilst the latter is about such things as concern for relationships, trust, and respect for the ideas of others. Robbins and Judge (2008) state that the two 'different' dimensions identified by each study are closely related and that most leadership researchers use the terms synonymously. The task and people styles were not at this time seen to be on a continuum but as independent; they therefore did not represent the extremes of a continuum so a leader was seen to be able to stress one or both, so could be high or low on both styles. It is believed that people-oriented leadership results in higher job satisfaction for employees and lower grievances and turnover but that performance tends to be lower than for employees with task-oriented leaders. However, whilst task-oriented leadership results in comparatively productive workers, it also leads to lower job satisfaction, more grievances and higher turnover but greater team unity. The dichotomy of styles identified by the studies reflects the idea of autocratic and democratic leadership which, in turn, reflect the 'scientific' and 'human relations' management movements respectively. These ideas were embedded in the Grid Organization Development (also known as Leadership Grid) developed by Blake and Moulton (1969) as a consulting tool for applying the Ohio findings.

The studies attracted considerable attention at the time as they seemed to be a productive way forward from the narrow trait theories. Indeed they gave insights into leadership behaviour which are still valuable today. However, subsequently their findings were not generally supported by research and so did not achieve their objective of defining universal behaviour patterns for leaders. What this research revealed was the huge complexity of behaviour patterns of leaders in the vast range of organizational settings encountered, which in turn drew attention to the effect on their performance of the situations in which leaders had to operate, leading to the development of contingency approaches to leadership.

11.4 Democratic vs. autocratic leadership styles

The last section made reference to democratic and autocratic leadership styles and, as these terms crop up frequently in the discourses on leadership which follow, it may be worth reflecting on them. They represent the extreme end of a spectrum of leadership styles and arise predominantly from society more generally, particularly in the political milieu. Most developed countries have expectations of open, collaborative, consultative and free systems of government. In contrast, authoritative regimes are repressive and certainly not free. Parallels are perceived in the way organizations are 'ruled'. Fincham and Rhodes (2005) consider this and state that 'democratic-autocratic leadership is perhaps the most complex construct in the leadership literature. It refers to the way in which decisions are taken; whose needs in the organization are met; and what characterised the relations between leader and follower, for instance, how much coercion is present'. Leadership models are underpinned by notions of relationship-oriented (democratic) and task-oriented (autocratic) styles of leadership. Likert (1961) saw autocratic systems as emphasizing managerial authority and control with no consultation with subordinates. Democratic systems were seen as being based on trust, consultation and participation; that is by getting the best out of subordinates. Using a range of variables Likert was able to position an organization along the autocratic–democratic dimension. In the developed world today totally autocratic leadership would be hard to find as the nature of modern organizations increasingly requires and receives democratic leadership as the only way to cope with increasingly volatile environments. Whilst the terms autocratic and democratic have been substituted by others in the leadership lexicon, the basic scale from the extremes of autocratic to democratic is still a useful concept.

11.5 Contingency approach

In common with the direction taken by organization theory – moving towards a contingency theory rather than believing there was only one way to organize – leadership theory reconsidered the idea that there was one best way to lead. It became apparent that having only one leadership style for all circumstances was inappropriate. Tannenbaum and Schmidt (1958) made the early proposal that management style was on a continuum from autocratic (task-centred) to democratic (relationship-centred). They illustrated points in this continuum and the styles to be adopted as shown in Figure 11.1 which shows that the leadership style to be adopted depends on the amount of freedom allowed to subordinates. Implicit in this idea is that a leader needs sensitivity to the situation regarding the feelings of the group being lead and needs to be sufficiently flexible to recognize the variety of leadership styles available. The next step was to consider the context within which the leader was operating. This requires

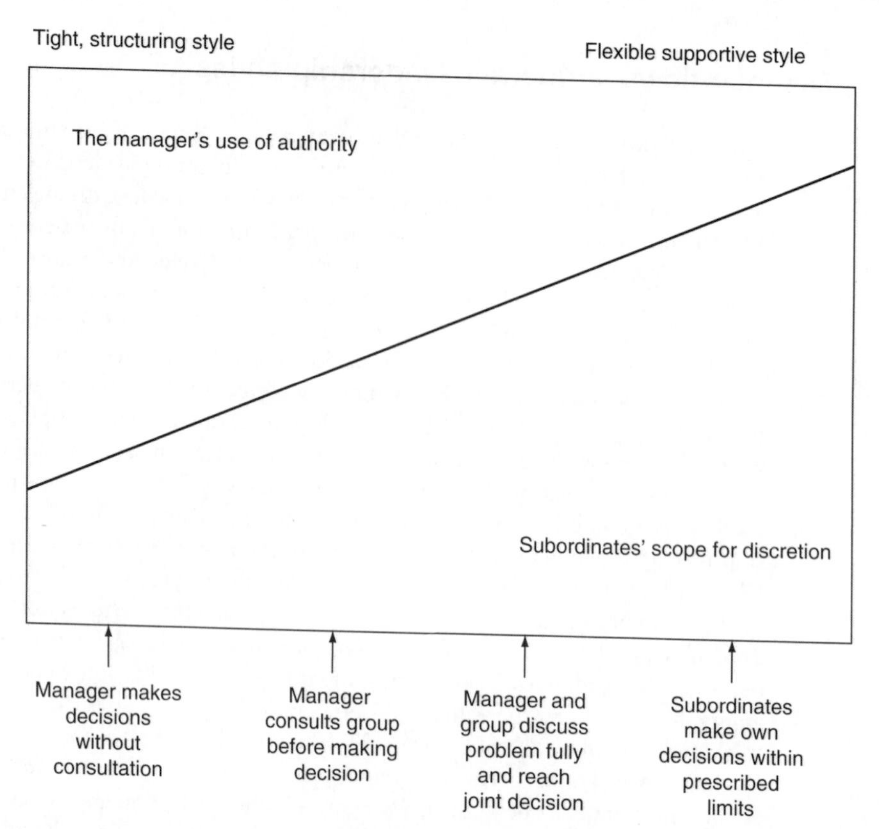

Figure 11.1 Some styles of leadership (adapted from Tannenbaum and Schmidt (1973) by Fryer (1985). By permission of Wiley).

taking into account the characteristics and abilities of the people being led, the nature of the task being undertaken, and the environment within which the task is being carried out. The idea is that with this knowledge it should be possible to tailor the leadership style to suit these requirements. This is known as the contingency theory of leadership and is analogous to the contingency theory of organization (Lawrence and Lorsch, 1967).

11.5.1 Fielder's model

Fielder's contingency model was the earliest contingency theory of leadership (Fielder, 1967). It states that leadership effectiveness is dependent on the matching of a leader's instinctive style to the situation being managed based on people-oriented and task-oriented styles. He used a new technique for measuring a leader's approach to managing people known as 'the leader's least preferred co-worker score'. Fielder's model has been found to be the most frequently used in researching leadership-related topics in construction, followed by Blake and Moulton's nine factor leadership grid (Toor and Ofori, 2008). Fielder's model suggests that the appropriate leadership style to be adopted depends on:

- Leader-follower relations: the extent to which followers trust and respect the leader and will accept guidance
- Degree of clarity of the task: that is, the extent to which the task is structured
- The leader's position power: the leader's formal, reward and coercive power over followers.

Huczynski and Buchanan (2007) describe the three sets of conditions which Fielder identifies as typical of those under which a leader may have to work which they condensed from the eight conditions which he originally identified:

Condition 1

- The task is highly structured
- The leader's position power is high
- Subordinates feel their relationships with the boss is good.

Task-oriented leaders get good results in these favourable situations. Relationship-orientated leaders waste time and energy working on relationships.

Condition 2

- The task is unstructured
- The leader's position power is low
- Subordinates feel that their relationships with the boss are moderately good.

Task-oriented leaders will ignore deteriorating relationships and as they are needed for ambiguous tasks, the outcome is likely to be poor. Relationship-oriented leaders get better results in these moderately favourable circumstances where the maintenance of good relations is important to getting the task done.

Condition 3

- The task is unstructured
- The leader's position power is low
- Subordinates feel that their relationships with the boss are poor.

Task-oriented leaders get better results in very unfavourable conditions.

Task-oriented leaders become impatient, try to structure the situation, ignore resistance from subordinates, reduce ambiguity and achieve good performance.

Relationship-oriented leaders will not exert the discipline necessary to complete the task.

Huczynski and Buchanan (2007) point out that the problems which exist with Fielder's model are that: task structure, power and relationships are difficult to assess; the 'least preferred co-worker' concept used to measure a leader's basic approach is confusing as it is not clear what it measures; the framework does not take into account the needs of subordinates; and the need for a leader to have technical competence is ignored. They identify two strengths: the model confirms the importance of contextual factors in determining leader behaviour and effectiveness and reinforces that there is no one ideal set of traits or best behaviour pattern; and it provides a systematic framework for developing the self-awareness of managers concerned about their leadership style.

The lack of concensus on the part of OB academics and writers on leadership is demonstrated by McShane and Von Glinow (2003) who comment that 'Fielder's contingency model may have become a historical footnote...' whilst Fincham and Rhodes (2005) state: 'Fielder's contingency theory is still the most widely cited theory of leadership', also 'And odd as it may seem almost forty years on, no one to date has done any better at predicting and explaining the way the situation a leader finds him/herself in interacts with his or her style to determine their likely effectiveness.'

11.5.2 Situational leadership

Situational leadership is a contingency approach to leadership developed by Hersey and Blanchard (1988). It draws heavily on the work of early leadership researchers such as Stogdill and Fielder. Hersey and Blanchard simplified earlier work, extended and popularized it. Their main contibution was the way in which they conceptualized the context or, in their terms, the situation. They believed the key issues to be: the competence; motivation; willingness and ability to take responsibility; education; and experience and achievement of subordinates, which they operationalized as subordinates' 'readiness' or 'maturity'. By plotting relationship behaviour on one axis and task behaviour on the other they defined four leadership styles: delegating, participating, selling and telling. The range spans from mechanistic instructions at one end in which subordinates have absolutely no discretion, to the other end which gives total discretion to subordinates. They then incorporated a continuum of the readiness/maturity of subordinates to perform a key task (also referred to as a scale of maturity of subordinates), all as shown on Figure 11.2, resulting in a readiness or maturity curve giving a basis for selecting an effective leadership style.

It is suggested that leadership behaviour should relate to high/low task and high/low relationship permutations as illustrated in Table 11.1.

An important difference between Fielder's and Hersey and Blanchard's views is that Fielder did not believe that leaders could readily change their leadership style whilst Hersey and Blanchard's situational leadership theory requires that leaders have the ability to change their style. The requirements of Hersey and Blanchard's model is that leaders diagnose the readiness (or maturity) level of

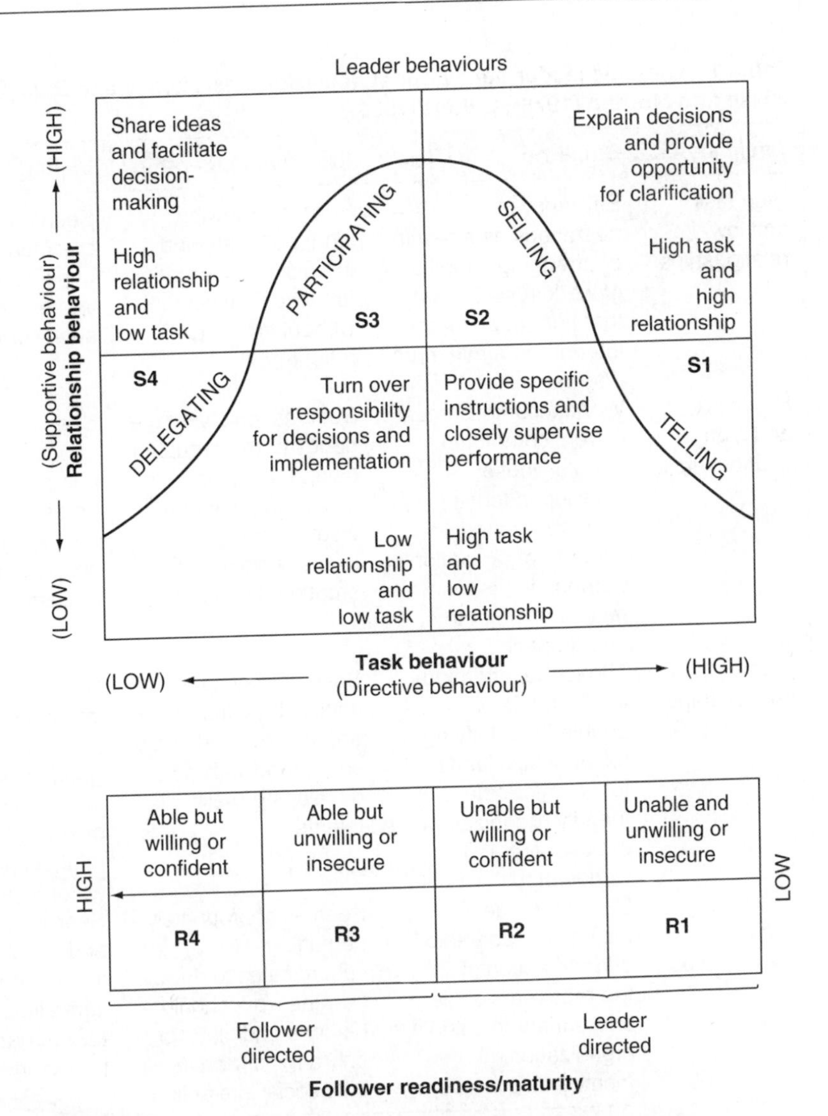

Figure 11.2 Hersey and Blanchard's situational leadership. (Fincham and Rhodes, 2005. By permission of Oxford University Press.)

their subordinates to undertake a specific task and then adopt the appropriate leadership style. The problem is that leaders may not be able to diagnose readiness (maturity) of their subordinates and also that they are unable to adjust their leadership style in the way required, as any of the behavioural aspects (which lead to irrationality) dealt with in other chapters may be manifest in the process. However, with the increasing need for leaders to be adaptable to

Table 11.1 Basic leader behaviour styles. (*Sources*: Hersey and Blanchard (1972); Blake and Mouton (1978); Gilbert (1983)).

Basic styles	Situation	Effective leadership	Ineffective leadership
High task and low relationships	Efficiency in operations as a result of arranging conditions of work in such a way that human elements interfere to a minimum degree	Seen as knowing what he wants and imposing his methods for accomplishing this without creating resentment	Seen as having no confidence in others, unpleasant and interested only in short-term output
High task and high relationships	Work accomplishment is from committed people whose interdependence through a 'common stake' in organization purpose leads to relationships of trust and respect	Seen as satisfying the needs of the group for setting goals and organizing work but also providing high levels of socio-emotional support	Seen as initiating more structure than is needed by the group and spends more time on socio-emotional support than necessary
High relationships and low task	Thoughtful attention to the needs of people for satisfying relationships leads to a comfortable friendly organization atmosphere and work tempo	Seen as having implicit trust in people and as being primarily concerned with developing their talents	Seen as primarily interested in harmony and being seen as 'a good person' and unwilling to risk disruption of a relationship to accomplish a task
Low task and low relationships	Exertion of the minimum leadership effort to accomplish the required work is appropriate to sustain organization membership	Seen as appropriately permitting subordinates to decide how the work should be done and playing only a minor part in their social interaction	Seen as uninvolved and passive, as a 'paper shuffler', who cares little about the task at hand or the people involved

rapidly changing business and political environments, the view that leaders cannot readily change their leadership styles creates difficulties for the effectiveness of leaders so Fielder proposed that situations should be changed to suit a leader's instinctive leadership style (Fielder and Chermers, 1984). Few leaders are likely to be able to find and remain in positions that suit their inflexible leadership styles and in which they perform well, or for that matter move readily to such positions, although it may be more possible in construction in which tasks, hence the situational factors, are more consistent. Increasingly today's organizations have flexible structures and the markets they seek to serve are flexible and adaptable such that similar characteristics are required of their leaders. In reponse to changing conditions it is now necessary for leaders to learn to change their style to suit changing circumstances. Leaders with such abilities are likely to be in increasing demand.

11.6 Path-goal theory

Path-goal theory of leadership (House and Mitchell, 1974) arises from the expectancy theory of motivation (see Chapter 5) which states that an individual's attitude and behaviour can be predicted from the extent to which someone believes that their performance will lead to various outcomes (expectancy) and the value of those outcomes (valences) to the individual. Hence, followers are motivated by their leaders' behaviours to the extent that their leaders' behaviours influence the followers' expectancies. Thus, leaders affect followers' performances by clarifying the behaviours (paths) that will lead to desired performances which lead to rewards (goals).

Four leadership behaviours are identified: the directive leader specifies what is expected of subordinates, schedules work, and gives specific guidance; the supportive leader shows concern for subordinates and is friendly towards them; the participative leader consults subordinates, asks for their suggestions and takes them into account before making a decision; the achievement-oriented leader sets challenging goals and expects subordinates to perform at the highest level and shows confidence in them that they will do so. Path-goal theory assumes that leaders are flexible and can adopt any of these behaviour patterns.

The theory proposes two types of contingency variable that moderate the relationship between a leader's style and effectiveness: subordinates' characteristics, and characteristics of the subordinates' work environment. The important personal characteristics are seen as locus of control and perceived ability and experience. Locus of control is the extent to which individuals believe that what happens to them is their responsibility (internal) or arises from external causes. Perceived ability and experience is how they view their capability regarding the task. Characteristics of the work environment are task structure, formal authority and the work group. Leader behaviour which helps subordinates to cope with environmental uncertainty caused by such factors will motivate subordinates. Robbins and Judge (2008) give the following examples of predictions arising from path-goal theory:

- Directive leadership leads to greater satisfaction when tasks are ambiguous or stressful than when they are highly structured and well laid out
- Supportive leadership results in high employee performance and satisfaction when employees are performing structured tasks
- Directive leadership is likely to be perceived as redundant among employees with high perceived ability or with considerable experience

- Employees with an internal locus of control will be more satisfied with a participative style
- Achievement-oriented leadership will increase employees' expectancies that effort will lead to high performance when tasks are ambiguously structured

In its inception, path-goal theory applied to relations between a leader and a subordinate but it has been extended to apply to relations between leaders and groups and organizations by adding new leadership styles of networking and value-based leadership (House and Mitchell, 1974). Networking recognizes that leaders play a significant political role and represent groups and organizations in this activity. Value-based leadership is about leaders having a vision; articulating and promoting their vision; and having the self-confidence to see it through.

McShane and Von Glinow (2003) identify path-goal theory as advocating servant leadership (McGill and Slocum, 1998) and state: 'Servant leaders do not view leadership as a position of power; rather they are coaches, stewards and facilitators. Leadership is an obligation to understand employee needs and to facilitate employee work performance. Servant leaders ask, "How can I help you?" rather than to expect employees to serve them.' Howell and Costley (2006) say that servant leaders think the best of others, seek moral excellence and model ethical behaviour for followers and use Abraham Lincoln and Mother Theresa as illustrations. This conception of leadership is not new; as the ancient Chinese Taoist philosopher Lao-tse said, 'To lead the people walk behind them' and also 'As for the best leaders, the people do not notice their existence. The next best, the people honour and praise. The next, the people fear; and the next the people hate ... When the best leader's work is done the people say "we did it ourselves".

In common with other approaches, path-goal theory has generated a range of views on its validity in developing leadership theory, but it is accepted that it reinforces the need for leaders to be insightful and flexible, with a whole range of leadership styles available to them in order to determine the situation in which they are acting and then select the appropriate leadership style, however difficult it may be to overcome their preferred style of leadership. The range of views on its validity encompasses, for example, McShane and Von Glinow (2003) who believe that it 'has withstood scientific critique better than others' and Robbins and Judge (2008) who say that 'a review of the evidence suggests mixed support' and that 'the jury is still out regarding the validity of path-goal theory. Because it is so complex to test, that may remain the case for some time to come.'

11.7 Vroom's decision tree approach

Vroom's decision tree approach (Vroom, 2000) again attempts to define leadership styles for given situations. It also assumes that leaders are capable of adopting different leadership styles. Its distinguishing feature is that it is concerned only with the degree to which subordinates should be encouraged to participate in decision-making as a function of the decision-making situation. Vroom proposes two decision trees: (1) a time driven model for making a decision as quickly as possible; and (2) a development-driven model for use when the leader wishes to develop the decision-making capabilities of others. Situational factors given are: decision significance; importance of commitment of followers; leader expertise; likelihood of commitment by followers; group support; group expertise and group competence. Using the appropriate decision tree, which has a funnelling effect and which incorporates five decision styles, the appropriate style is arrived at. The decision styles are:

Decide: The leader makes the decision.
Delegate: The group defines the nature of the problem and develops a solution.
Consults (group): The leader presents the problem to the group, receives their suggestions and then decides.
Consults (individuals): As last but with individuals.
Facilitates: The leader presents the problem to the group and facilitates discussion as the group makes the decision.

Vroom's approach has not been fully tested but reseach that has been done has been generally supportive.

11.8 Adopting different leadership styles

As stated earlier when reviewing Fielder's and Hersey and Blanchard's work, the former did not believe that leaders could change their instinctive leadership style while the latter believed that it was possible for leaders to do so. However, most contingency theories of leadership now incorporate the idea that leaders need to adopt different management styles, essentially along a spectrum from authoritarian to democratic (although different terms are usually used, for example, 'task-oriented' or 'employee, people or relationship-oriented') as a function of the situation in which leadership has to be exercised. Put simply, if starkly, at one end of the spectrum the authoritarian leader issues orders without consultation and at the other end the democratic leader allows the group to take the decision after having fully discussed the issues. Much of the research suggests that the situation to be managed should determine at which point on the spectrum the leader should be and that leaders should change their style according to the needs of the situation.

However, the practical application of these ideas does create problems. It presupposes that either the person who selects the leader has the ability to analyse situations sufficiently well to match the type of leader needed to each situation, or that, if a situation changes, can change the leader to suit the new situation. Alternatively, leaders need to be able to change their leadership style. OB writers appear to believe that leaders can learn from experience to adjust their leadership styles according to the context and that this ability is vital with the growth in demand for flexibility and adaptability in modern organizations. Nevertheless, it is extremely difficult for an individual to change management styles and for a group, assuming its constitution does not change, to accept different leadership styles from the same leader. By the time someone reaches a high leadership level, their character and personality are likely to already have been established, they have settled into a position on the autocratic/democratic spectrum, and they have a relatively short span of flexibility. Also leaders have difficulty in analysing situations so that they fully understand the context within which they are managing given the relatively vague nature of the situation elements within the different theories. In addition, contingency theories fail to consider other key dimensions of context, such as organizational culture; degree of change; levels of stress; working conditions; external economic factors; organizational design and technology. The huge number of variables inherent in individual behaviours and organizational settings lead to the extraordinary complexity of leadership.

Notwithstanding the deficiencies of the contingency theories of leadership, they are probably the most relevant to the leadership of projects in construction due to the wide range of different types of task to be carried out, from the truly creative to the rigidly technical. This may not be so pronounced in the leadership of the firms that contribute to projects to which other leadership approaches may more readily apply. Both applications are discussed later.

11.9 Leader-member exchange theory

Leader-member exchange theory (LMX) is quite different from other theories already discussed. Its focus is the differential relationships that exist between leaders and individual subordinates (Dansereau et al., 1975). Leaders are said to establish especially close relationships with certain subordinates who are known as the 'in-group'. The in-group receives special treatment such as responsibilities, more autonomy, and a disproportionate amount of the leader's attention and usually receives special privileges. Those not forming the in-group are known as the 'out-group' and receive no special treatment. It is not known how leaders allocate members to each group but it has been suggested that in-group members are chosen because they have personal compatibility with the leader or have a higher level of competence. Hardly surprisingly, research has found that in-group members have a better level of satisfaction and performance and lower turnover than out-group members. Robbins and Judge (2008) make an important point that whilst few would want to be outside a leader's inner circle, there is a danger that if a leader is fired, particularly one at a high position in the hierarchy, then their close followers are also likely to be lose status. They also comment on the point that research to test LMX has been generally supportive by stating: 'These positive findings for in-group member shouldn't be totally surprising given our knowledge of the self-fulfilling prophesy. Leaders invest their resources with those they expect to perform best. And with the belief that in-group members are the most competent, leaders treat them as such and unwittingly fulfill their prophecies.'

11.10 Inspirational approaches

Two contemporary aspects constitute the inspirational approach to leadership: charismatic leadership and transformational leadership. The difference between them (they are closely related) has caused some confusion and, whilst occasionally writers use them interchangably, they do differ. Charisma is interpersonal attraction through which followers develop an unthinking belief, respect and trust in an individual. Although seen as a contemporary approach, it contains many of the elements of traits discussed previously. Transformational leadership is mainly about behaviours that people use to lead the change process, essentially by inspiring employees to transcend their own self-interest.

11.10.1 Charismatic leadership

Fincham and Rhodes (2005) point out that 'The concept of charisma has its roots in the New Testament: St. Paul describing the gifts of the Holy Spirit. And, as Wheen (2004) argues, charismatic CEOs like religious leaders are expected to have the "gift of tongues" – to inspire employees to work harder and follow their vision to the promised land'. Robbins and Judge (2008) cite Max Weber (1947 trans.) as 'the first scholar to discuss charismatic leadership'. More than a century ago, he defined charisma (from the Greek for 'gift'): 'a certain quality of an individual's personality, by virtue of which he or she is set apart from ordinary people and treated as endowed with supernatural, superhuman, or at least specially exceptional powers or qualities. These are not accessible to the ordinary person, but are regarded as of divine origin or as exemplary, and on the basis of them the individual concerned is treated as a leader.'

House (1976) believes that charismatic leaders have high self-confidence and self-belief and high ideals and a strong need to influence people. They express high expectations and confidence in their followers. At the extreme, followers believe in the charismatic leader without question and fully accept whatever the leader says. But in many cases what is promised by charismatic leaders may not come to fruition as they may act in their own interests rather than in the interests of their organizations or their followers. Their egos can blur the lines between their personal and their organization's interests, leading to unethical and, at worst, illegal behaviour not only in themselves but also in their followers. Examples in the political milieu are many worldwide and business is also littered with examples. But charismatic leadership accompanied by honour, political/business skills and vision can represent a powerful combination for good.

11.10.2 Transformational (and transactional) leadership

The contingency approaches discussed earlier are known as transactional leadership models and differ from transformational leadership approaches. Transactional leadership is essentially about achieving current objectives more effectively whereas transformational leadership is concerned with vision and change. The latter inspires followers to overcome their own self-interest for the benefit of their organizations in the rapidly changing environments in which many businesses find themselves. Transformational leaders:

- Pay attention to the developmental needs of individual followers
- Help followers to look at old problems in new ways
- Inspire followers to make extra efforts to achieve goals
- Lay out the vision of the organization.

The over-riding distinction of transformational leadership is *vision* which is achieved through the following routes:

Creating a strategic vision: A vision in this context is seen as much more than a mission statement. Rather it is seen as the grand view of the future of the organization. It is intended to represent the reason for existence for the organization and is often expressed in abstract terms subsequently made real through discussion, decisions and events. It is intended to bring employees together and motivate them to follow the vision and may be envisaged in almost evangelical terms.

Communicating the vision: This is seen as a major and necessary role of transformational leaders based on emotional appeal to motivate employees. Symbols, metaphors and stories are used to project the vision and inspire employees to sign up for the vision. The objective is to persuade employees to accept the vision as their own.

Modelling the vision: Transformational leaders are required to enact the vision through the manner in which they conduct themselves. They need to be consistent and persistent in acting out the vision and changing how the organization conducts its business in large and small ways in order that it is consistent with the vision. Consistency between the words and actions of leaders is vital.

Transformational leaders are themselves creative and innovative and encourage their followers to be more so. They have ambitious goals generated by their vision and also encourage their followers to have ambitious goals. Whilst charisma is not essential to transformational leadership, charismatic transformational leaders are seen to wield enormous influence; no doubt they often achieve their visions but 'full blown' charisma may not be essential. Nevertheless, Burns (1978), the first researcher to write about transformational leadership, incorporated charisma but what is probably more important is the beauty and realism of the vision, as a highly charismatic leader with a flawed vision will not be successful in the long term. Tellingly, Huczynski and Buchanan (2007) state: 'It is tempting to regard these novel terms [transformational, charisma], and this shift in emphasis, as fresh developments in leadership theory. However, the identification of super, new, transformational leaders and associated behaviour represents a return to trait-spotting ('hunt the visionary'), overlooking much of what is known about the influence of context on leadership effectiveness.'

Robbins and Judge (2008) make the point that the best leaders are both transactional and transformational leaders. They should not be viewed as opposing approaches to getting things done. Transformational leadership builds on top of transactional leadership and produces levels of followers' effort and performance that go beyond that which would occur with a transactional approach. Too many leaders become enmeshed in transactional leadership activities so they either do not recognize the need for transformational leadership or lose touch with this important aspect of leadership.

Whilst there is evidence that transformational leadership has been found to be related to the motivation and satisfaction of followers and to the higher performance and perceived effectiveness of leaders, it appears to be seen as universal rather than contingency-oriented and as such overshadows the value of contingency theories. That transformational leadership is no panacea is illustrated by a number of failures of high profile companies led by transformational leaders.

11.11 Authentic leaders

To add to the mix of what constitutes an effective leader Robbins and Judge (2008) introduce the important components of ethics and trust referred to as authentic leaders (cf. May et al., 2003; Gardner and Schermerhorn, 2004). Authentic leaders make their beliefs and values explicit and incorporate them in both their goals and the means of achieving them. Those being lead by authentic leaders will recognise that they are ethical and that they generate trust in their emplyees.

11.12 More widely-based leadership

Huczynski and Buchanan (2007) generate an argument that 'Alongside this focus on visionary super-leaders sits the recognition that leadership acts can be observed at all levels of an organization.' Citing Buchanan et al. (1999) they find that the vast majority of managers do not believe that change management should be left to full-time professionals but that the skill and knowledge required are relevant to all managers. They suggest that leadership is a widespread phenomenon capable of being undertaken by those at all levels of the organization who have the interest, knowledge, skills and motivation to perform them effectively and point to self-managing teams as a manifestation of this idea. They therefore describe leadership as comprising both visionary leadership and widely-dispersed leadership which will induce changes in leadership characteristics. This view of leadership is not of an autocratic or a wildly charismatic figure but of someone, male or female, who enthuses, inspires, coaches and facilitates subordinates. Such an approach seems to be in tune with the increasingly knowledge-based industries and learning organizations and, although it may be thought of as a new direction, reflects servant leadership and the sayings of Lao-tse referred to earlier in this chapter. Such developments in leadership were anticipated by Bennis and Nanus (1985) over 25 years ago. They believed then that rather than organizations having few top leaders and many managers who would direct and supervise, there would be leaders at every level reflecting flat collegiate organizations – not hierarchies – that would empower, inspire and facilitate; that leading by vision would replace leading by goal setting; leaders would act as change agents so that change would be creative and anticipated, not generated by reacting to the organization's environment; information would not be the preserve of a few but shared by many decision makers; essentially leaders would be coaches creating learning organizations and developing future leaders. Twenty-five years on such views may still be largely no more than aspirational but the vision continues to need to be followed.

They also identify an important perspective opposing the current conventional image of formal leadership as being indispensable, charismatic and transformational which goes so far as to call it potentially dangerous. Citing Morgan (2001) and Huy (2001), they argue that rather than relying on high profile forceful charismatic leaders to force through dramatic change, organizations would be better changing incrementally and in a controlled manner and that it is actually middle managers who achieve the necessary balance between continuity and change. They also cite Myerson (2001) who supports the contribution of middle managers in initiating, leading and driving change and Baradacco (2001) who advocates a 'quiet approach to leadership' and describes effective leadership as 'unglamorous, not heroic.'

11.13 Leadership in perspective

There is no doubt that there is no agreement on the relative significance of the various leadership theories amongst OB academics and practitioners. They seem to go through phases of fashion without, it would appear, any method of comparing or validating their universality. Rather, each approach has some merit and provides a range of insights that are useful to practising leaders without providing the 'holy grail' which all but those with gigantic self-belief seek. An outcome from this state of affairs seems to be that the merits of each newly-developed idea tends to be greatly overstated, compounded by leaders being given too much credit or blame for their performance. Perspectives that challenge widely accepted beliefs about leadership are as follows.

11.13.1 *Attributing leadership*

Attribution theory is about how people make sense out of cause and effect relationships. People want to believe leaders make a difference as this belief simplifies organizational events. In leadership terms this causes subordinates to believe that what happens is due to leadership abilities rather than to environment forces as this avoids the need to analyse possibly complex causes. High-ranking leaders take advantage of good outcomes by claiming credit even when they had nothing to do with the success and vice versa, resulting in their vulnerability. This may be particularly so for charismatic leaders who give the impression of invulnerability but are brought down by events outside their control.

Also subordinates may have stereotypes of what effective leaders should be like and how they should behave, e.g., intelligent, outgoing, articulate, aggressive. Leaders matching their subordinates' view of what an effective leader should be like will be seen as effective, in spite of poor performance, until perhaps that becomes sufficiently dire to contradict the subordinates' ingrained view. Arising from these ideas is that leadership is actually the perception of followers, so what is important is giving the appearance of being a leader rather than focusing on achievements. Robbins and Judge (2008) sum up: 'Aspiring leaders can attempt to shape the perception that they're smart, personable, verbally adept, aggressive, hardworking, and consistent in their style. And by doing so, they increase the probability that their bosses, colleagues, and employees will view them as effective leaders', which is reinforced by the statement that: 'We distort reality and attribute events to leaders because we feel more comfortable believing that a competent individual is at the organization's helm' (McShane and Von Glinow 2003).

Sitting alongside is Fincham and Rhodes' (2005) observation: 'Leaders represent the return of the "primal father" with whom, like the father of our early childhood, is easy to identify.' They quote Freud (1951) who suggested, 'We know that in the mass of mankind there is a powerful need for authority who can be admired, before whom one bows down, by whom one is ruled and perhaps even ill treated' and Kets de Vries (1993) who points out, 'leaders are partly defined by the desires of their followers. A great potential for distortion exists'. They continue: 'Leaders are therefore from a Freudian perspective, a prime target for the process of transference – transferring the feelings we have about those who have taken care of us in our past onto organizational leaders.'

11.13.2 *Leadership substitutes and neutralizers*

Leadership is generally seen as a necessary part of organizational life to give direction to the achievement of goals. However, the 'theory of leadership substitutes' suggests that in some situations leadership may be unimportant; that is, employees can do their job without direction such that hierarchical leadership is unimportant. Leadership is neutralized in cases when it is impossible for leader behaviour to make any difference and leadership is substituted when conditions make leadership unnecessary. One set of circumstances that illustrates this phenomenon is when employees have the knowledge, ability, experience, professionalism and high need for independence which neutralizes leadership. Highly-skilled professionals fall into this category which will include construction-related professionals. But this may only be the case when they are undertaking specific professional tasks, when they are working in project teams, leadership will be necessary to maintain coordination. Several other conditions have been identified. Leadership may become neutralized when employees are indifferent to organizational rewards for whatever reason. When a task is routine with a high degree of structure and intrinsic satisfaction, subordinates may not need leadership. Similarly, clearly formalized plans, goals, rules and procedures together with a rigid reward system may substitute for leadership. Also, strictly performance-based reward systems keep employees focused and also reduce or eliminate the need for leadership. This is subject to continuing research; much is still unclear because the topic is complicated and controversial with a wide range of potential substitutes and neutralizers.

11.14 Leaders' perceptions

It is not really possible to separate leaders' instinctive (preferred) leadership styles from their perceptions of the people they are leading. As a result, the degree of flexibility of styles of which leaders are capable may be inhibited by the fundamental assumptions they make about the motivation of the people they are leading. The early work on understanding people's attitude to work (and hence being led) was undertaken by Maslow (1954) who developed his famous 'hierarchy of people's needs'. This was complemented by McGregor (1960), who discovered that leaders' thinking and approaches were based on two different sets of assumptions about people: that people are lazy and wish to avoid responsibility or that people are self motivated and wish to achieve and enjoy responsibility. Hertzberg's (1968) subsequent work on motivation advanced previous work but reflected the same perceptions. This early work referred to above, whilst still useful, was the forerunner of the behaviour–performance–outcome relationship approach to motivation (e.g., Vroom, 1964; Porter and Lawler, 1968; Campbell et al., 1970) which believes that the attractiveness of outcomes and individuals' expectation of success determine the amount of effort that they are willing to apply to achieve goals and that individuals evaluate their own performances against the expected outcomes, which is akin to the path-goal theory discussed earlier.

Leader' perceptions of what motivates the people they are leading will be influenced by the motivation theories outlined above (for a fuller account see Chapter 5), which in turn will strongly influence the position of their instinctive leadership style on the task/relationship spectrum. It is therefore important for leaders to have knowledge and understanding of theories of motivation, hence the behavioural issues that drive their subordinates. Such understanding should help leaders evaluate their subordinates more effectively so that they are better able to adopt the leadership style suited to their followers rather than just relying on their instinctive style which will be based on their perceptions rather than reality. Whilst the argument for leaders to be flexible in their leadership styles is now generally accepted, is there not also an argument – at the level of professional employees – for those being led to adopt flexibility in their expectations of their leaders? Professional-level employees could be expected to understand the problems of leadership and management and adapt to situations occuring at any given time so that leaders are relieved of some pressure and leaders and followers work in partnership. But is this too much to expect?

An example of an application of the later motivation theories to construction is given by Liu (1999) who models the effect of project complexity and goal commitment on project outcomes. However, Maslow's ideas continue to provide inspiration for research in construction (Shoura and Singh, 1998) even though the field has moved on considerably. There is little doubt that they are more easily digested and applied by practising leaders.

11.15 Women as leaders

Until relatively recently it was generally assumed that leaders were men and a woman in a leadership position was cause for surprise and comment. Whilst this is less so now the assumption still exists, particularly in some more traditional businesses. However, the changing nature of economies in developed countries leading to new organizational forms has presented more opportunities for women to take up leadership roles to which their particular talents are well suited. Many modern organizations require organizational structures to be flatter, less hierarchical and more decentralized. This is seen to require skills which are more often found in women than men, such as communicating, participating and consensus making; that is, essentially interpersonal relationship-oriented. Such skills are very necessary in leading teams and team-building and, as team-working increases, so does the need for the skills in using teams and leading them.

McShane and Von Glinow (2003) point out that these ideas reflect sex role stereotypes; that is, men tend to be task-oriented and women more people-oriented. They ask is it true that women adopt more people-oriented and participative leadership styles? They answer that 'male and female leaders are equally people-oriented, but female leaders do tend to be more participative than their male counterparts', and that studies have 'generally found that male and female leaders do not differ in their levels of task-oriented or people-oriented leadership. The main explanation why men and women do not differ on these styles is that real world jobs require similar behaviour from male and female job incumbents'. McShane and Von Glinow (2003) also suggest that women are more participative as their upbringing makes them more egalitarian and less status conscious and also that they have better interpersonal skills.

Also, subordinates' expectations of women is that they will be more participative due to their sex role stereotypes.

These factors have led commentators to suggest that women will tend to excel in transformational leadership positions, whilst men will be oriented towards transactional roles. Studies by Metcalfe and Metcalfe (2002, 2003) confirmed this view and also found that women were better than men at being decisive, focusing effort, mentoring, managing change, inspiring others, and showing openness to others. However, generalizations are all too easy and whether men or women are the best leaders depend on the person and the leadership circumstances.

Of course, the single factor which dominates whether women can fully generate a leadership career is the demands of family life: whether they have the energy and resilience to persevere in both roles; whether they are prepared to sacrifice family life for their career or whether their partner will. Hewlett (2002) found that in the USA high and ultra-achieving women were less likely to be married than their male colleagues. Many organizations expect long hours of work followed up by socializing after work by drinking and other male-dominated activities to the disadvantage of women with family responsibilities.

11.16 Culture and leadership

Culture manifests itself on leadership from both an organizational culture and national culture perspective. The scope of this book does not allow room for a full consideration of leadership approaches and styles in other national cultures, which would be a book or more in itself, except to say that national culture can be expected to have a profound effect on leadership styles and may modify the theories discussed in this chapter. Fincham and Rhodes (2005) examine the studies undertaken and find that this is so. For example, they refer to a study conducted by the Cranfield School of Management and based on Kakabadse et al. (1997) which found that the predominent leadership style in UK, Ireland and Spain was 'Leading from the front', in Sweden and Finland as 'Consensus', in France as 'Managing from a distance' and in Germany and Austria as 'Towards a common goal'. However, an examination of the leadership styles of Chinese and expatriate construction project managers in Hong Kong found that, despite their cultural background, they did not differ significantly in terms of leadership perceptions and power relationships (Wong et al., 2007). Based on there being only minor differences in these respects, the authors proposed two possible reasons: first, that the Western expatriates were trying to adapt to host-national culture in order to be successful in their international management; and second, that Hong Kong Chinese managers tend to adapt to the Western model of business and leadership style. Although not mentioned by the authors, it should be said that, due to Hong Kong's colonial history, there is a long experience of Western (particularly British) and Chinese managers working together which may have led to a long-term adaptation.

Whilst organizational culture is seen as an important feature of organizations, as dicussed in Chapter 8, it has not been enduringly linked to leadership even though 'culture has always been on the agenda of management theorists' (Fincham and Rhodes, 2005). They stress that Barnard and Drucker from the 1930s through the 1950s to the present, 'emphasised the centrality of *values* in the management of enterprises'. But the link is not strongly emphasized in modern management literature; for example, Huczynski and Buchanan (2007) comment that it is surprising that organizational culture has not featured as a variable in the context of leadership styles in relation to contingency theories. Leadership is central to change management and a significant part of this process is likely to be to change the culture of an organization.

Changing an organization's culture is very difficult as it can be expected to be deeply entrenched and resistant to change but, as described in Chapter 8, conditions can prevail that make cultural change possible. If leaders are to effect organizational change, a change in organizational culture to foster desired values is likely to be a necessity and will need to be within the capabilities of leaders, of which transformational leaders can be expected to be the most adept. This requires leaders to be aware of the symbolism of their actions and that by communicating and reinforcing their vision for the enterprise through, for example, internal media, e.g., company magazines, they embed it into the organization's culture and by injecting their own stories to become company folk-lore and so develop shared meaning. However, Fincham and Rhodes (2005) cite Morgan (1997) in stating the reality: 'Managers [leaders] can influence the evolution of culture by being aware of the symbolic consequences of their actions and by attempting to foster desired values. But they can never control culture in the sense that management writers advocate ....'

11.17 Leadership in construction

11.17.1 Leadership of construction-related organizations

The companies related to front-line construction industry activities can be broadly classified into contracting firms, professional consultancy firms and clients. The leadership needs differ between them but with substantial overlap. In the case of the leadership of contracting firms and professional consultancy firms, common issues relate to leading the company and to leading in-house project teams. The distinction is that contracting firms also need leadership of more mechanistic on-site tasks, but where some of the tasks may also require innovation and creativity. In-house project teams' members will in all cases be highly-qualified professionals such as architects, engineers, quantity surveyors and other specialists and, for certain contractual arrangements, builders. In addition, the influence of the leadership of client organizations will be of great importance to the construction-related organizations that have to collaborate with them, as the often delicate relationships can have a great impact on the success of a project. Overlying these situations is the leadership of the interdisciplinary client's project teams led by project leaders who may be members of the client's organization, members of a separate project management company or members of one of the contracting/consultancy firms (dealt with separately in the next section).

Such complex leadership requirements generate a myriad of leadership issues, which together with the range and complexity of variables within leadership concepts and theories make it hardly surprising that leadership research in construction has not been conclusive. The lack of meaningful leadership research in construction has not been aided by construction researchers only basing their research on a limited range of mainstream management theories. Through their review of empirical research on leadership in the construction industry, Toor and Ofori (2009) find that the mainstream management leadership constructs which have not been used in empirical studies in construction include charismatic, servant, spiritual, self, political, shared, authentic and aesthetic leadership. They also discovered that in most studies researchers used their own questionnaires and semi-structured interviews or case studies with the majority of studies using quantitative methods based on questionnaire surveys with very few alternative methods of data collection and analysis such as interviews and case studies. Hence, the conclusions of the empirical studies on leadership in the construction industry are mostly based on the perceptions of respondents without providing objective measurement of outcomes of the leadership process. A paper by Limsila and Ogunlana (2008) examines the effect of transactional and transformational styles of leadership by project managers on the work performance and organizational commitment of subordinates in the Thai construction industry. They found that 'the transformational leadership style has a positive association with work performance and organizational commitment of subordinates more than the transactional style. Transformational leaders produce higher leadership outcomes as well'. Also they claimed: 'Contrary to the hypotheses set in this study, the results show that the leadership style most adopted and proving to be most suitable to Thai people is the transformational style.' They also point out that their finding is contrary to earlier research which 'confirmed that autocratic or task-structure are the preferred leadership styles in Thailand', although there have been other results that showed that a relationship-oriented leadership style is considered to be more important, all of which is less than conclusive. However, as said previously, transactional and transformational leadership shouldn't be viewed as opposing approaches to getting things done. They complement each other but that doesn't mean that they're equally important. A further example of difficulty of dealing with the complexity and inconclusiveness of leadership studies when researching leadership in construction is Cheung et al. (2001) who developed a leadership behaviour model for design consultants in Hong Kong. Whilst interesting, their focus was on inspirational approaches which were not fully placed within the context of broader leadership studies and their lack of consensus.

So the complexity of leadership and the difficulty of obtaining results that lead to categorical conclusions other than 'there are no categorical conclusions' pervades most leadership studies.

The various approaches to leadership find relevance to different aspects of the construction milieu. At the highest level of leadership in consultancy and contracting firms, transformational leadership would appear to have a role to play even though the fundamental objectives of professional and contracting firms are often said to differ in their orientation towards profit and service to the client. The turbulent economic, social and technical environment of today, together with green and sustainability issues, requires leaders to have vision and to be adaptable. That many have been so is shown by the amalgamations of firms to form multidisciplinary companies more able to deal with the increasingly complex issues of construction and development. Amalgamation of contracting and consultancy firms to provide an even broader provision of capability and with the ability to invest in new initiatives and techniques such as relational contracting, design and build and build-operate-transfer in its various forms, is a result of leaders having a vision of the future. But the vision of such leaders should also extend to having a vision of how to deal with an environment that requires entrenchment in difficult times. Firms that do not have leaders with vision face the possibility of being absorbed by others. Nevertheless, rapid change may not be the answer for many professional firms; stability and constancy may be what a client seeks for the control of their construction project which often forms the largest investment a client's company ever makes and, in the case of public organizations, on which their reputation rests. Hence professional firms may see it as being in their best interests not to change too rapidly unless it proves to be essential, as by nature they tend to be conservative in their outlook becauses the actual operational needs of development and construction generally require a transactional style of leadership. But it should be recognized that transactional leadership does not deny change; rather, it is not defined by change, and change, if necessary, takes place in a more controlled manner and probably without such a high profile. The way in which they change, if at all, and their rate of change is finely balanced. Development and construction companies tend to be less conservative, more commercial and more aggressive in terms of change and are likely to see benefit in a transformational style of leadership. Nevertheless, all organizations in construction will see value in both transactional and transformational styles of leadership.

The type of leadership issues outlined here are also likely to be present in client organizations and need to be understood by construction-related organizations if they are to relate to their clients and help to bring their clients' visions to fruition. The range of businesses and public sector organizations from which the construction industry's clients are drawn is vast so the profile of their leadership can be based on any of the theories which have been discussed. An understanding of these theories by construction-related organizations is therefore necessary if they are to better interpret what their clients require.

Leaders in construction also need to be able to put their visions into practice. They therefore need to ensure, first and foremost, that appropriately qualified staff are appointed and that the management of the company is effective or those visions will not be realized. Charismatic leaders of construction-related organizations are perhaps not widespread as such firms frequently see themselves as practical and 'down to earth' but examples can be found, particularly amongst architectural practices, through such people as Norman Foster and Richard Rogers. But in construction charismatic leaders need to have more to offer than just charisma, they need high professional abilities which generate respect from their peers.

As a construction-related firm's business is practically totally devoted to projects, and as projects proceed through a series of stages, each of which consists of a number of tasks that vary from structured through various degrees of being unstructured, the leadership style required is essentially transactional as epitomized by contingency theories, although transformational leadership has a part to play in leading the team creating design solutions. There are essentially two types of project team: one type is 'in-house' to each contributing firm as they develop their contribution (design, engineering, etc. which is dealt with here), and the other type is the client's interdisciplinary project teams comprising members from all the contributing firms and which coordinates their input. The leadership of the latter is particularly complex and is dealt with in the next section. The in-house project teams will work up their contribution to the project before it is amalgamated with the contribution of the other firms through the client's interdisciplinary project team. The types of issue being dealt with by in-house project teams will be diverse but can be classified simply as:

- technical for which knowledge is available
- creative which requires original thought, and
- problem-solving which will include definition of the problem as well as its solution.

In-house project teams from each of the different contributors are likely to have a preponderance of issues of the type reflecting their discipline, for example, contractors will have mainly technical problems and architects mainly creative problems. All contributors will have a range of each type in varying degrees as each participant's work will impinge on the others, so they are all required to have a sound understanding of the tasks of each of the others.

According to contingency theories of leadership, project team leaders need the ability to vary their leadership style depending on the type of issue and its context. The applications which follow relate to the leadership of in-house teams of professionally qualified employees. Some applications of Fielder's model (referenced to the Conditions referred to earlier) could be:

- *A conventional structural solution (engineering consultancy)*: Condition 1; task highly structured, leader's power high (senior partner or director), relationship with boss high. Task-oriented leader gets good results.

- *A challenging design problem (architectural practice)*: Condition 2; task unstructured, leader's power low (new senior appointment); relationship with boss moderate. Relationship-oriented leader better.
- *Feasibility of project in doubt (cost consultancy)*: Condition 3; task unstructured, leader's power low, relationship with boss poor. Task-oriented leader better.
- *Re-programming construction work (during construction) after a major variation is issued whilst still being required to meet the original completion date (main contractor)*: A variation on Condition 1; task unstructured, leader's power high, relationship with boss high. Task-oriented leader obtains better results.

Although these examples illustrate the useful guidance of Fielder's model, they also show that the three components (structure, power and relationship) are ends of spectrums and situations do not generally occur in practice in such a stark manner. Rather, situations lie somewhere between the extremes. For example, in the applications illustrated above, tasks are rarely either totally structured or unstructured; they contain components of each, all of which have to be dealt with before a task is able to be completed. Similarly, a leader's power is on a scale from high to low, as are relationships.

Hersey and Blanchard's model stresses the 'maturity' of subordinates, which is taken to be essentially their ability and experience which, when applied to construction, places the construction professionals at the most mature end of the maturity scale. This results in a low task (because the professionals are highly proficient) / low relationship (because they are members of the same organization, familiar with relationships and not in need of support). The high skills of subordinates mean that generally they require minimum (if any) supervision of their work and generally enjoy harmony in their relationships. This combination requires a delegating leadership style characterized by an implicit trust in people and a primary concern with developing their talents.

Path-goal theory identifies directive leaders, supportive leaders, participative leaders and achievement-oriented leaders. Two types seem appropriate: participative leaders, because employees in professional firms have an internal high locus of control as they readily accept responsibility for their work and expect consultation from leaders, not unilateral decisions; achievement-oriented leaders, as employees are frequently faced with ambiguous professional tasks and are expected to be able to solve them. Vroom's approach identifies a wider range of more detailed leadership styles. Apart from 'decide', all other styles may be used by leaders of in-house project teams both sequentially and in parallel as a project develops dependent on the situational factors discussed earlier.

Although all contingency theories, except Fielder's, require leaders to be able to vary their leadership styles depending on the situation and other factors, leaders of in-house project teams would not be expected to need to adopt a wide range of leadership styles as directing professionals does not require detailed attention to their work as they are generally self-starting, self-directing individuals who expect democratic type leadership and are likely to react unfavourably to impositions by leaders. However, leaders of clients' interdisciplinary project teams may have to draw on a wider range of leadership styles, as discussed in the next section. A construction firm's in-house project team differs very distinctly as it also has to manage the actual construction process. This will generally need a directive (task-oriented) leadership style as instructions have to be given to site workers about decisions that have been taken by their leaders and the professions involved. However, the construction process is not entirely mechanistic as technical 'solutions' may not work in practice and more participatory (relationship-oriented) methods of leadership involving site staff may be necessary; similarly, unexpected problems can occur, for example in excavation and drainage, which will require site staff to be involved in their solution.

Leader–member exchange theory is as likely to occur in construction-related firms as in any other and is just as potentially dangerous. It can occur in the leadership of the organization as a whole and in in-house project teams. In terms of the organization, it is the prerogative of leaders to listen and be influenced by whom they choose, but they need to be conscious of the divisions and envy that may occur between staff, which are the breeding ground for organizational politics. In the case of the project leader, favouring advice from some members of the in-house project team over others can distort the team's input to the interdisciplinary project team, to the detriment of the project outcome and hence the client.

11.17.2 Leadership and the client's interdisciplinary project team leader

The term 'project team leader' (often called project manager) always needs qualifying by stating its context. In the last section a project team leader was described as 'in-house' as the term referred to project team leaders within a single skill, construction-related organization. In this section, the client's project team leader (either in-house to the client's organization or not) is the focus and will be referred to as such. The client's project team leader leads the team with responsibilty for developing and delivering the project, not just one element of it as dealt with in the last section. The team therefore comprises members drawn from all the professional and contracting firms working on the project. Frequently these organizations are independent firms but can be in a variety of associations and may contribute one or more element. The client's project team leader can be from an independent project management firm, or from one of the other independent firms that also provides another specialism, or from the client organization itself. There could also be cases where there are two project team leaders: one from the client's organization (whose team is drawn from the client organization) and one from the professional and contractor group.

Whichever arrangement is used, the position of a client's project team leader is particularly challenging. Management research on leadership has focused on general leadership situations and in-house-type teams, not on the position of a client's project team leader who has to deal with unique features such as the relationship with the client and the interorganizational nature of the team and process. However, many of the leadership features discussed in the last section on in-house project team leaders are relevant to the client's project team leaders. A client's project team leader is usually in the position of leading contributors from a large number of commercially-independent organizations. Client project team leaders appointed from an independent professional firm are likely to have only limited authority over the project team members. However, project team leaders drawn directly from the client organization will have much greater authority and power, usually including contractual authority. A significant outcome of this situation is that in the emerging leadership matrix each team member will be subject to leadership from both the client's project team leader and their own leader in the organization (Rowlinson, 2001). In effect, the leaders are in 'competition'. As long as the leaders have the same objectives there should be no problem, but if they disagree the potential for leaders from the team members' organization to distort the client's project team leader's leadership is ever present, as is pressure on team members.

A client's project team leaders will be leading a group of mature, experienced professionals and are often only slightly elevated over them in terms of authority (though more so if an employee of the client's organization). Team members may even resent having project team leaders in this position. Consequently, the client's project team leader's leadership style will tend to be relationship-oriented and rely on influence and persuasion rather than positional authority.

The development and realization of a construction project consists of a series of stages (subsystems) (Walker, 2007) which terminate in a significant decision. Project team leaders are continually in receipt of information from their teams and must interact with them, constantly exchanging large volumes of information of a creative, technical and financial nature. The nature of the work to be undertaken within each stage should determine the leadership style most appropriate to arriving at the best outcome for each stage. However, in each case the decision will effectively be made by the client's project team leader following input from all specialist contributors rather than the project team as a whole, although in the case of the most fundamental (key) decisions it is likely to be in the form of a recommendation to the client who will make the final decision. So whilst for some stages the client's project team leader should adopt a relationship-oriented form of leadership as ideas are formed and developed, at the decision point at the end of each stage it is likely to be that the client's project team leader will act 'autocratically' in making or recommending a decision. Whilst it will be the client's project team leader's responsibility to act in this way, in practice it may not appear to be autocratic as in many cases the decision will be arrived at by consensus. However, many clients' project team leaders might not wish to appear in such a high profile decision-making role and may seek to have decisions seen as team decisions. There are degrees of democracy and the actual position on the scale taken by client's project team leader will depend on a number of factors such as the urgency of the decision, their personal characteristics and the experience of the team being led.

The stages that lend themselves to a relationship-oriented style of leadership are those concerned with developing initial designs and evaluations, outline design strategies and selecting contractual arrangements. Other stages may lend themselves to a leadership style which is more towards the task-oriented end of the scale. These will include developing working drawings, preparing contract documentation, cost control during construction and much of the construction stage. However, should problems occur on any of the mechanistic stages it may be that a more relationship-oriented style of leadership is necessary to enable the team to solve the problem. Conversely, within a stage being led in a relationship-oriented style, it may be necessary on occasions to take a task-oriented position to bring it to a conclusion in order to make progress. The underlying work for each stage will have been undertaken by one or more teams in one or more of the contributing specialist firms led by in-house project team leaders. These leaders may not necessarily have used the same style of leadership as the client's project team leader, as the former may have had to explore alternatives using a relationship-oriented leadership style to identify their input to the client's project leader's team's work. If the matter, when put to the client's project leader's team, is seen by the client's project team leader to be straightforward, the client's project team leader may adopt a task-oriented style in order to come to a swift decision.

Client's project team leaders have the taxing job of varying their leadership styles as required by most contingency theories depending on the needs of the particular stage in the process. Anderson (1992) confirmed this requirement and found that leadership skills had the highest frequency of significance for the eight project management functions examined yet, surprisingly, project managers were perceived as having only average or less than average managerial skills. However, it should be recognized that a client's project team leaders' need for flexibility is extremely high, not only because of the nature of the process being led, but also because of the maturity and organizational independence of the contributors. In such a context, client's project team leaders' position on the scale will tend to lie predominantly towards the relationship end as he or she seeks to cope with what is inevitably a challenging leadership position.

The need for a client's project team leaders to vary their leadership styles to suit the situation is readily apparent. For example:

- At the early stages of a project, the client's project team leader has to weld together the range of professional specialists involved in the project and lead them in balancing the conflicting objectives which will no doubt have emerged from the client organization. Developing a viable solution that balances function, cost and time requires the client's project team leader to create a climate in which everyone is free to speak up, make suggestions and criticize. To achieve this, the client's project team leader must adopt a relationship-oriented style of leadership. Once the proposed solution has been defined, the process of developing the solution into working documents becomes a much more structured process requiring a more task-oriented leadership style.

- During the process of developing working drawings, contract documentation, etc., which requires a task-oriented approach, it may become apparent that a major problem occurs in translating the proposed design solution into a buildable solution. In this case, the client's project team leader is unlikely to be able to solve the problem alone. He or she will need to bring together professional specialists to discuss and resolve the problem and his/her style will switch from a task- to a relationship-oriented style until the routine work is recommenced.
- The client's project team leader's authority affects the style adopted. A client's project team leader given authority under the contract during the construction stage of a project under conventional contractual arrangements will normally have to adopt a directive- or task-oriented style as the contract places specific obligations upon him or her and on the contractor. Alternatively, the architect may be given authority under the contract in which case the client's project manager will become a member of the project team but with referent power which gives great influence. Both conditions make it difficult to adopt a relationship-oriented style even when it appears to be the best way of approaching a problem. Partnering and other relational arrangements are approaches that seek to allow a more relationship-oriented style which could be particularly effective in the construction stage.
- The degree of authority can also affect the client's project team leader's style during the pre-contract stage. A client's project team leader who is a direct employee of the client organization (or, if not, is given clear authority by the client), has a greater opportunity to be autocratic, which is not the case his or her authority is ambiguous. However, paradoxically, the former gives the project leader a greater scope to use a range of leadership styles if he/she is capable of exercising such flexibility as he/she can fall back on his or her authority if this is felt to be appropriate in difficult situations but care needs to be taken not to do so inappropriately.

Whilst the demands of the client's project team leadership may be rather different from general management, and even in-house project team leadership, and even though leadership research is far from coherent, it is useful to a client's project team leaders to review the previously described research models. This will enable them to ask themselves questions about how they lead project teams and help them to focus on the variables that have been seen to have an effect on leadership. The comments made in the last section for in-house project managers are generally relevant to client's project team leaders and are complemented by the further comments here, adapted for client's project team leaders.

Fielder's (1967) work asks project team leaders to consider how well liked they are by their teams and also to evaluate the degree of uncertainty of the task being undertaken. In construction, the latter is very variable, ranging often from practically 'a blank sheet of paper' in the early stages of some projects to what can be a well-defined task of producing a standard building type. It also asks project team leaders to assess where they see themselves on a scale from task- to relationship-oriented leader and the degree of flexibility which they consider themselves to have. Unfortunately, these are questions which we can not easily answer about ourselves nor are they opinions we like to hear from others. Herein lies one of the greatest dilemmas in improving leadership abilities – that of recognizing honestly one's own abilities.

The low task and low relationship behaviour requiring a delegating leadership style identified by Hersey and Blanchard's model is clearly appropriate on many construction projects for which the project team leader is dealing with highly educated, well trained and emotionally mature people and has been suggested previously as appropriate for in-house project team leaders. However, a participating leadership style, that reflects low task/high relationship behaviour may be more appropriate to clients' project team leaders as relationships may not be so familiar and may be more difficult to generate because team members are from independent firms. Therefore, clients' project team leaders may have to work harder at these relationships. Hersey and Blanchard's model also demonstrates other modes that may be necessary in specific circumstances, particularly if a crisis situation occurs. However, such situations are rare and the client's project team leader is more likely to have to switch to a delegating leadership style to reflect low task/low relationship behaviour once good relationships have been established, or to a selling leadership style to reflect high task/high relationship mode if leading a somewhat less mature group, particularly in times of high construction activity when less experienced staff may occupy more senior positions.

Path-goal theory identifies four leadership behaviours: directive, supportive, participative and achievement-oriented. Contingency variables are locus of control and perceived ability and experience. Client project team members generally accept responsibility and are highly competent, so leaders should adopt a participative or achievement-oriented leadership style. Whilst proponents of the various contingency theories will point to differences between them, their general steer is essentially in the same direction and consistent for clients' project team leaders.

Organization culture has been referred to as an element of the situation in which leaders have to work. The problem with this concept in relation to clients' project team leaders is that analysis of organizational culture of clients' project teams does not appear yet to have been undertaken so its impact on leadership is not known. The concept of organizational culture in relation to temporary organization structures such as clients' project teams, rather than permanent evolving business organizations, is likely to need special consideration as no sooner will a culture begin to grow than the clients' project team will be disbanded and the culture will dissipate. One of the major claims for the benefit of relational contracting is that a culture which is seen to be valuable can be preserved and developed.

Leader–member exchange theory highlights a particular danger for clients' project team leadership. The essence of clients' project team leadership is achieving balance between the inputs of the specialists in seeking to achieve clients' objectives. If the input of some specialists is favoured because they are part of an influence group, distortion of the outcome of the project is likely to adversely affect clients' satisfaction.

Transformational leadership has generally been associated with business and public organizations rather than project organizations. Clarke (2010) has identified studies that have shown no significant impact of transformational leadership within project situations. Keegan and Den Hartog's (2004) study of project managers (not construction) who undertook both line and project management found that transformational leadership showed no significant correlation with commitment and motivation for project teams but did so for line management. Whilst the leaders behaved the same way in both situations, the effect was opposite for the two types of team. Sang's (2005) case study of four project leaders found that transformational leadership was not always needed to produce good project results. He also observes that for projects which exhibit high uncertainty 'the importance of transformational leadership, and therefore its association with emotional intelligence, may be of less significance'. However, the associated characteristic of charisma can have a high impact on the standing of clients' project team leaders in the eyes of team members. A leader's standing is enhanced by referent power which arises from the client's project team leader having led teams to success in completing complex, high profile, prestigious projects to clients' satisfaction. The construction industry is tightly knit and the grapevine is quick to pass around stories of success and failure so is a breeding ground for establishing reputations, good and bad.

11.17.3 Project leadership qualities

Clients' project team leaders have probably the most demanding role in the construction industry. Essentially, they manage by influence, sometimes with little authority, but the nature of the job requires it to be that way. The demands on in-house project team leaders may not be so high but some may argue that dealing with in-house colleagues is equally difficult as dealing with colleagues from other firms. What then are the qualities that project leaders should possess? What are the qualities that allow project team leaders to be sensitive to the different situations which arise on projects and be sufficiently flexible to use an appropriate leadership style?

These qualities can be split into characteristics and skills. Project team leaders' characteristics will in many cases determine how well they employ their skills. Examples of the characteristics that help to form good leaders in construction project management are seen to be:

- Integrity
- Preferred leadership style (towards a relationship-orientation)
- Self-confidence
- Ability to delegate and trust others
- Ability to cope with stress
- Decisiveness
- Judgement
- Consistency and stability
- Personal motivation and dedication
- Determination
- Positive thinking
- Excellent health
- Openness and the ability to hear what others say
- Ease in social interactions with many types of people.

In terms of skills, the following are seen to be important:

- Persuasive ability
- Negotiation skills
- Commercial expertise
- 'Political' awareness
- Breadth of vision
- Integrative skills
- Ability to set clear objectives
- Communication skills
- Management of meetings
- Early warning antennae
- Skills of diplomacy
- The skills of discriminating important information.

The context of these qualities is the experience of the project team leader. A broadly-based experience is required of all phases of a project. The client's project team leader will require an appreciation of all the specialist areas whilst not needing to be a specialist in any. Nevertheless, generally, certain aspects will be more significant to a client's project team leader such as contract strategies, cost and time control, money management including project finance and capital and revenue relationships. This type of experience can give clients' project team leaders the confidence needed, but equally importantly is the way in which it affects the team's perception of its leader. If project managers are seen to have the right experience or, more importantly, they have a successful record of achievement, be it by good leadership or good luck, the team will more readily accept them as leader; a reflection of attribution theory.

It is generally accepted that there is no leader equally suited for all situations. The variables involved in leadership behaviour and success are wide and often unquantifiable. It is only possible to identify them and recognize the kinds of situations in which different combinations of talents may be most beneficial. It is necessary that leaders have the ability to vary the way in which they amalgamate their attributes depending on the situation in which they find themselves or, failing leaders' ability to vary their leadership style, it would be necessary to change the leader when the situation changes. The OB literature accepts that the former is possible but just what proportion of leaders is able to do this is not known. The latter, a transfer of leadership, is for most organizations unrealistic which in any case could be extremely disruptive if allowed to happen frequently. Hence, the ability to select appropriate leaders consistently against such a backcloth must be in question. Respected leaders have an amalgam of characteristics, skills and experience which are recognized and respected by those they lead. Leadership is granted by the people being led and not by an organizational position. The effective leader influences not just subordinates but superiors and peers and in construction project management has the additional need to exercise such influence over people in other companies, which requires leadership characteristics of a very high order.

In trying to answer the question of how one spots a leader it is worth turning to Townsend (2007): 'They [leaders] come in all ages, shapes and sizes and conditions. Some are poor administrators, some are not overly bright. One clue: since most people *per se* are mediocre, the true leader can be recognized because, somehow or other, his people consistently turn in superior performances.'

11.18 Learning organizations and organizational change

Learning organizations and organizational change are fashionable but controversial topics in OB and are closely linked to leadership. Leaders will be responsible for instigating those initiatives that are more likely to be instigated by transformational leaders. Learning and change are also closely linked as learning organizations are in many ways in readiness for change should external forces determine it to be necessary.

11.19 The learning organization

A learning organization is seen as one that has developed the ability to modify itself and change in response to shifting environmental conditions. To achieve this, organizations have been seen to learn in the same way as individuals. Wilson (2004) says, 'Organizational learning has been seen as the aggregate of individual learning in an organizational context. This view now appears to have been superseded by organizations being seen as collective entities.' But this concept that organizations themselves can actually learn is challenged by Fincham and Rhodes (2005) who believe that the learning organization is pure metaphor. They continue that 'Learning in human beings is about being able to detect changing circumstances, modifying past behaviour that has been unsuccessful, and on this basis building a repertoire of complex responses and skills' and also that '... learning is quintessentially a human capacity and we need to resist attempts to reify the organization in any way'. Organizational learning is seen to be the embodiment of these ideas by allowing these human characteristics to flourish. This lack of clarity is reflected in Huczynski and Buchanan (2007) comment that 'the literature of this topic is preoccupied with refining the way in which learning, knowledge and the learning organization can be conceptualized'.

To achieve a learning organization people are seen to have to discard traditional ways of thinking; learn to be open with each other; fully understand how their organization actually works; have a vision that everyone can agree on; and work together to achieve it. Obstacles to be overcome include: bureaucratic tendencies; reactionary attitudes; fragmentation producing barriers; and competition between segments of the organization. Interestingly, Robbins and Judge (2008) see the learning organization as an ideal which no company has successfully achieved and one to be aspired to rather than a realistic description of structured activity. They also point out that learning organizations reflect many aspects of OB such as: culture, which needs developing to support and sustain organizational learning rather than inhibiting change; creative constructive criticism which stimulates new ideas; and transformational leadership which should aim to generate conditions for the characteristics of organizational learning to flourish. All of this can lead to developments such as new governance initiatives, new management methods, and ways in which people carry out their work. Transformational leadership is cited as encouraging the development of learning organizations but so also can transactional leadership. Leaders of any persuasion who provide employees with learning opportunities create conditions where employees' learning can feed in to create organizational learning.

Construction-related firms should be learning organizations by instinct and those that are not could be expected to be relatively unsuccessful. They are increasingly challenged by dynamic environments in which techniques, materials and methods are continually evolving. Not to adopt an organizational learning culture would inhibit solutions, particularly those requiring creativity which would result in stagnation and a shrinking reputation. An example of construction-related organizations adopting organizational learning is illustrated by their take-up of total quality management's commitment to continuous improvement, although Love et al. (2000) do not believe that this has been as extensive as it may have been. Also, the fact that construction-related organizations are generally staffed by professionally qualified members who are required by their professional institutions to undertake continuing professional development should mean that they continue to learn over their working life. Although much of this learning is on profession-specific topics, many members take the opportunity to attend business-orientated activities, all of which contribute to the sum of knowledge of their organization and provide a basis for adaptation and change. Fu et al. (2006) identified that professional engineers who need a wide domain of knowledge learn actively through their 'learning networks'. They see learning networks in construction as those that provide practitioners with physical and virtual platforms where collective learning takes place. The way in which construction-related organizations learn is illustrated by their ability to adapt to the emergence of different building types as economic and political forces shift, for example, increases in the number and value of public sector projects with emphasis on, say, health-related projects. They need to anticipate such developments and adapt to them, alongside which they need to develop their learning base to be able to handle such projects, which may also require them to adapt their organizational arrangements. The literature on organizational learning tends to deal with organizations comprising a wide range of levels of employees and to point out that organizations can learn from employees at all levels, whereas professional construction-related organizations, with the exception of contractors, are dealing with a much more unified level of employee so the organizational learning literature is only relevant in part.

A study of organizational learning related to construction by Chan et al. (2005) draws attention to the lack of understanding of what is meant by a learning organization and hence the lack of a solid theoretical and empirical base for taking it forward, as pointed out earlier. They do not focus on the organizational learning of firms but rather point out that the difficulty of understanding organizational learning in construction is exacerbated by the industry being largely project-based. In particular, they emphasize the need to consider organizational learning in construction beyond strategic partnering alliances, on which construction-related studies into organizational learning have been centred, and shift the focus towards viewing projects as learning networks.

The features of a learning company as identified by Pedler et al. (1997) are, briefly:

A learning approach to strategy: Creative thinking to generate improvements and to modify strategy.
Participative policy-making: All members involved in strategy, decisions, values and resolving conflict.
Informative: Information available to everyone.
Accounting and control: People are assisted to understand organization finance.
Internal exchange: Company is seen as an internal supply chain with groups learning from each other.
Reward flexibility: Flexible reward system with financial and non-financial elements.
Enabling structures: Structures are temporary and can be changed to suit the task.

Environmental scanners: All members dealing directly with external contacts are seen as good information sources.
Inter-company learning: Learning through association with other organizations.
A learning climate: A leader's task is to generate learning in others.
Self-development opportunities: Opportunities for learning made available for all employees who are responsible for taking the opportunities offered.

This list reinforces the notion presented earlier that a learning organization is one to aspire to, as the chances of achieving such an organization in reality is extremely remote.

Senge (1990) identified five 'learning disciplines' which are seen as necessary in order to construct a learning organization:

Personal mastery: Individuals aspiring to achieve the best for themselves.
Mental models: People constantly refining their awareness of their organizational environment.
Shared vision: A sense of common purpose and how to achieve it.
Team learning: Working effectively within teams.
Systems thinking: This is the 'fifth discipline' which was the title of his book, and relates to understanding the differentiation, interdependency and complexity that binds the parts of a company together.

Fincham and Rhodes (2005) remark that, 'Some of these are rather exotic "deep" skills' and the expectation may be that all or the majority of organizational members may see them as unrealistic but members of construction-related organizations will recognize these 'disciplines' as familiar requirements for developing and constructing projects, particularly"systems thinking", which is central to project management (Walker, 2007) and confirms the comment made earlier that many construction-related organizations have the characteristics of learning organizations.

In the early development of the concept of the learning organization Argyris and Schon (1974) distinguished between single- and double-loop learning. Single-loop learning is a mechanical-type response borrowed from cybernetics in which corrections are made to deviations within given parameters of the system using feedback to bring it back on course. In this sense it 'learns' to return the system to the norm but cannot recognize whether the norm is too high or too low. It has been argued that single-loop learning is not really learning and certainly not learning to learn which is what is claimed for double-loop learning. Double-loop learning is about fundamental challenges about how systems work; it challenges assumptions, beliefs, norms and decision-making rather than accepting what is given and working within its limitations as does single-loop learning. To create truly learning organizations double-loop learning should not be the preserve solely of the top echelons of management but should permeate the whole organization so that all employees are able to instigate and cope with change.

Knowledge management is often linked with the idea of learning organizations and is complementary to it. Whereas organizational learning is developing the ability to change in response to forces acting on the organization, knowledge management has been said to be concerned with 'turning individual learning into organizational learning' (Huczynski and Buchanan, 2007) or managing the intellectual capital (the sum of an organization's human, structural and relational capital) of an organization (Stewart, 1997), which appear to be somewhat different perceptions. The latter is probably more recognizable to construction-related organizations for which intellectual capital is of paramount importance. The connection between the two is that an organization's intellectual knowledge is the spur to signalling the need for an organization to adapt to changing circumstances due to technical, social, economic and political forces, the knowledge of which is held within an organization's intellectual capital. Whilst they do not appear to draw this distinction, Chinowsky and Carrillo (2007) examine the process from knowledge management to learning organizations through case studies of four engineering-construction organizations which showed that successful organizations found that the success of the process strengthened both knowledge management and the learning organization.

Human capital is the knowledge that employees have, including their skills and experience. In construction, such knowledge is the most significant asset of the professions. Structural capital is knowledge retained in an organization's data-retaining systems in documentation and data sets. Whilst construction organizations have vast documentation of projects, including drawings, specifications, contract documentation and records, it is impossible to retain all knowledge in this form and much remains in the minds of employees, particularly in relation to reasons for decisions. Relationship capital is knowledge of the contacts which members of the organization have and the quality of their relations with them. Construction relies on networks of contacts in firms with which they work, public authorities, clients, subcontractors, etc. These knowledge bases allow an organization to cope with the future and, particularly in construction, to keep moving forward aesthetically, technologically and procedurally. The critical aspect of knowledge management is communicating the knowledge in the organization's intellectual capital bank to the right people at the time they need it. Collecting, updating, storing and distributing knowledge is an expensive but essential task but worthless if employees do not know what is available.

Intellectual capital can be seen to be an organization's memory, and loss of memory can occur. Whilst documenting helps to avoid this, not all knowledge can be retained in this way. Tacit knowledge, which includes insights, intuition and judgements, is personal to the individual and when a person leaves the organization this type of knowledge also leaves. This is the same with explicit knowledge which has not been saved even though it was possible to do so. An attempt to acquire and document knowledge from leavers before they go seems sensible but probably unrealistic.

The jury would appear to be still out on the validity of an effective learning organization. Huczynski and Buchanan (2007) examine the positives and negatives and say that 'it will be interesting to observe whether the learning organization remains fashionable'. In particular they are concerned that it

- is a complex and diffuse concept
- uses dated concepts
- encourages compliance with management directives and so strengthens management control
- is technologically-dependent which ignores how people use knowledge.

On the other hand it can be seen as a new set of rich challenging multi-dimensional concepts; an innovative approach to learning, investment in intellectual capital, organization, management, staff development and technology.

11.20 Organizational change

The ability of organizations to be flexible and to adapt and hence change is seen to be vital to survival and continued development in today's turbulent economic, technological, social and political environment. Yet organizations, particularly large ones which tend towards being bureaucratic, struggle to change as a result of organization inertia which is often the product of employees' fear of uncertainty and perceived threats to their positions leading to avoidance behaviour and evasion. Fincham and Rhodes (2005) point out that much of the management literature treats change as an objective process consisting of a series of steps with the assumption that if they are followed change will take place without question or problems and 'are the kinds of models and frameworks beloved of management consultants.' Huczynski and Buchanan (2007) make the same point saying that such approaches have 'attracted much criticism as organizations rarely operate in such a tidy and predictable manner, particularly with respect to strategic (major, messy radical) change'. So achieving change is not so simple; it is highly problematic.

It is as Collins (1998) sees it: 'a social activity, involving people from diverse social groups who will tend to interpret issues and situations in different often quite divergent ways.' He points out that the beginning of change is embedded in what exists, that is in the wider networks of the organization; that change takes time and should be seen as an emergent process; and that oversimplified views of harmony need to be questioned, which may result in conflict (which may be creative) as groups re-form. A contingency approach is proposed by Burnes (2000) who says that there is 'no one best way' or simple recipe for change but that implementation of change should draw on the wide range of approaches available. Measuring how an organization is performing is not easy; hence knowing that change is needed is often not clear and, if it is clear, how to agree the necessary response? Leadership of what could become a maelstrom demands high quality leaders who have the capability to gain acceptance from the workforce by involving them in the change process and so recognizes the saying that 'change means changing people not things'. It is argued that learning organizations should be in greater readiness to change and the process of guiding change is frequently referred to as change management, although it would be better called 'change leadership'.

Change management in construction firms has not figured strongly in the construction-related academic research literature until relatively recently, although there have been many government and industry sponsored initiatives for change (Murray and Langford, 2003) aimed at improving industry methods. Fernie et al. (2006) call these initiatives 'the reform movement' and believe that 'the reform movement's allegiance to approaches based on best practice is … dismissed as contextual, unreflective and insufficient in providing wholly reliable explanations for the relationship between practice and performance. Indeed, it is unclear why the reform movement has failed to engage in an exercise to understand the current legitimacy of managerial practice in the construction sector over repeated attempts to describe the sector as "ill" and in need of reform and change'. Essentially they are saying that the reform movement should recognize the organizational change field of study before attempting to prescribe the changes it believes should take place in the industry in order to understand how the industry's firms change themselves in response to their business environment. They argue that many firms' managerial practices which are criticized by the reform movement are legitimate actions within the context of the industry's setting. Bresnen et al. (2005) carried out case studies to examine barriers to construction firms implementing 'best practice' types of change called for by government- and industry-sponsored initiatives. They found that the well-established ways of working by project managers (vital for achieving project objectives and the levels of performance demanded of them), together with the distribution of power resources and the autonomy of project managers created major barriers to wider organizational change which was intended to implement government- and industry-sponsored initiatives as reflected in Fernie et al. (2006). Case studies were also used by Price and Chahal (2006) in their quest to identify the key steps that could improve the management of change. Perhaps reflecting the lack of change management in construction firms, they had to rely for their case studies on two firms on the fringe of the industry and one which, whilst not a mainstream construction company, was at least a subsidiary of one. It should be pointed out here that in the construction industry the term 'change management' is sometimes used to mean the management of changes in the design and detail of construction projects during construction (cf. Senaratne and Sexton, 2009) but in the previous reference and in this text the term is used in the sense attributed to it in the mainstream management literature.

Huczynski and Buchanan (2007) draw on Whittington and Mayer (2002), saying that their research shows that the results of major changes in organizations were often disappointing leading to poor financial results and lower staff morale and retention. In the change process people were often forgotten, with employee morale and retention being the lowest-ranked aims of change with little employee participation in the design and implementation of change. They again cite Whittington and Mayer (2002) who argue: 'for change to be successful, the "soft" human issues need to be integrated with "hard" structures and systems. This requires skilled change agents, and organizational characteristics such as a culture which welcomes change, appropriate management styles and supportive human resource policies'.

Whilst recognizing that change has been the buzzword of business over recent years, they continue by making the argument for less and more controlled change, citing Mintzberg (1994) who has argued that our preoccupation with change exaggerates the need for change. Not everything needs to change and changes that do take place need to be allowed to mature and take effect before another change is imposed. Change needs to be sustained and the benefits of improvements allowed to accrue before they are wasted by being overtaken by new changes: a process known as *initiative decay*. Also, constant change produces *initiative fatigue* which leads to lack of enthusiasm and lack of commitment to more change. They also cite Abrahamson's (2000) argument for 'painless change' based on 'dynamic stability' which, rather than dramatic large-scale transformation, involves mixing small and larger changes to working methods and structures in incremental steps to meet the organization's vision. Organizations should resist changing continually because this leads to cynicism and burnout.

Changes are seen to be achieved by 'tinkering' and 'kludging'. Tinkering is fiddling with what already exists in order to make creative improvements. Kludging means tinkering on a larger scale which may involve new organizational structures to cope with a changing sector of the firm's market but not a root and branch transformation. Morgan (2001) who argues for 'a quieter, more evolutionary approach to change' is cited in support. He argues for reducing the number of change initiatives, abandoning the preoccupation with large-scale transformation and focusing on incremental improvements. Such an approach to change is likely to appeal to construction-related organizations. They are generally learning organizations and can be expected to be aware of developments in their environment (economically, professional and political) and so be ready to implement incremental changes as they are demanded rather than leave things as they are until dramatic changes are required. For example, significant changes have taken place in the types of clients for construction work as a result of economic and other forces. There have been increases in the number of clients from communication industries, tourism, leisure and health services and a decline in those from more traditional industries together with the rise of partnering and other relational contracting arrangements but adaptation and reorganization has generally been achieved without great upheaval.

The 'holy grail' is an organization which is flexible enough to adapt to changes but stable enough not to be destabilized by them but rather be able to sustain them. Huczynski and Buchanan (2007) refer to Stace and Dunphy (2001) who identify organizations which they describe as 'prudent mechanistics', which have retained traditional structures, avoided the 'organizational fashion show', survived and performed well.

11.20.1 Change agents

Change agents are responsible for managing the change process. They can be drawn from the organization subject to change, normally from the management team but this does not have to be so. The main criteria are interpersonal and management skills, particularly communication skills, persuasiveness and political awareness. Change agents should be transformational by nature. It may be necessary in certain situations that they have high technical skills in the field. Alternatively, organizations may hire an external change agent in a consultancy capacity. The argument for an external agent is that they can offer an objective perspective as they have no allegiances within the organization; on the other hand, they are disadvantaged by not having insights which an internal appointee would possess. Also, they walk away from the job on completion no matter what the long-term outcome and are not held responsible. Internal change agents carry the responsibility for their work and have to live with the consequences. If internal change agents are high calibre employees held in high regard within their organizations then their work is more likely to be accepted than that of external change agents who can be criticized (rightly or wrongly) for not understanding the organization.

11.20.2 Resistance to change

Resistance to change to some extent on the part of employees at all levels seems inevitable but is likely to be less, or less severe, in learning organizations. Nevertheless, readiness for change can be tested by asking: is there pressure for change; a shared vision; liaison and trust; the will and power to act; enough capable people; suitable rewards and defined accountability; known first steps; and has the organization a capacity to learn (Eccles, 1994)? If resistance does occur, approaches to overcoming it include:

- *Communication and support*: explain why change is necessary and its rationale, alleviate fear and suspicion
- *Participation*: including employees in the decision-making process increases commitment to change
- *Recruit people who accept change*: not always possible but beneficial; they influence others
- *Negotiation*: compromise to reach agreement if viable
- *Manipulation*: play one group off against another
- *Coercion*: the last resort when there is no chance of compromise, use financial and promotion sanctions.

11.20.3 Business process re-engineering

Put simply, the approach of business process re-engineering (BPR) advocates a completely new beginning to organizational and operational design; that is, a re-design from a blank sheet of paper. Past history and current practice are discarded. A totally new process is then designed to deliver the product or service. BPR is an example of change management which was championed in the 1990s as a revolutionary way forward for the construction industry (Betts and Wood-Harper, 1994). However, although it may have informed other initiatives, e.g., lean construction, supply chain management and partnering, it has more recently been claimed by Green and May (2003) in their 'overtly critical perspective' to be 'impossible to define in terms of substantive content and is best understood as a rhetorical label'.

Courtney and Winch (2003) reported on research sponsored by the International Council for Research and Innovation in Building and Construction (CIB) and found that re-engineering not only implied radical change but also that improvement came from incremental change, which CIB should facilitate. They also made the point that re-engineering had to be concerned not only with the process of delivery, but also with the specification of what was to be delivered in order to truly represent eventual value to the client. They considered the concept of re-engineering as restrictive and proposed the idea of *re-valuing construction* as a broader alternative. In justifying the term 're-valuing construction', they say that in order to meet the desire for change in the construction industry it is necessary to find a way that meets the distinctive needs of the industry and, in particular, captures the role of the client in the construction process. Continuing this theme, they believe that improving construction performance concerns not technological but organizational and behavioural issues, which point to the contribution that the social sciences could make to re-engineering construction.

The argument in favour of BPR is that rapid and radical change is the only way for firms to deal with increasingly volatile environments, and that some organizations have reported significant improvements in performance. Others dismiss the approach as futile and irrelevant repackaging of traditional management methods. Research suggests that re-engineering has a high failure rate. In America, BPR quickly earned a 'slash and burn' reputation for the job losses or 'downsizing' that applications typically caused (Huczynski and Buchanan, 2007). However, as summarized by Ozcelik (2010), 'more recently literature suggests that the first generation of BPR . . . is evolving into modest process management, which is softened by the lessons learned from successes and failures in the course of implementations. The contemporary definition of BPR, therefore, encompasses a continuum of approaches to process transformation that may include both radical and incremental improvements, depending on the nature the problem. In fact many studies have been published in the literature in order to explain and promote this new approach to BPR, . . . Nevertheless, even the recent literature is rife with anecdotal evidence and short on empirical evidence of performance impacts of BPR projects.'

11.20.4 Processual/contextual perspectives

Processual/contextual theory is not a single theory but applies to a range of processual perspectives (Huczynski and Buchanan, 2007) and is in complete contrast to BPR. They cite Pettigrew (1985, 1987a, 1987b), saying that the theory does not accept a simple explanation for change and that 'He points instead to the many related factors – individual, group, organizational, social and political – that influence the nature and outcome of change'. Pettigrew argues that 'change is a complex and "untidy cocktail" of rational decisions, mixed with competing individual perceptions, stimulated by visionary leadership, spiced with "power plays" and attempts to recruit support and build coalitions behind particular ideas' and that the unit of analysis should be 'the process of change in context'. Whilst reminiscent of an organization's relationship with its environment in contingency theories, it claims to be more complex with three dimensions: the internal context (structure and culture); the external context (as the environment above); and past and current events.

This approach was further developed by Dawson (2003a) who argued that to understand the change process we need to take into account the past, present and future context of the organization; the type and nature of the change; the transition process needed; political activity, both internal and external; and the interaction between these factors. Dawson (2003b) identifies lessons for change as:

- There are no rules for how best to manage change
- Sensitivity is needed to people and context
- Change takes time
- Change impacts on people in different ways
- Learning arises from both unsuccessful as well as successful change
- Training in new methods is essential
- Communication is vital
- Change strategies must fit the substance and context
- Change is a political process
- Contradiction often arises from complex interactions.

A major drawback often given for this approach is that it does not give practical directions for implementing change, but this is the nature of change; it cannot be achieved by following a set of rules. What it does provide is guidance regarding the issues which are significant.

11.20.5 Organizational development

Organizational development (OD) is yet another concept in OB which is described as not easily defined or singular. Robbins and Judge (2008) believe it to be 'a term used to encompass a collection of planned-change interventions built on humanistic-democratic (behavioural science) principles that seek to improve organizational effectiveness and employee wellbeing'. They also see the underlying values in most OD efforts as:

- Respect for people
- Trust and support
- Power equalization which de-emphasises authority
- Confrontation openly confronted
- Participation of those affected by change.

OD techniques and interventions are employed for bringing about change. Many have been introduced in earlier chapters such as job enrichment, team building and participative management. The first step is to identify the problem before establishing the treatment. The intervention tool kit includes the following strategies.

Sensitivity training (also known as laboratory training, encounter groups and T-groups). Members meet in an open environment where they discuss themselves and their interactions. Professional behavioural scientists facilitate conditions for members to freely express their ideas, beliefs and attitudes without constraints and formality, hence allowing them to gain increased awareness of themselves. The objectives are to allow members to: better empathize with others; improve listening skills; be more open; increase tolerance; and develop conflict resolution skills. It is said to be one way to develop emotional intelligence.

Change of structure. Changing the organization structure can include job rotation, enlargement and enrichment and autonomous team working and decentralization to enhance empowerment.

Survey feedback uses survey methods to obtain feedback (anonymously) in order to identify discrepancies in perceptions amongst and between managers and employees. The findings are used as a basis for discussing and resolving any problems identified, usually through group discussions.

Team building is used to develop cohesion and effectiveness of teams. Survey feedback techniques can be used within the team to identify areas that need attention.

Inter-group development aims to change the attitudes that different groups within an organization have towards each other as a result of their perceptions. In construction-related organizations it is not uncommon for professional groups to perceive administrative departments as a hindrance to their work. One approach is for each group to meet independently to draw up a description of how it perceives both itself and the other group. The groups exchange descriptions and discuss their perceptions. Differences are clearly identified and reasons and possible answers established which will aid understanding and relationships. A similar approach can be adopted by two individuals whose relationship is ineffective.

Appreciative enquiry differs from the approaches discussed above as it highlights successes rather than problems and their solution. It aims to identify the distinctive elements of the organization in order to build on its unique qualities and special strengths to improve performance. Such an approach enables the identification of competitive advantage rather than creating defensiveness, as may occur with other approaches. Appreciative enquiry is carried out through large group meetings over a two or three day period, breaking down into smaller groups as necessary.

References

Abbott, V. (2009) Enormous white elephant. www.bhamweekly.com/2009/07/29/, accessed 17 August, 2009.

Abrahamson, E. (2000) Change without pain. *Harvard Business Review,* **78**, 4.

Abrahamson, E. and Fombrun, C. (1994) Macrocultures: Determinants and consequences. *Academy of Management Review*, **19**, 4.

Ackroyd, S. and Thompson, P. (1999) *Organizational Misbehaviour.* London: Sage.

Adams, J. (1963) Towards an understanding of inequality. *Journal of Abnormal and Social Psychology*, **67**, 4.

Adler, N. (2002) *International Dimensions of Organisational Behaviour.* London: International Thompson.

Adler, N. and Jelinek, M. (1986) Is organisational culture "Culture Bound"? *Human Resource Management*, **25**, 2.

Aitken, A. and Crawford, L. (2007) Coping with stress: Dispositional coping strategies of project managers. *International Journal of Project Management*, **25**, 666–673.

Ajzen, I. (2001) Nature and operations of attitudes. *Review of Psychology*, **52**, 27–58.

Akintoye, A., Macintosh, G. and Fitzgerald, E. (2000) A survey of supply chaincollaboration and management in the UK construction industry. *European Journal of Purchasing and Supply Management*, **6**, 159–168.

Alderfer, C. (1972) *Human Needs in Organizational Settings.* New York: Free Press.

Anderson, S. (1992) Project quality and project managers. *International Journal of Project Management*, **10**, 138–144.

Ankrah, N. and Langford, D. (2005) Architects and contractors: A comparative study of organisational cultures. *Construction Management and Economics*, **23**, 595–607.

Annett, J. and Stanton, N. (2000) Editorial: Team work: A problem for ergonomics? *Ergonomics*, **2**, 8.

Applebaum, D. and Lawton, S. (1990) *Ethics and the Professions.* Englewood Cliffs: Prentice-Hall.

Argyris, C. and Schon, D. (1974) *Theory in Practice.* San Francisco: Jossey-Bass.

Asch, S. (1951) Effects of group pressure upon the modification and distortion of judgements. In H. Guetzkow (Ed.) *Groups, Leadership and Men.* New York: Carnegie Press.

Ashforth, B. (1985) Climate formations: Issues and extensions. *Academy of Management Review,* **10**, 4.

Ashkanasy, N., Hartel, C. and Daus, C. (2002) Diversity and emotion: The new frontiers in organizational behaviour. *Journal of Management*, **28**, 3.

Baiden, B., Price, A. and Dainty, A. (2006) The extent of team integration within construction projects. *International Journal of Project Management*, **24**, 13–23.

Organizational Behaviour in Construction, First Edition. Anthony Walker.

Bales, R. (1950) *Interaction Process Analysis: A Method for the Study of Small Groups.* Cambridge, MA: Addison-Wesley Press.

Bales, R. (1955) How people interact in conferences. *Scientific American*, **192**, 31–50.

Baltes, B., Dickson, M., Sherman, M., Bauer, C. and La Ganke, J. (2002) Computer-mediated communication and group decision making: A meta-analysis. *Organisational Behaviour and Human Decision Processes*, January, 156–79.

Baradacco, J. (2001) We don't need another hero. *Harvard Business Review*, **79**, 8.

Barnard, C. (1938) *The Functions of the Executive.* Cambridge, MA: Harvard University Press.

Barney, J. (1991) Firm resources and sustained competitive advantage. *Journal of Management*, **17**, 1.

Barrett, M. and Stanley, C. (1999) *Better Construction Briefing.* Oxford: Blackwell Science.

Barthorpe, S. (2002) The origins and organisational perspectives of culture. In R. Fellows and D. Seymore (Eds) *Perspectives on Culture in Construction*, CIB Publication No. 275, Rotterdam: CIB.

Baskin, O. and Aronoff, C. (1980) *Interpersonal Communications in Organizations.* Santa Monica, CA: Goodyear.

Bayles, M. (1988) The professional–client relationship. In J. Callahan (Ed.) *Ethical Issues in Professional Life.* Oxford: OUP.

Behling, O. and Eckel, N. (1991) Making sense out of intuition. *Academy of Management Executive*, 5, February.

Belbin, M. (1993) *Team Roles at Work.* London: Butterworth-Heinemann.

Belbin, M. (1996) *The Coming Shape of Organization.* London: Butterworth-Heinemann.

Belbin, M. (2000) *Beyond the Team.* Oxford: Butterworth-Heinemann.

Bell, J. (2001) Patterns of interactions in multidisciplinary child protection teams in New Jersey. *Child Abuse and Neglect*, **25**, 65–80.

Bem, D. (1972) Self perception theory. In L. Berkowski (Ed.) *Advances in Experimental Social Psychology*, Vol. 6. New York: Academic Press.

Benne, K. and Sheats, P. (1948) Functional roles of group members. *Journal of Social Issues*, **4**, 41–49.

Bennis, W. (1959) Leadership theory and administrative behaviour. *Administrative Science Quarterly*, **4**, 259–301.

Bennis, W. and Nanus, B. (1985) *Leaders: The Strategies for Taking Charge.* New York: Harper Collins.

Berggren, C. (1993) *The Volvo Experience: Alternatives to Lean Production in the Swedish Auto Industry.* London: Macmillan.

Bess, T. and Harvey, R. (2002) Binomial score distributions and the Myers-Briggs Type Indicator: Fact or Artefact? *Journal of Personality Assessment,* Feb., 176–186.

Betts, M. and Wood-Harper, T. (1994) Re-engineering construction: A new management research agenda. *Construction Management and Economics*, **12**, 551–556.

Blake, R. and Moulton, J. (1969) *Building a Dynamic Corporation through Grid Organization Development.* Reading, MA: Addison-Wesley.

Blake, R.R. and Moulton, J.S. (1978) *The New Managerial Grid.* Houston: Gulf Publishing.

Blockley, D. and Godfrey, P. (2000) *Doing it Differently: Systems for Re-thinking Construction.* London: Thomas Telford.

Bowen, P., Pearl, R. and Akintoye, A. (2007) Professional ethics in the South African construction industry. *Building Research & Information*, **32**, 2.

Bowen, P., Cattell, K. and Distiller, G. (2008) Job satisfaction of South African quantity surveyors: An empirical study. *Construction Management and Economics*, **26**, 765–780.

Bowley, M. (1966) *The British Building Industry*. Cambridge: Cambridge University Press.

Bramel, D. and Friend, R. (1981) Hawthorne and the myth of the docile worker andclass base in society. *American Psychologist*, **38**, 8.

Breckler, S. (1984) Empirical validation of affect, behavior, and cognition asdistinct components of attitude. *Journal of Personality and Social Psychology*, **47**, 1191–1205.

Bresnen, M.J. (1991) Construction contracting in theory and practice: A case study. *Construction Management and Economics*, **9**, 247–63.

Bresnen, M. and Marshall, N. (2000a) Building partnerships: Case studies of client–contractor collaboration in the UK construction industry. *Construction Management and Economics*, **18**, 819–832.

Bresnen, M. and Marshall, N. (2000b) Motivation, commitment and the use of incentives in partnerships and alliances. *Construction Management and Economics*, **18**, 587–598.

Bresnen, M. and Marshall, N. (2000c) Partnering in construction: A critical review of issues, problems and dilemmas. *Construction Management and Economics*, **18**, 229–237.

Bresnen, M., Goussevskaia, A. and Swan, J. (2005) Implementing change in construction project organizations: Exploring the interplay between structure and agency. *Building Research and Information*, **33**, 6.

Briscoe, G. (2005) Women and minority groups in UK construction: recent trends. *Construction Management and Economics*, **23**, 10.

Brochner, J., Josephson, P. and Kadfors, A. (2002) Swedish construction culture, quality management and collaborative practice. *Building Research and Information*, **30**, 6.

Brown, A. (1998) *Organisational Culture*. London: Financial Times Management.

Buchanan, D. (2000) An eager and enduring embrace: The ongoing rediscovery of teamworking as a management idea. In S. Procter and F. Mueller (Eds) *Teamworking*. London: Macmillan.

Buchanan, D., Claydon, T. and Doyle, M. (1999) Organization development and change: The legacy of the nineties. *Human Resource Journal*, **9**, 2.

Building Industry Communication (1966) *Interdependency and Uncertainty: A Study of the Building Industry*. London: Tavistock.

Buller, P., Kohls, J. and Anderson, K. (1997) A model for addressing cross-cultural ethical conflicts. *Business and Society*, **26**, 169–193.

Burnes, B. (2000) *Managing Change: A Strategic Approach to Organizational Dynamics*. Harlow: Financial Times/Prentice Hall.

Burns, J. (1978) *Leadership*. New York: Harper and Row.

Burt, R. and Knez, M. (1996) Trust and third party gossip. In M. Kramer and T. Tyler (Eds.) *Trust in Organizations: Frontiers of Theory and Research*. Thousand Oaks, CA: Sage.

Butler, C. and Chinowsky, P. (2005) Emotional intelligence and leadership behaviour in construction executives. *Journal of Management in Engineering*, **22**, 3.

Butler, T. and Waldroop, J. (1999) Job sculpturing: The art of retaining your best people. *Harvard Business Review*, **77**, 5.

Campbell, J.P., Dunnette, M.D., Lawler, E.E. and Weick, K.E. (1970) *Managerial Behaviour, Performance and Effectiveness*. New York: McGraw-Hill.

Capraro, R. and Capraro, M. (2002) Myers-Briggs Type Indicator Score reliability across studies: A meta-analytic reliability generalization study. *Educational & Psychological Measurement*, **62**, 590–602.

Carlson, D. and Perrewe, P. (1995) Institutionalisation of organisational ethics through transformative leadership. *Journal of Business Ethics*, **14**, 10.

Carlsson, B., Josephson, P.E. and Larson, B. (2001) Communication in building projects: Empirical results and future needs. Paper HPT29, Proceedings of CIB World Building Congress, Wellington, New Zealand.

Carr, P., de la Garza, J. and Vorster, M. (2002) Relationship between personality traits and performance for engineering and architectural professions providing design services. *Journal of Management in Engineering*, **18**, 4.

Cartwright, S. and Cooper, C. (1993) The role of cultural compatibility in successful organisational marriage. *Academy of Management Executive*, **7**, 2.

Chan, P., Cooper, R. and Tzortzopoulos, P. (2005) Organisational learning: Conceptual challenges from a project perspective. *Construction Management and Economics*, **23**, 747–756.

Chatman, J. and Jehn, J. (1994) Assessing the relationship between industry characteristics and organisational culture: How different can you be? *Academy of Management Journal*, **37**, 3.

Chau, K.W., Raftery, J. and Walker, A. (1998) The baby and the bathwater: Research methods in construction management. *Construction Management and Economics*, **16**, 1.

Cherns, A.B. and Bryant, D.T. (1984) Studying the client's role in construction management. *Construction Management and Economics*, **2**, 177–184.

Cheung, S., Ng, T., Lam, K. and Yue, W. (2001) A satisfying leadership behaviour model for design consultants. *International Journal of Project Management*, **19**, 421–429.

Chinowsky, P. and Carrillo, P. (2007) Knowledge management to learning organisation connection. *Journal of Management in Engineering*, **23**, 3.

Clarke, N. (2010) Emotional Intelligence and its relationship to transformational leadership and key project management competences. *Project Management Journal*, **41**, 2.

Claydon, T. and Doyle, M. (1996) Trusting me, trusting you: The ethics of employee empowerment. *Personnel Review*, **25**, 6.

Cleland, D.I. and King, W.R. (1972) *Management: A Systems Approach*. New York: McGraw-Hill.

Coch, L. and French, J. (1948) Overcoming resistance to change. *Human Relations*, **1**, 512–532.

Collins, D. (1998) *Organizational Change: Sociological Perspectives*. London: Routledge.

Conger, J.A. and Kanungo, R.N. (1988) The empowerment process: Integrating theory and practice. *Academy of Management Review*, **13**, 471–482.

Cornick, T. and Mather, J. (1999) *Construction Project Teams: Making Them Work Profitably*. London: Thomas Telford.

Costa, P. and McCrae, R. (1997) Longitudinal stability and adult personality In R. Hogan, J. Johnson and S. Briggs (Eds) *Handbook of Personality Psychology*. San Diego: Academic Press.

Costa, A. (2003) Work team trust and effectiveness. *Personnel Review*, **32**, 5.

Courtney, R. and Winch, G. (2003) Re-engineering construction: The role of research and implementation. *Building Research and Information*, **31**, 2.

Cray, D. and Mallory, G. (1998) *Making Sense of Managing Culture.* London: International Thomson Business Press.

Culp, G. and Smith, A. (2001) Understanding psychological type to improve project team performance. *Journal of Management in Engineering*, **17**, 1.

Dainty, A., Bryman, A., Price, A., Geasley, K., Soetanto, R. and King, N. (2005) Project Affinity: The role of emotional attachment in construction projects. *Construction Management and Economics*, **23**, 241–244.

Dainty, A., Moore, D. and Murray, M. (2006) *Communication in Construction: Theory and Practice.* Abingdon: Taylor and Francis.

Dansereau, F., Graen, G. and Haga, W. (1975) A vertical dyad linkage approach to leadership within formal organizations: A longitudinal investigations of the role-making process. *Organizational Behavior and Human Performance*, **15**, 46–78.

Dawson, P. (2003a) *Reshaping Change: A Processual Approach.* London: Routledge.

Dawson, P. (2003b) *Understanding Organizational Change: The Contemporary Experience of People at Work.* London: Sage Publications.

De Graft-Johnson, A., Manley, S. and Greed, C. (2005) Diversity or the lack of it in the architectural profession. *Construction Management and Economics*, **23**, 10.

De Grada, E., Kruglandski, A., Mannetti, L. and Pierro, A. (1999) Motivation cognition and group interaction: Need for closure affects the contents and processes of collective negotiations. *Journal of Experimental Social Psychology*, **35**, 346–365.

Deal, T. and Kennedy, A. (1982) *Corporate Cultures: The Rights and Rituals of Organizational Life.* Reading MA: Addison-Wesley.

Denison, D. (1996) What is the difference between organizational culture and organizational climate? A native's point of view on a decade of paradigm wars. *Academy of Management Review*, **21**, 3.

Dewsbury, D. (1978) *Comparative Animal Behaviour.* New York: Mc Graw-Hill.

Dolfi, J. and Andrews, E. (2007). The sublime characteristics of project managers: An exploratory study of optimism overcoming challenge in the project management work environment. *International Journal of Project Management*, **25**, 674–682.

Dornbusch, S.M. and Scott, W.R. (1975) *Evaluation and the Exercise of Authority.* San Francisco: Jossey-Bass.

Druskat, V. and Wolff, S. (2001) Building emotional intelligence in groups. In *Harvard Business Review of Teams That Succeed.* Boston, MA: Harvard Business Review Publishing Corporation.

Dunham, R., Grube, J. and Castaneda, M. (1994). Organizational Commitment: The Utility of an Integrative Definition. *Journal of Applied Psychology*, **88**, 3.

Eccles, T. (1994) *Succeeding with Change: Implementing Action-Driven Strategies.* London: McGraw-Hill.

Eilion, S. (1979) *Aspects of Management.* Oxford: Pergamon Press.

Emerson, R.M. (1962) Power-dependence relations. *American Sociological Review*, **27**, 31–40.

Emery, F. (1959) *Characteristics of Socio-Technological Systems.* Tavistock Document 527. London: Tavistock Publications.

Emery, F. and Trist, E. (1960) Socio-technical systems. In C. Churchman and M. Verhulst (Eds) *Management Science, Models and Techniques, vol. 2.* London: Pergamon Press.

Emmitt, S. and Gorse, C. (2003) *Construction Communications.* Oxford: Blackwell.

Emmitt, S. and Gorse, C. (2007a) *Communication in Construction Teams.* Abingdon: Taylor and Francis.

Emmitt, S. and Gorse, C. (2007b) Communication behaviour during management and design team meetings: A comparison of group interaction. *Construction Management and Economics*, **25**, 1195–1211.

Eskrod, P. and Blichfeldt, B. (2005) Managing team entries and withdrawals during the project life cycle. *International Journal of Project Management*, **23**, 7.

Evans, G. and Johnson, D. (2000). Stress and open-office noise. *Journal of Applied Psychology*, **85**, 5.

Ezzamel, M. and Wilmott, H. (1998) Accounting for teamwork: A critical study of group-based systems of organisational control. *Administrative Science Quarterly*, **43**, 358–396.

Fan, L., Ho, C. and Mg, V. (2001) A study of quantity surveyor's ethical behaviour. *Construction Management and Economics*, **19**, 19–36.

Farmer, G. and Radford, A. (2010) Building with uncertain ethics. *Building Research and Information*, **38**, 4.

Faure, C. (2004) Beyond brainstorming: Effect of different group procedures on selection of ideas and satisfaction with the process. *Journal of Creative Behaviour*, **38**, 113–134.

Fayol, H. (1949 trans.) *General and Industrial Management*. London: Pitman (firstpublished in 1919).

Fellows, R. (2006) Culture. In D. Lowe and R. Leiringer (Eds) *Commercial Management of Projects: Defining the Discipline*. Oxford: Blackwell.

Fernie, S., Leiringer, R. and Thorpe, T. (2006) Change in construction: A critical sperspective. *Building Research and Information*, **34**, 2.

Fielder, F. (1967) *A Theory of Leadership Effectiveness*. New York: McGraw-Hill.

Fielder, F. and Chemers, M. (1984) *Improving Leadership Effectiveness: The Leader Match Concept*. Chichester: John Wiley & Sons Ltd.

Fincham, R. and Rhodes, P. (2005) *Principles of Organizational Behaviour*. Oxford: OUP.

Fineman, S. (1995). Stress, Emotion and Intervention. In T. Newton (Ed.) *Managing Stress: Emotion and Power at Work*. London: Sage.

Fineman, S., Sims, D. and Yannis, G. (2005) *Organizing and Organizations*. London: Sage.

Finkelstein, S. (1992) Power in top management teams: Dimensions, measurement and validation. *Academy of Management Journal*, **35**, 505–538.

Folkman, S., Lazarus, R. Dunkel-Scetter, C. Delongis, A. and Gruen, R. (1986) Dynamics of a stressful encounter: Cognitive appraisal coping and encounter outcome. *Journal of Community Psychology*, **1**, 113–137.

Foucault, M. (1977) *Discipline and Punishment: The Birth of Prison*. Harmondsworth: Penguin.

Fox, M., Dwyer, D. and Ganster, D. (1993) Effect of stressful job demands and control on psychological and attitudinal outcomes in a hospital setting. *Academy of Management Journal*, **36**, 289–318.

Fox, P. (2003) *Construction Industry Development: Analysis and Synthesis of Contributing Factors*, unpublished PhD thesis. Brisbane: Queensland University of Technology.

Fox, P. (2007) The culture of the construction industry. In A. Dainty, S. Green and B. Bagilhole (Eds) *People and Culture in Construction*. Abingdon: Taylor and Francis.

Francesco, A. and Gold, B. (1998) *International Organizational Behaviour: Text, Readings, Cases and Skills*. Upper Saddle River, NJ: Prentice-Hall.

French, W. and Bell, C. (1990) *Organization Development*. Englewood Cliffs, NJ: Prentice-Hall.

Freud, S. (1951) *Moses and Monotheism,* standard edition, vol. xxiii. London: Hogarth Press.

Frey, B. (2000). The impact of moral intensity on decision making in a business context. *Journal of Business Ethics*, **26**, 181–195.

Friedman, M. and Rosenmam, R. (1974) *Type A Behaviour and Your Heart.* New York: Knopf.

Fryer, B. (1985) *The Practice of Construction Management.* Oxford: Blackwell Science.

Fu, W., Lo, H. and Drew, D. (2006) Collective learning, collective knowledge and learning networks in construction. *Construction Management and Economics*, **24**, 1019–1028.

Gallie, D., White, M., Cheng, Y. and Tomlinson. M. (1998) *Restructuring the Employment Relationship.* Oxford: Clarendon Press.

Gambetta, D. (1988) *Trust: Making and Breaking Co-operative Relations.* Oxford: Blackwell.

Gambetta, D. and Hamill, H. (2005) *Streetwise.* New York: Russell Sage Foundation.

Gameson, R. (1992) *An investigation into the interaction between potential clients and construction professionals.* PhD. thesis, University of Reading.

Ganzach, Y. (2003) Intelligence, education and facets of job satisfaction. *Work and Occupations*, **30**, 1.

Gardner, W. and Martinko, M. (1996) Using the Myers-Briggs Type Indicator to study managers: A literature review and research agenda. *Journal of Management*, **22**, 1.

Gardner, W. and Schermerhorn Jr., J. (2004) Performance gains through positive organizational behaviour and authentic leadership. *Organizational Dynamics*, August, 270–281.

Garvin, D. and Roberto, M. (2001) What you don't know about making decisions. *Harvard Business Review*, **79**, 8.

Geier, J. (1967) A trait approach to the study of leadership in small groups. *Journal of Communications,* December, 316–23.

Gilbert, G. (1983) Styles of project management. *International Journal of Project Management*, **1**, 189–193.

Glass, J. and Symonds, M. (2007) "Considerate construction": Case studies of current practice. *Engineering, Construction and Architectural Management*, **14**, 2.

Glassop, L. (2002) The organisational benefits of teams. *Human Relations,* February, 225–250.

Goleman, D. (1995) *Emotional Intelligence: Why it can matter more than IQ.* London: Bloomsbury.

Gordon, G. (1991) Industry determinants of organisational culture. *Academy of Management Review*, **16**, 2.

Gorse, C. (2002) Effective interpersonal communication and group interaction during construction management and design team meetings. PhD. Thesis, University of Leicester.

Gorse, C. and Emmitt, S. (2003) Investigating interpersonal communications during construction progress meetings: Challenges and opportunities. *Engineering, Construction and Architectural Management*, **10**, 4.

Gorse, C. and Emmitt S. (2009) Informal interaction in construction progress meetings. *Construction Management and Economics*, **27**, 983–993.

Gorse, C., Emmitt, S. and Lowis, M. (1999) *Problem and appropriate communication medium.* Proceedings of ARCOM, 15th Annual Conference. Liverpool: Liverpool John Moore's University.

Gouldner, A. (1957) Cosmopolitans and locals: Towards an analysis of latent roles. *Administratve Science Quarterly*, **2**, 3.
Gray, R. (2001) Organisational climate and project success. *International Journal of Project Management*, **19**, 103–109.
Greasley, K., Bryman, A., Dainty, A., Price, A. and Soetanto, R. (2005) Employee perceptions of empowerment. *Employee Relations*, **27**, 354–368.
Greasley, K., Bryman, A., Dainty, A., Price, A. and Soetanto, R. (2008) Understanding empowerment from an employee perspective: What does it mean and what do they want? *Team Performance Management*, **14**, 39–55.
Green, S. (1994) Sociological paradigms and building procurement. In S.M. Rowlinson (Ed.) *Proceedings of CIB W92 Symposium: East Meets West: Procurement Systems*. CIB Publications No. 175. Hong Kong: Department of Surveying, University of Hong Kong.
Green, S. (1998) The technocratic totalitarianism of construction process improvement: A critical perspective. *Engineering, Construction and Architectural Management*, **5**, 4.
Green, S. and May, S. (2003) Re-engineering construction: Going against the grain. *Building Research and Information*, **31**, 2.
Greenacre, T. (2008) White elephant, white elephants on parade! *Building, QS Online*, www.buiding.co.uk, 23 Sept. 2008.
Grint, K. (1995) *Management: A Sociological Introduction*. Cambridge: Polity.
Grint, K. (2005) *Leadership Limits and Possibilities*. Basingstoke: Palgrave Macmillan.
Guirdham, M. (1995) *Interpersonal Skills at Work*. Hemel Hempstead: Prentice Hall.
Hackman, J. and Oldham, G. (1974) The job diagnostic survey: An instrument for the diagnosis of jobs and the evaluation of job redesign projects. *Technical Report No. 4, Department of Administrative Sciences*. New Haven, CT: Yale University.
Hackman, J. and Oldham, G. (1980) *Work Redesign*. Upper Saddle River, NJ: Pearson Educational.
Hackman, J., Oldham, G. and Purdy, K. (1975) A new strategy for job enrichment. *California Management Review*, **17**, 4.
Hammuda, I. and Dulaimi, M.F. (1997) The theory and application of empowerment in construction: A comparative study of the different approaches to empowerment in construction, service and manufacturing industries. *International Journal of Project Management*, **15**, 5.
Hampton, M. (1999) Work groups. In Y. Gabriel (Ed.) *Organisations in Depth*. London: Sage Publications.
Handy, C. (1993) *Understanding Organisations*. Oxford: Oxford Polity Press.
Hartman, F. (2003) Ten commandments of better contracting: A practical guide to adding value to an enterprise through more effective SMART contracting. *ASCE Press*, 235–260.
Harter, J., Schmit, F. and Hayes, T. (2002) Business-unit level relationship,between employee satisfaction, employee engagement, and businessoutcomes: A meta-analysis. *Journal of Applied Psychology*, **87**, 268–279.
Hayes, N. (1997) *Successful Team Management*. London: International Thompson Business Press.
Haynes, N. and Love, P. (2004) Psychological adjustment and coping among construction project managers. *Construction Management and Economics*, **22**, 129–140.
Hegedus, M. and Rasmussen, R. (1986) Task effectiveness and interaction process of a modified nominal group technique in solving an evaluation problem. *Journal of Management*, **12**, 545–560.

Hersey, P. and Blanchard, K.H. (1972) *Management of Organization Behavior*, 2nd edition. New Jersey: Prentice-Hall.

Hersey, P. and Blanchard, K.H. (1988) *Management of Organization Behavior*, 5th edition. New Jersey: Prentice-Hall.

Hertzberg, F. (1968) *Work and the Nature of Man*. London: Staples Press.

Hertzberg, F., Mausner, B. and Snyderman, B. (1959) *The Motivation to Work*. New York: John Wiley & Sons Inc.

Hewlett, S. (2002) Executive women and the myth of having it all. *Harvard Business Review*, **80**, 44.

Higgins, G. and Jessop, N. (1965) *Communications in the Building Industry*. London: Tavistock Publications.

Higgins, G. and Jessop, N. (1966) *Interdependence and Uncertainty: A Study of the Building Industry*. London: Tavistock Publications.

Hilmer, F. and Donaldson, L. (1996) *Management Redeemed: Debunking the Fads that Undermine Corporate Performance*. New York: Free Press.

HKEDC (1996) *Ethics for Professionals (Architecture, Engineering and Surveying): A Resource Portfolio for Hong Kong Universities*. Hong Kong: Hong Kong Ethics and Development Centre.

Ho, C. (2010) A critique of corporate ethics codes in Hong Kong construction. *Building Research and Information*, **38**, 4.

Ho, C. and Ng, V. (2003) Quantity surveyors' background and training, and their ethical concepts, conceptions and interests considerations. *Construction Management and Economics*, **21**, 43–67.

Hofstede, G. (1991) *Cultures and Organisations*. London: McGraw-Hill.

Hofstede, G. (1993) Cultural constraints in management theories. *Academy of Management Executive*, **7**, 1.

Hofestede, G. (2001) *Culture's Consequences: International Differences in Work-related Values*. London: Sage Publications.

Hofestede, G. and Bond, M. (1988) The confucian connection: From cultural roots to economic growth. *Organizational Dynamics*, **16**, 4.

Hofstede, G., Neuijen, B., Ohayv, D. and Sanders, G. (1990) Measuring organisational cultures: A qualitative and quantitative study across twenty cases. *Administrative Science Quarterly*, **35**, 286–316.

Holland, J. (1997) *Making Vocational Choices: A Theory of Vocational Personalitiesand Work Environments*. Odessa, FL: Psychological Assessment Resources.

Holmes, T. and Rahe, R. (1967) The social adjustment rating scale. *Journal of Psychosomatic Research*, **11**, 2.

Hood, N. (2003) The relationship of leadership style and CEO values to ethical practices in organisations. *Journal of Business Ethics*, **43**, 263–273.

House, R. (1976) A 1976 theory of charismatic leadership. In J. Hunt and L. Larson (Eds) *Leadership the Cutting Edge*. Carbondale Il: Southern Illinois University.

House, R. and Mitchell, T. (1974) Path-Goal Theory of Leadership. *Journal of Contemporary Business*, Autumn, 81–98.

Howell, J. and Costley, L. (2006). *Understanding Behaviours for Effective Leadership*. Upper Saddle River, NJ: Pearson Educational.

Huczynski, A. (2004) *Influencing within Organisations: Getting In, Rising Up and Moving On*. London: Routledge.

Huczynski, A. and Buchanan, D. (2007) *Organizational Behaviour: An Introductory Text*, 6th edition. Harlow, England; New York: Prentice Hall/Financial Times.

Huy, Q. (2001) In praise of middle manager. *Harvard Business Review*, **79**, 8.

Ive, G. and Gruneberg, S. (2000) *The Economics of the Modern Construction Sector.* Basingstoke: Macmillan.

Jackall, R. (1988) *Moral Mazes: The World of Corporate Managers.* New York: OUP.

Janis, I. (1972) *Victims of Groupthink.* Boston: Houghtom Mifflin.

Janis, I. (1982) *Groupthink,* 2nd ed. Boston: Houghton Mifflin.

Jin, X. and Ling, F. (2005) Constructing a framework for building relationships andtrust in project organizations: Two case studies of building projects in China. *Construction Management and Economics*, **23**, 685–696.

Johnson, D. (1991) *Ethical Issues in Engineering.* Englewood Cliffs, NJ: Prentice Hall.

Johnson-George, C. and Swap, W. (1982) Measurement of specific personal trust: Construction and validation of a scale to assess trust in a specific other. *Journal of Personality and Social Psychology*, **43**, 1306–1317.

Jones, J. (1973) Model of group development. *The Annual Handbook for Group Facilitators.* San Francisco, CA: Pfeiffer/Jossey-Bass.

Judge, T. and Church, A. (2000) Job satisfaction: Research and practice. In C. Cooper and E. Locke (Eds) *Industrial and Organizational Psychology: Linking Theory with Practice.* Oxford: Blackwell.

Judge, T., Heller, D. and Mount, M. (2002) Five-Factor Model of Personality and Job Satisfaction. *Journal of Applied Psychology*, **87**, 530–541.

Judge, T., Piccolo, R., Podsakoff, N., Shaw, J. and Rich, B. (2005) Can happiness be 'earned'? The relationship between pay and job satisfaction. Working paper, University of Florida.

Judge, T., Thoresen J., Bono, J. and Patton, G. (2001) The job satisfaction–job performance relationship: A qualitative and quantitative review. *Psycholgical Bulletin*, **127**, 3.

Kadfors, A. (2004) Trust in project relationships – inside the black box. *International Journal of Project Management*, **22**, 175–182.

Kadfors, A. (2005) Fairness in interorganisational project relations: Norms and strategies. *Construction Management and Economics*, **23**, 871–878.

Kakabadse, A., Myers, A., Mcmahon, T. and Spony, G. (1997) Top management styles in Europe: Implications for business and cross national teams. In R. Grint (Ed) *Leadership: Classical, Contemporary and Critical Approaches.* Oxford: Oxford University Press.

Kakabadse, A., Bank, J. and Vinnecombe, S. (2004) *Working in Organisations.* Aldershot: Gower.

Kanter, R.M. (1977) *Men and Women of the Corporation*. New York: Basic Books.

Kanter, R.M. (1983) *The Change Masters.* London: Unwin.

Karasek, R. (1979) Job demands, job decision latitude and mental strain. *Administrative Science Quarterly*, **24**, 129–44.

Katzenbach, J. and Smith, D. (1993) *The Wisdom of Teams: Creating the High Performance Organisation.* Boston: Harvard Business Press.

Keegan, A. and Den Hartog, D. (2004) Transformational leadership in a project based environment: A comparative study of the leadership styles of project managers and line managers. *International Journal of Project Management*, **22**, 609–618.

Kets de Vries, M. (1993) *Leaders, Fools and Imposters: Essays on the Psychology of Leadership.* San Francisco: Jossey-Bass.

Kirkman, B. and Rosen, B. (1999) Beyond self-management: Antecedents and consequences of team empowerment. *Academy of Management Journal*, **42**, 1.

Kleinginna, P. and Kleinginna, A. (1981). A categorized list of motivation definitions with a suggestion for a consensual definition. *Motivation and Emotions*, **5**, 263–292.

Knights, D. and McCabe, D. (2000) Bewitched, bothered and bewildered: The meaning and experience of teamworking for employees in an automobile company. *Human Relations*, **53**, 11.

Knights, D. and Willmott, H. (1987) Organisational culture as corporate strategy. *International Studies of Management and Organisation*, **17**, 3.

Kobasa, S. (1979) Stressful life events, personality and health: An inquiry into hardness. *Journal of Personality and Social Psychology*, **Jan.**, 1–11.

Kohn, A. (1993) Why incentive plans cannot work. *Harvard Business Review*, **71**, 5.

Kotter, J. (1982) *The General Managers.* New York: Free Press.

Kotter, J. and Heskett, J. (1992) *Corporate Culture and Performance.* New York: Free Press.

Kramer, R. (2003) The virtues of prudent trust. In R. Westwood and S. Clegg (Eds) *Debating Organizations: Point–Counterpoint in Organizational Studies.* Oxford: Blackwell Publishing.

Kramer, R. and Cook, K. (2004) *Trust and Distrust in Organizations: Dilemmas and Approaches.* New York, NY: Russell Sage Foundation.

Kumaraswamy, M., Rowlinson, S., Rahman, M. and Phua, F. (2002) Strategies for triggering the required 'cultural revolution' in the construction industry. In R.Fellows and D. Seymore (Eds) *Perspectives on Culture in Construction.* CIB Publication No. 275, Rotterdam: CIB.

Kunda, G. (1982) *Engineering Culture.* Philadelphia: Temple University Press.

Lamm, H. (1988) A review of our research on group polarization: Eleven experiments on the effects of group discussion on risk acceptance, probability estimation and negotiation positions. *Psychogical Reports*, **62**, 807–813.

Lancaster, L. and Stillman, D. (2002) *When Generations Collide.* San Francisco: Jossey-Bass.

Laurent, A. (1983) The cultural diversity of western conceptions of management. *International Studies of Management and Organisation*, **13**, 1–2.

Lavers, A. (1992) Communication and clarification between designer and client: Good practice and legal obligation. In M. Nicholson (Ed.) *Architectural Management*, London: Spon.

Lawler, E. (1973) *Motivation in Work Organizations.* New York: Brooks-Cole Publishing.

Lawrence, P.C. and Lorsch, J.W. (1967) *Organization and Environment: Managing Differentiation and Integration.* Boston: Graduate School of Business Administration, Harvard University.

Lazar, F. (2000) Project partnering: Improving the likelihood of win/win outcomes. *Journal of Management in Engineering*, **16**, 2.

Leavitt, H. (2005) *Top Down: Why Hierarchies are Here to Stay and How to ManageThem More Effectively.* Boston, MA: Harvard Business School Press.

Lee, R. and Ashforth, B. (1996) A meta-analytic examination of the correlates of three dimensions of job burnout. *Journal of Applied Psychology*, **81**, 2.

Lenway, S. and Rehbein, K. (1991) Leaders, followers and free riders: An empirical test of variation in corporate political involvement. *Academy of Management Journal*, **34**, 4.

Leung, M., Liu, A. and Wong, M. (2006) Impact of stress-coping behaviour on estimation performance. *Construction Management and Economics*, **24**, 55–67.

Leung, M., Ng, T. and Cheung, S. (2004) Measuring construction project participant satisfaction. *Construction Management and Economics*, **22**, 319–331.

Leung, M., Olomolaiye, P., Chong, A. and Lam, C. (2005) Impacts of stress on estimation performance in Hong Kong. *Construction Management and Economics*, **23**, 891–903.

Likert, R. (1961) *New Patterns of Management*. New York: McGraw-Hill.

Limsila, K. and Ogulana, S. (2008) Performance and leadership outcome correlates of leadership styles and subordinate commitment. *Engineering, Construction and Architectural Management*, **15**, 2.

Ling, F., Ofori, G. and Low, S. (2000) Importance of design consultants' soft skills in design-build projects. *Engineering, Construction and Architectural Management*, **7**, 4.

Lingard, H. (2003) The impact of individual and job characteristics on 'burnout' amongst civil engineers in Australia and the implications for employer turnover. *Construction Management and Economics*, **21**, 69–80.

Lingard, H. and Francis, V. (2004) The work–life experiences of office and site-based employees in the Australian construction industry. *Construction Management and Economics*, **22**, 991–1002.

Lingard, H. and Francis, V. (2006) Does a supportive work environment moderate the relationship between work–life conflict and burnout among construction professionals? *Construction Management and Economics*, **24**, 185–196.

Littlepage, G. and Silbiger, H. (1992) Recognition of expertise in decision making groups: Effect of group size and participation patterns. *Small Group Research*, **23**, 344–355.

Liu, A. (1999) A research model of project complexity and goal commitment effects on project outcome. *Engineering, Construction and Architectural Management*, **6**, 2.

Liu, A. and Fellows, R. (2001) An Eastern perspective on partnering. *Engineering, Construction and Architectural Management*, **8**, 1.

Liu, A. and Fellows, R. (2008) Behaviour of quantity surveyors as organizational citizens. *Construction Management and Economics*, **26**, 1271–1282.

Liu, A. and Walker, A. (1998) Evaluation of project outcomes *Construction Management and Economics*, **16**, 209–219.

Liu, A., Fellows, R. and Fang, Z. (2003) The power paradigm of project leadership. *Construction Management and Economics*, **21**, 819–829.

Liu, A., Chiu, W. and Fellows, R. (2007) Enhancing commitment through work empowerment. *Engineering, Construction and Architectural Management*, **14**, 568–580.

Locke, E. (1968) Towards a theory of task performance. *Organizational Behaviour and Human Performance*, **3**, 157–189.

Loosemore, M. and Waters, T. (2004) Gender differences in occupational stress among professionals in the construction industry. *Journal of Management in Engineering*, **20**, 3.

Love, P. and Edwards, D. (2005) Taking the pulse of UK construction project managers' health. *Engineering, Construction and Architectural Management*, **12**, 1.

Love, P., Li, H., Irani, Z. and Faniran, O. (2000) Total quality management and the learning organisation: A dialogue for change in construction. *Construction Management and Economics*, **18**, 321–331.

Lovell, R.J. (1993) Power and the project manager. *International Journal of Project Management*, **11**, 73–78.

Lubatkin, M., Calori, R., Very, J. and Veiga, J. (1998) Managing mergers across borders: A two nation explanation of a nationally bound administrative heritage. *Organizational Science*, **9** (6).

Luhmann, N. (1988) Familiarity, confidence, trust: Problems and alternatives. In D. Gambetta (Ed.) *Trust: Making and Breaking Co-operative Relations*. Oxford: Blackwell.

Lukes, S. (1975) *Power: A Radical View*. London: Macmillan.

Maslow, A. (1954) *Motivation and Personality.* New York: Harper.

May, D., Chan, A., Hodges, T. and Avolio, B. (2003) Developing the moral component of authentic leadership. *Organizational Dynamics,* August, 247–260.

Mayer, R., Davis, J. and Schoorman, F. (1995) An integrative model of organisational trust. *Academy of Management Review,* **20**, 3.

Mayo, E. (1949) *The Social Problems of an Industrial Civilization.* New York: Routledge.

McClelland, D. (1961) *The Achieving Society.* New York: Van Nostrand Reinhold.

McClelland, D. and Burnham, D. (1995) Power is the great motivator. *Harvard Business Review,* **73**, 1.

McDermott, P., Malik, M. and Swann, W. (2005) Trust in construction projects. *Journal of Financial Management of Property and Construction,* **10**, 1.

McGill, M. and Slocum Jr., J. (1998) A little leadership please? *Organizational Dynamics,* **26**, 3.

McGregor, D. (1960) *The Human Side of Enterprise.* New York: McGraw-Hill.

McLuhan, M. (1964) *Understanding Media: The Extensions of Man.* New York: McGraw-Hill.

McShane, S. and Von Glinow, M. (2003) *Organizational Behavior.* New York: McGraw-Hill.

Metcalfe, B. and Metcalfe, J. (2002) The great and the good. *People Management,* **8**, 11.

Metcalfe, B. and Metcalfe, J. (2003) Under the influence. *People Management,* **9**, 5.

Meyer, J., Allen, N. and Smith, C. (1993) Commitment to organizations and occupations: Extension and test of a three-component conceptualisation. *Journal of Applied Psychology,* **78**, 4.

Miller, E.J. and Rice, A.K. (1967) *Systems of Organisation: The Control of Task and Sentient Boundaries.* London: Tavistock Publications.

Mintzberg, H. (1973) *The Nature of Managerial Work.* New York: Harper and Row.

Mintzberg, H. (1979) *The Structure of Organizations.* Englewood Cliffs, NJ: Prentice-Hall.

Mintzberg, H. (1989) *Mintzberg on Management: Inside Our Strange World of Organisations.* New York: Free Press.

Mintzberg, H. (1994) That's not "turbulence", Chicken Little, it's real opportunity. *Planning Review,* **22**, 6.

Moodley, K., Smith, N. and Preece, C. (2008) Stakeholder matrix for ethical relationships in the construction industry. *Construction Management and Economics,* **26**, 625–632.

Moore, D. (2001) William of Sen to Bob the Builder: Cultural perceptions. *Engineering, Construction and Architectural Management,* **8**, 3.

Moore, M. (2008) Shanghai's World Finance Center may prove to be white elephant. www.telegraph.co.uk, 1 September 2008, accessed 17 August 2009.

Moorhead, G. and Griffin, R. (2001) *Organizational Behaviour: Managing People and Organisations.* Boston MA: Houghton Miffin.

Moreno, J. (1953) *Who Shall Survive?* New York: Beacon Press.

Morgan, G. (1997) *Images of Organization.* Thousand Oaks, CA: Sage.

Morgan, N. (2001) How to overcome 'change fatigue'. *Harvard Management Update,* July, 1–3.

Morse, N. and Reimer, E. (1956) The experimental change of a major organizational variable. *Journal of Abnormal and Social Psychology,* January, 120–129.

Moscovici, S. (1980) Towards a theory of conversion behaviour. *Advances in Experimental Social Psychology,* **13**, 209–239.

Munns, A. (1995) Potential influence of trust on the successful completion of a project. *International Journal of Project Management*, **13**, 1.

Murnighan, J., Malhotra, D. and Weber, J. (2004) *Paradoxes of trust: Empirical and theoretical departures from a traditional Model.* In R. Kramer and K. Cook (Eds) *Trust and Distrust in Organizations: Dilemmas and Approaches*. New York, NY: Russell Sage Foundation.

Murray, M. and Langford, D. (2003) *Construction Reports 1944–98*. Oxford: Blackwell Science.

Myerson, D. (2001) Radical change, the quiet way. *Harvard Business Review*, **79**, 9.

National Economic Development Office (NEDO) (1990) *The Innovation Toolkit*. London: Her Majesty's Stationary Office.

Naylor, J., Pritchard, R.D. and Ilgen, D.R. (1980) *A Theory of Behaviour in Organisations*. New York: Academic Press.

Needle, D. (2000) Culture at the level of the firm: Organizational and corporate perspectives. In J. Barry, J. Chandler, H. Clark, R. Johnston and D. Needle (Eds) *Organisation and Management: A Critical Text*. Thomson Learning Business Press.

Neilsen, E. (1986) Empowerment strategies: Balancing authority and responsibility. In S. Sirwastra (Ed.) *Executive Power*. San Francisco: Jossey-Bass.

Nemeth, C. (1986) Differential contributions of majority and minority influence. *Psychological Review*, **93**, 1.

Newcombe, R. (1994) Procurement paths – a power paradigm. In S.M. Rowlinson (Ed.) *Proceedings of CIB W92 Symposium: East Meets West: Procurement Systems*. CIB Publication No. 175. Hong Kong: Department of Surveying, University of Hong Kong.

Newcombe, R. (1996) Empowering the construction project team. *International Journal of Project Management*, **14**, 2.

Newcombe, R. (1997) Procurement paths – A cultural/political perspective. In C.H. Davidson and T.A. Meguid (Eds) *Procurement – A Key to Innovation*. Proceedings of CIB W92, Montreal: IF Research Corporation.

Newton, T. and Keenan, A. (1985) Coping with work related stress. *Human Relations*, **38**, 107–126.

Ng, T., Skitmore, M. and Leung, T. (2005) Manageability of stress among construction project participants. *Engineering, Construction and ArchitecturalManagement*, **12**, 3.

Nicolini, D. (2002) In search of 'project chemistry'. *Construction Management and Economics*, **20**, 167–177.

Ofori, G. and Toor, S. (2009) Research on cross-cultural leadership and management in construction: A review and directions for future research. *Construction Management and Economics*, **27**, 119–133.

Ogbonna, E. and Harris, L.C. (1998) Managing organisational culture: Compliance or genuine change? *British Journal of Management*, **9**, 273–288.

O'Reilly, C. and Pondy, L. (1979) Organizational communication. In S. Kerr (Ed.) *Organizational Behavior*. Columbus, Ohio: Grid.

Organ, D. (1988) *Organizational Citizen Behavior: The Good Soldier Syndrome*. Lexington, MA: Lexington Books.

Ouchi, W. (1981) *Theory Z*. Reading MA: Addison-Wesley.

Ozcelik, Y. (2010) Do business process reengineering projects payoff? Evidence from the United States. *International Journal of Project Management*, **28**, 7–13.

Palmer, I. and Hardy, C. (2000) *Thinking about Management: Implications and Organisational Debates for Practice*. London: Sage.

Pascale, R. and Athos, A. (1982) *The Art of Japanese Management.* Harmondworth: Penquin.

Pease, A. (1997) *Body Language: How to Read Others' Thoughts by their Gestures.* London: Sheldon Press.

Pedler, M., Burgoyne, J. and Boydell, T. (1997) *The Learning Company: A Strategy for Sustainable Development.* London: McGraw-Hill.

Peter, L.J. and Hull, R. (1969) *The Peter Principle.* New York: Morrow.

Peters, T. and Waterman, R. (1982) *In Search of Excellence: Lessons from America's Best Run Companies.* New York, NY: Harper and Row.

Petrovic-Lazarevic, S. (2008) The development of corporate social responsibility in the Australian construction industry. *Construction Management and Economics,* **26**, 193–101.

Pettigrew, A. (1985) *The Awakening Giant: Continuity and Change in ICI.* Oxford: Basil Blackwell.

Pettigrew, A. (1987a) Context and action in the transformation of the firm. *Journal of Management Studies,* **24**, 6.

Pettigrew, A. (Ed.) (1987b) *The Management of Strategic Change.* Oxford: Basil Blackwell.

Pfeffer, J. (1992) *Managing with Power: Politics and Influence in Organisations.* Boston, MA: Harvard Business School Press.

Philips, M. (1994) Industry mindsets: Exploring the cultures of two micro-organisational settings. *Organisational Science,* **5**, 3.

Phua, F. and Rowlinson, S. (2003) Cultural differences as an explanatory variable for adversarial attitudes in the construction industry: The case of Hong Kong. *Construction Management and Economics,* **21**, 777–785.

Phua, F. and Rowlinson, S. (2004) Operationalizing culture in construction management research: A social identity perspective in the Hong Kong context. *Construction Management and Economics,* **22**, 913–925.

Pietroforte, R. (1997) Communication and governance in the building process. *Construction Management and Economics,* **15**, 1.

Pinto, J., Slevin, D. and English, B. (2009) Trust in projects: An empirical assessment of owner/contractor relationships. *International Journal of Project Management,* **27**, 638–648.

Plous, S. (1993) *The Psychology of Judgement and Decision Making.* New York: McGraw-Hill.

Podsakoff, P., MacKenzie, S., Paine, J. and Bachrach, D. (2000) Organizational citizen behaviors: A critical review of the theoretical and empirical literature and suggestions for future research. *Journal of Management,* **26**, 3.

Poirot, J.W. (1991) Organising for quality: matrix organisation. *Journal of Management in Engineering,* **7**, 178–186.

Porter, L. and Lawler, E. (1968) *Managerial Attitudes and Performance,* Homewood Il.: Irwin.

Posner, B. and Schmidt, W. (1992) Values and the American manager: An update updated. *California Management Review,* **34**, 3.

Price, A. and Chahal, K. (2006) A strategic framework for change management. *Construction Management and Economics,* **24**, 237–251.

Putnam, L. and Mumby, D. (1993) Organizations, emotion and the myth of rationality. In S. Fineman (Ed.) *Emotion in Organizations.* Thousand Oaks, CA: Sage.

Quenk, N. (2000) *Essentials of Myers-Briggs Type Indicator Assessment.* New York: John Wiley & Sons Inc.

Reason, J. (1987) The Chernobyl errors. *Bulletin of the Psychology Society*, **40**, 201–206.

Ree, M., Carretta, T. and Steindl, J. (2001) Cognative ability. In N. Anderson, D. Ones, H. Sinangil and C. Viswesvaran (Eds) *Handbook of Industrial, Work and Organizational Psychology, Vol. 1.* Malden, MA :Blackwell.

Ries, C.J. (1964) *The Management of Defense.* Baltimore: John Hopkins Press.

Riketta, M. (2002) Attitudinal organizational commitment and job performance: A meta-analysis. *Journal of Organizational Behavior*, **23**, 257–266.

Riley, M. and Clare-Brown, D. (2001) Comparison of cultures in construction and manufacturing industries. *Journal of Management in Engineering*, July, 149–158.

Robbins, S. and Judge, T. (2008) *Essentials of Organisational Behavior.* Upper Saddle River, NJ: Pearson Educational.

Robertson, I. (2001) Undue diligence. *People Management*, **7**, 23.

Robinson, J. and Torvik, R. (2005) White elephants. *Journal of Public Economics*, **89**, 2–3.

Roethlisberger, F. (1997) *The Elusive Phenomenon: An Autobiographical Account of My Work in the Field of Organisational Behaviour at the Harvard Business School.* Cambridge MA: Harvard University Press.

Rokeach, M. (1973) *The Nature of Human Values.* New York: The Free Press.

Rooke, J., Seymore, D. and Fellows, R. (2003) The claims culture: A taxonomy of attitudes in the industry. *Construction Management and Economics*, **21**, 167–174.

Rose, M. (1994) Skill and Samuel Smiles: Changing the British work ethic. In R. Penn, M. Rose and J. Rubery (Eds) *Skill and Occupational Change.* Oxford: OUP.

Rose, R., Jenkins, C. and Hurst, M. (1978) Air traffic controller health change study: A prospective investigation of physical, psychological and work related changes. Galveston: University of Texas (mimeo).

Rose, R. (1988) Organisational as multiple cultures: A rules theory analysis. *Human Relations*, **41**, 2.

Rosnow, R. and Fine, G. (1976) *Rumor and Gossip: The Social Psychology of Hearsay.* New York: Elsevier.

Rotter, J. (1966) Generalized expectancies for internal versus external control of reinforcement. *Psychological Monographs*, **80**, 609.

Rousseau, D. (1990) Assessing organisational culture: The case for multiple methods. In B. Schneider (Ed.) *Organisational Climate and Culture.* San Francisco, CA: Jossey-Bass.

Rousseau, D., Sitkin, S., Burt, R. and Camerer, C. (1998) Not so different after all: A cross-discipline view of trust. *Academy of Management Review*, **23**, 3.

Rowlinson, S. (2001) Matrix organisational structure, culture and commitment: A Hong Kong public sector case study for change. *Construction Management and Economics*, **19**, 669–673.

Royer, I. (2005) Why bad projects are hard to kill. In *Harvard Business Review on Managing Projects.* Boston, MA: Harvard Business Review Publishing Corporation.

Rudolph, H. and Peluchette, J. (1993) The power gap: Is sharing or accumulating power the answer? *Journal of Applied Business Research*, **9**, 3.

Ruthankoon, R. and Ogunlana, S. (2003) Testing Herzberg's two-factor theory in the Thai construction industry. *Engineering, Construction and Architectural Management*, **10**, 5.

Sackmann, S. (1992) Culture and subcultures: An analysis of organisational knowledge. *Administrative Science Quarterly*, **37**, 140–161.

Sang, K. (2005) Examining effective and ineffective transformation project leadership. *Team Performance Management*, **11**, 3/4.

Sang, K., Dainty, A. and Ison, S. (2007) Gender: A risk factor for occupational stressin the architectural profession. *Construction Management and Economics*, **25**, 1305–1317.

Sang, K., Ison, S. and Dainty, R. (2009) The job satisfaction of UK architects and relationships with work–life balance and turnover intentions. *Engineering, Construction and Architectural Management*, **16**, 3.

Sathe, V. (1983) Implications of corporate culture: A manger's guide to action. *Organisational Dynamics*, **12**, 2.

Schein, E. (1985) *Organisational Culture and Leadership.* San Francisco, CA: Jossey-Bass.

Schermerhorn, J., Jr., Hunt, G. and Osborn, R. (2004) *Core Concepts of Organisation Behaviour,* Hoboken, NJ: John Wiley & Sons Inc.

Schleicher, D., Watt, J. and Greguras, G. (2004) Re-examining the job satisfaction–performance relationship: The complexity of attitudes. *Journal of Applied Psychology*, **89**, 1.

Scott, W. (1992) *Organizations: Rational, Natural and Open Systems,* 3rd edition. London: Sage.

Scott, W. (1998) *Organizations: Rational, Natural and Open Systems,* 4th edition. London: Sage.

Schwartz, M. (2001) The nature of the relationship between corporate codes of ethics and behaviour. *Journal of Business Ethics*, **32**, 247–262.

Selye, H. (1936) A syndrome produced by nocuous agents. *Nature*, **138**, 32.

Senaratne, S. and Sexton, M. (2009) Role of knowledge in managing construction project change. *Engineering, Construction and Architectural Management*, **16**, 2.

Senge, P. (1990) *The Fifth Discipline: The Art and Practice of the Learning Organisation.* New York: Doubleday Currency.

Sheldon, K. and Bettencourt, B. (2002) Psychological need-satisfaction and subjective well-being within social groups. *British Journal of Social Psychology*, **41**, 1.

Shohet, I. and Frydman, S. (2003) Communication patterns in construction at construction manager level. *Journal of Construction Engineering and Management*, **129**, 5.

Shoura, M.M. and Singh, A. (1998) Motivation parameters for engineering managers Using Maslow's Theory. *Journal of Management in Engineering*, **15**, 5.

Shriberg, A., Shriberg, D. and Kumari, R. (2005) *Practicing Leadership*. Hoboken, NJ: John Wiley & Sons Inc.

Sievers, B. (2003) *Fool'd with hope men favour the Deceit, or, Can we trust in trust?* In R. Westwood and S. Clegg (Eds) *Debating Organizations: Point–Counterpoint in Organizational Studies.* Oxford: Blackwell Publishing.

Simon, H. (1957) *Administrative Behaviour.* New York: Macmillan.

Simon, H. (1960) *The New Science of Management Decision.* Harper: New York.

Singh, A. (2002) Behavioural perceptions of design and construction engineers. *Engineering, Construction and Architectural Management*, **9**, 2.

Skitmore, M. and Vee, C. (2003) Professional ethics in the construction industry. *Engineering, Construction and Architectural Management*, **10**, 2.

Smith, F. (1977) Work attitudes as predictors of attendance on a specific day. *Journal of Applied Psychology*, **62**, 1.

So, Y. and Walker, A. (2006) *Explaining Guanxi: The Chinese Business Network*. London: Routledge.

Soetanto, R. and Proverbs, D. (2002) Modelling the satisfaction of contractors: The impact of client performance. *Engineering, Construction and Architectural Management*, **9**, 5/6.

Sparks, J. and Hunt, S. (1998) Marketing researcher ethical sensitivity: Conceptualisation, measurement and exploratory investigation. *Journal of Marketing*, **62**, 92–109.

Spreitzer, G. and Quinn, R. (2001) *A Company of Leaders: Five Disciplines for Unleashing the Power in Your Workforce*. San Francisco, CA: Jossey-Bass.

Stace, D. and Dunphy, D. (2001) *Beyond the Boundaries: Leading and Re-creating the Successful Enterprise*. Sydney: McGraw-Hill.

Stacey, R., Griffin, D. and Shaw, P. (2000) *Complexity and Management*. London: Routledge.

Strang, K. (2005) Examining effective and ineffective transformation project leadership. *Team Performance Management*, **11**, 3/4.

Staw, B. (1981) The escalation of commitment to a course of action. *Academy of Management Review*, **6**, 4.

Staw, B. and Ross, J. (2005) Knowing when to pull the plug. In *Harvard Business Review on Managing Projects*. Boston, MA: Harvard Business School.

Steiner, I. (1972) *Group Process and Productivity*. New York: Academic Press.

Stewart, T. (1997) *Intellectual Capital: The New Wealth of Organisations*. New York: Doubleday.

Stogdill, R. (1950) Leadership, membership and organization. *Psychological Bulletin*, **47**, 1–14.

Strategic Forum for Construction (2009) *Strategic Forum for Construction. www.strategicforum.org.uk, accessed on 24 April 2009.*

Sundstrum, E., De Meuse, K. and Futrell, D. (1990) Work teams. *American Psychologist*, **45**, 2.

Sunindijo, R., Hadikusomo, H. and Ogunlana, S. (2007) Emotional intelligence and leadership styles in construction project management. *Journal of Management in Engineering*, **23**, 4.

Symes, M., Eley, J. and Seidel, A. (1995) *Architects and their Practices: A Changing Profession*. Oxford: Butterworth Architecture.

Tajful, H. and Turner, J. (1986) The social identity theory on inter-group behaviour, In S. Worchel and W. Austin (Eds) *Psychology of Inter-group Relations*. Chicago Ill: Nelson-Hall.

Tannen, D. (1991) *You Just Don't Understand: Women and Men in Conversation*. New York, NY: Ballantine Books.

Tannen, D. (1995) *Talking from 9 to 5*. New York: William Morrow.

Tannenbaum, R. and Schmidt, W.H. (1973) How to choose a leadership pattern. *Harvard Business Review*, **51**, 162–80 (originally in vol. 37, March–April, 1958).

Tavistock Institute (1966) *Interdependency and Uncertainty*. London: Tavistock Publications.

Taylor, A. (2002) Job satisfaction among early school leavers working in the trades and the influence of vocational education in schools. *Journal of Youth Studies*, **5**, 3.

Taylor, D., Berry, P. and Bloch, C. (1958) Does group participation when using brainstorming techniques facilitate or inhibit creative thinking? *Administrative Science Quarterly*, **3**, 1.

Taylor, F. (1911) *The Principles of Scientific Management*. New York: Harper.

Taylor, L. (1972) The significance of interpretation of replies to motivational questions: The case of sex offenders. *Sociology*, **6**, 1.

Taylor, L. and Walton, P. (1971) Industrial sabotage: Motives and meanings. In S. Cohen (Ed.) *Images of Deviance*. Harmondsworth: Penguin.

Thompson, J. (1967) *Organisation in Action*. New York: McGraw-Hill.

Thompson, P. and Findlay, P. (1999) Changing the people: Social engineering in the contemporary workplace. In I. Ray and A. Sayer (Eds) *Culture and Economy after the Cultural Turn*. London: Sage Publications.

Tinsley, C. (1998) Models of conflict resolution in Japanese, German and American cultures. *Journal of Applied Psychology*, **83**, 2.

Tolman, E. (1932) *Purposive Behaviour in Animals*. New York: Appleton-Century-Crofts.

Toor, S. and Ofori, G. (2008) Taking leadership research into the future: A review of empirical studies and new directions for research. *Engineering, Constructionand Architectural Management*, **15**, 4.

Townsend, R. (2007) *Up The Organization* (Commemorative ed.). San Francisco: Jossey-Bass.

Trist, E. and Bamforth, K. (1951) Some social and psychological consequences of the longwall method of coal getting. *Human Relations*, **4**, 1.

Trompenaars, F. and Wooliams, P. (2002) Model behaviour. *People Management*, **8**, 24.

Tuckman, B. (1965) Development sequences in small groups. *Psychological Bulletin*, **63**, 6.

Tuckman, B. and Jensen, M. (1977) Stages of small group development revisited. *Group and Organisational Studies*, **2**, 4.

Tuuli, M. and Rowlinson, S. (2009a) Empowerment in project teams: A multilevel examination of the job performance implications. *Construction Managementand Economics*, **27**, 5.

Tuuli, M. and Rowlinson, S. (2009b) Performance consequences of psychological empowerment. *Journal of Construction Engineering and Management*, **135**, 12.

Tuuli, M. and Rowlinson, S. (2010) What empowers individuals and teams inproject settings? A critical incident analysis. *Engineering, Construction and Architectural Management*, **17**, 1.

Tylor, E. (1871) *Primitive Culture: Researches into the Development of Mythology, Philosophy, Religion, Language, Art and Custom*. London; Murray.

Verquer, T., Beehr, T. and Wagner, S. (2003) A meta-analysis of relations between person–organization fit and work attitudes. *Journal of Vocational Behavior*, **63**, 3.

Vroom, V. (1964) *Work and Motivation*. New York: John Wiley & Sons Inc.

Vroom, V. (2000) Leadership and the decision-making process. *Organizational Dynamics*, **28**, 4.

Walker, A. (1994) *Building the Future – The Story of the Controversial Construction of the Campus of the Hong Kong University of Science and Technology*. Hong Kong: Longman Asia Ltd.

Walker, A. (2007) *Project Management in Construction*. Oxford: Blackwell.

Walker, A. and Newcombe, R. (2000) The positive use of power on a major construction project. *Construction Management and Economics*, **18**, 37–44.

Walker, A. and Kalinowski, M. (1994) An anatomy of a Hong Kong project – organisation, environment and leadership. *Construction Management and Economics*, **12**, 191–202.

Walker, D. and Rowlinson, S. (2008) *Procurement Systems: A Cross-industry Project Perspective*. Abingdon: Taylor and Francis.

Walker, D., Segon, M. and Rowlinson, S. (2008) *Business ethics and corporate citizenship*. In D. Walker and S. Rowlinson (Eds) *Procurement Systems: A Cross-industry Project Perspective*. Abingdon: Taylor and Francis.

Weber, M. (1947 trans.) *The Theory of Social and Economic Organisation*. In A.H. Henderson and T. Parsons (Eds) *The Theory of Social and Economic Organization*. Glencoe, IL: Free Press (first published in 1924).

Weber, M. (1968 trans.) *Economy and Society: An Interpretive Sociology*, 3 vols. (Eds. Guenther Roth & Claus Wittich). New York: Bedminster Press (first published in 1924).

Weihrich, H. and Koontz, H. (1993) *Management: A Global Perspective*. New York: McGraw-Hill.

Westwood, R. and Clegg, S. (2003) *Debating Organizations: Point–Counterpoint in Organizational Studies*. Oxford: Blackwell Publishing.

Wheen, F. (2004) *How Mumbo-Jumbo Conquered the World*. London: Fourth Estate.

Whittington, R. and Mayer, M. (2002) *Organizing for Success in the Twenty-First Century: A Starting Point for Change*. London: Chartered Institute of Personnel and Development.

Whittock, M. (2002) Women's experiences of non-traditional employment: Is gender equality in this area possible? *Construction Management and Economics*, **20**, 5.

Wilkins, A. (1983) The culture audit: A tool for understanding organisations. *Organisational Dynamics*, **12**, 2.

Williams, R. (1983) *Keywords*. London: Fontana.

Williams, T.M. (1997) Empowerment vs risk management? *International Journal of Project Management*, **15**, 4.

Williamson, O. (1975) *Markets and Hierarchies: Analysis and Antitrust Implications*. New York, NY: The Free Press.

Williamson, O. (1985) *The Economic Institutions of Capitalism*. New York, NY: The Free Press.

Williamson, O. (1990) Chester Barnard and the incipient science of organisation. In: O. Williamson (1996) *The Mechanisms of Governance*. New York, NY: Oxford University Press.

Wilson, F. (1999) *Organizational Behaviour and Work: A Critical Introduction*. Oxford: Oxford University Press.

Wilson, F. (2004) *Organizational Behaviour and Work: A Critical Introduction*, 2nd edition. Oxford: Oxford University Press.

Wong, E., Then, D. and Skitmore, M. (2000) Antecedents of trust in intra-organisational relationships within three Singapore public sector constructionproject management agencies. *Construction Management and Economics*, **18**, 797–806.

Wong, P. and Cheung, S. (2004) Trust in construction partnering: Views from parties of the partnering dance. *International Journal of Project Management*, **22**, 437–446.

Wong, W., Cheung, S., Tak, W. and Hoi, Y. (2008) A framework for trust in construction contracting. *International Journal of Project Management*, **26**, 821–829.

Wong, W., Wong, P. and Heng, L. (2007) An investigation of leadership styles and relationship cultures in Chinese and expatriate managers in multinationalconstruction companies in Hong Kong. *Construction Management and Economics*, **25**, 95–106.

Wood, G. and McDermott, P. (1999) Searching for trust in the UK construction industry: An interim view. In *Proceedings of CIB W92 International Procurement Systems Conference*. Rotterdam: CIB.

Wood, G. and Ellis, R. (2005) Main contractor experiences of partnering relationship on UK construction projects. *Construction Management and Economics*, **23**, 317–325.

Wood, J. (1995) Mastering management: Organisational behaviour. *Financial Times*, supplement (part 2 of 20).

Woodruffe, C. (2001) Promotional intelligence. *People Management*, **7**, 1.

Yadong, L. (2007) *Global Activities of Corporate Governance.* Oxford: Blackwell.

Yannis, G. (1999) *Organisations in Depth.* London: Sage Publications.

Yanow, D. (1993) Review: Controlling cultural engineering? *Journal of Management Inquiry*, **2**, 2.

Yates, D. (1998) Conflict dispute resolution in the Hong Kong construction industry: A transaction cost economics perspective. M.Phil Thesis, Hong Kong: University of Hong Kong.

Yip, B., Rowlinson, S. and Oi Ling Siu (2008) Coping strategies as moderators in the relationship between role overload and burnout. *Construction Managementand Economics*, **26**, 871–882.

Yip, B. and Rowlinson, S. (2009a) Job burnout amongst construction engineers working within consulting and contracting organisations. *Journal of Management in Engineering*, **25**, 3.

Yip, B. and Rowlinson, S. (2009b) Job re-design as an intervention strategy of burnout: An organisational perspective. *Construction Engineering and Management*, **135**, 8.

Yoshida, R., Fentond, K. and Maxwell, J. (1978) Group decision making in the planning team process: Myth or reality? *Journal of School Psychology*, **16**, 237–244.

Zaghloul, R. and Hartman, F. (2000) Construction contracts: The cost of mistrust. *International Journal of Project Management*, **21**, 419–424.

Zhang, S. and Liu, A. (2006) Organisational culture profiles of construction enterprises in China. *Construction Management and Economics*, **24**, 817–828.

Zohar, D. (1997) Predicting burnout with a hassle-based measure of role demands. *Journal of Occupational Behaviour*, **18**, 101–115.

Zohar, D. (1999) When things go wrong: The effect of daily work hassles on effort,exertion and negative mood. *Journal of Occupational and Organizational Behaviour*, **72/3**, 265–84.

Zolin, R., Hinds, P., Fruchter, R. and Levit, R. (2004) Interpersonal trust in cross-functional, geographically distributed work: A longitudinal study. *Information and Organisation*, **14**, 1.

Index

Organizational Behaviour in Construction, First Edition. Anthony Walker.